Microfluidics for Biologists

Chandra K. Dixit • Ajeet Kaushik
Editors

Microfluidics for Biologists

Fundamentals and Applications

Springer

Editors
Chandra K. Dixit
Department of Chemistry
University of Connecticut
Storrs, CT, USA

Ajeet Kaushik
Department of Immunology
Center for Personalized Nanomedicine
Institute of NeuroImmune Pharmacology
Herbert Wertheim College of Medicine
Florida International University
Miami, FL, USA

ISBN 978-3-319-40035-8   ISBN 978-3-319-40036-5 (eBook)
DOI 10.1007/978-3-319-40036-5

Library of Congress Control Number: 2016952550

Printed on acid-free paper

This Springer imprint is published by Springer Nature
The registered company is Springer International Publishing AG Switzerland

Preface

Microfluidics has revolutionized the way we deal with biological samples and biological matrix. It has enabled us to understand how a single cell is completely different from the information we can obtain by current tools and techniques. It is because of the microfluidic technology we are able to study physiology of a single cell and to understand the heterogeneity in the cellular population of the same descent. This is just one example of how microfluidics has changed the way we perceive biological information. This technology has enormous applications in every field of life sciences, from basics to industrial to diagnostics. However, there is a big communication gap between biologists and microtechnologists, which is due to a lack of training in the fields other than theirs.

In the first chapter of this book, we have presented the fundamentals of physics that govern microfluidics. These principles are presented in such a way that biologists can easily understand what controls the fluidics and how to use those for studying biological phenomenon. The second chapter of this book is dedicated to acquaint biologists with an overview of tools, techniques, and applications of microfluidics. The following chapters will cover manufacturing methods for developing custom microfluidic tools including 3D printing. Valving for controlling fluids in fluidic tools is also explained. Surfaces, sensors, and their integration are described such that the layman can understand the concepts. In the following chapters, the application of microfluidics in the field of cell and molecular biology, single cell biology, and disease diagnostics are introduced with simplicity. All these chapters are discussed in relation to commercial technologies so biologists can better correlate functioning of these tools with applications they desire to employ. This book is an attempt to describe the need of novel microtechnologies and their integration strategies for developing a new class of assay systems to retrieve the desired health information of patients in real time. This book also describes the selection and integration of sensor components and of operational parameters for developing point-of-care (POC). System-on-a-Chip (SoC), Diagnostic-on-a-Chip (DoC), and Lab-on-a-Chip (LOC) are the core to the next-generation bioanalytical sciences; therefore, this book can be lab assistance for those who work with biology-microfluidics interface, thus helping them to understand these systems and allowing them to make educated decisions on selecting the nature and type of microtechnologies that suits best to their methods thereby enhancing the rate of translational research in the field.

Salient Features of This Book

- This book serves as a resource guide for biologists and chemists to understand the complex physics of microfluidics.
- Describes the preparatory methods for developing 3-dimensional microfluidic structure and their use for LOC designing.
- Explains the significance of miniaturization and integration of sensing components to develop wearable sensors for POC.
- Demonstrates the application of microfluidics in life sciences and analytical chemistry including disease diagnostics and separations.
- Motivates new ideas related to novel platforms, valving technology, miniaturized transduction methods, and device integration to develop next-generation sequencing platforms, future diagnostic systems, and platforms for single cell biology applications.
- Discusses the future prospects and challenges of the field of microfluidics in the areas of life.

Storrs, CT, USA Chandra K. Dixit
Miami, FL, USA Ajeet Kaushik

Contents

Chapter 1
Fundamentals of Fluidics

Chandra K. Dixit

1 Introduction

Microfluidics has had tremendous impact on miniaturization of biological experiments by reducing the reagent volumes, shortening the reaction times, and enabling multiplexed parallel operations by integrating an entire laboratory protocol onto a single chip (i.e., lab-on-a-chip or LOC). Best examples of microfluidic tools in biology are Gene chips, Capillary electrophoresis, CD-based inertial cell separation devices, integrated transcriptome analysis systems, and others. Along with miniaturization comes a tremendous opening at the microscale where slight manipulation in physics can provide unprecedented number of applications for each design. An understanding of the physical processes at microscale and their dynamics can allow biologists to leverage those for performing experiments that are practically not feasible at macroscale. Since microfluidics can allow new processes and experimental paradigms to emerge therefore, here we will focus on fundamentals that predominantly govern the processes at microscales and how we can manipulate those to address problems in the field of biology.

2 Microfluidic Physics

Dimension is the key in understanding the magnitude of a physical event taking place. Prior to discussing physics of microfluidic processes we must first understand that on what we are working. Few important symbols representing physical quantities and the microfluidic scales that are mainly relevant to biologists are mentioned in Tables 1.1 and 1.2, respectively.

C.K. Dixit (✉)
Department of Chemistry, University of Connecticut, Storrs, CT, USA
e-mail: chandra.kumar_dixit@uconn.edu

C.K. Dixit, A. Kaushik (eds.), *Microfluidics for Biologists*,
DOI 10.1007/978-3-319-40036-5_1

Table 1.1 Common symbols for physical parameters

Greek letter symbols			
α	alpha	λ	lambda
β	beta	μ	mu
γ	gamma	ν	nu
Δ	delta	Π	pi
ε	epsilon	ρ	rho
ζ	zeta	σ	sigma
η	eta	τ	tau
Θ	theta	ω	omega
κ	kappa		

Table 1.2 Length scales for common biological moieties

Sample matrix	Approximate scales
Distance between molecules in a liquid	0.1 nm
Distance between molecules in a gas	3 nm
Mean free path between collision in a gas, air at ambient pressure (λ)	61 nm
Sample	
Protein, lipid molecule of the membrane	1 nm
Virus	10 nm
Cells	1–20 μm

These are few illustrative sample matrices and sample types that are routinely employed in biological analysis. Given the sizes, our focus should be on the phenomenon that can be used to manipulate micron and sub-micron entities. Reagent mixing, reagent delivery, cell capture, and shear-free conditions for biological analysis are few typical applications that are sought by biologists. We will understand physical entities in this chapter with respect to these applications that will allow developing an understanding of microfluidics.

2.1 Hierarchy of Dimensions

Before advancing to the complex physics dominating the micron regimen, we must first review the basic concepts and their respective dimensions. Table 1.3 summarizes few of the most basic scaling entities.

Entity	Dimension
Size	$[l]$
Surface	$[l]^2$
Volume	$[l]^3$
Van der Waals	$[d]^{-3 \text{ to } -7}$
Various Forces	$[l]^{1 \text{ to } 3}$
l is size of an object,	
d is distance between two objects	

Table 1.3 Scaling laws: variation at changing length scales

Quantity	Scaling law
Time	$[l]^0$
Length	$[l]^1$
Area	$[l]^2$
Volume	$[l]^3$
Velocity	$[l]^1$
Acceleration	$[l]^1$
Density	$[l]^{-3}$
Viscosity	$[l]^{-2}$
Diffusion time	$[l]^2$
Reynolds number	$[l]^2$
Peclet number	$[l]^2$
Hydraulic resistance	$[l]^{-4}$

With our previous knowledge of physical processes, we can realize that size, shape, and volume have tremendous impact on the forces acting upon/between bodies. For example, let us consider the force exerted upon a body by earth. This force is called gravitational pull and is represented as the ratio of the product of masses of earth and ours to the squared distance between us. As we realize this force has dimensional dependence on the distance between the two bodies, which is $[l]^2$. Similarly, a body flowing through a water stream will experience some force exerted upon it by the flow. This is dependent on the size and surface of the body and is somewhat close to how biomolecules and cells will feel in the microfluidic channels. Therefore, we must now look few years back in high school physics, which is actually the foundation to our advanced understanding of microfluidics.

2.2 *Non-dimensionalization and Dimensionless Numbers*

This section is intended to introduce the concept and importance of non-dimensionalization because you will now know terms that will be commonly used throughout the text; if it is hard to understand at this point then these can be revisited once all the basics are learnt. Dimensions are critical in physical analysis as they draw boundaries around a physical quantity by defining them in dimensions. Their importance becomes predominant when we are working at structures in micrometer range where surface area increases drastically relative to volume. This characteristic dependence of physical processes on dimensions must be addressed in such a way that the process can be explained as a function of the intrinsic properties of the fluid rather than the dimensions of those properties. In other words, we must make equations governing these processes without any resultant dimensions. This can be achieved by carefully replacing quantities in those equations with others, such that their dimensions cancel out each and have no net dimensional dependence. These quantities may be constants and can be employed for understanding the relative importance of entities within the process itself. Thus, non-dimensionalization is known as removal of units from the mathematical expression of a phenomenon by substituting with appropriate variables. This is also termed as *scaling*.

Scaling reduces the dependence of the process on several variables and significantly contributes to understand the relative importance of the physical quantities in the process and to realize the variation in their dimensions. This certainly helps in neglecting the smaller terms from the equation, which simplifies the associated physics. Therefore, it allows understanding physics at smaller scales and thus, is very important in microfluidics.

We will not deal scaling in great detail as it is a complex method but generally non-dimensionalization can be achieved via following steps:

(a) Identify the unit for which scaling is required; developing a scaling law
(b) Identify all the variables dependent and independent to that unit
(c) Identify a set of physically-relevant dimensionless groups and plug them in
(d) Determine the scaling exponent for each one, and
(e) Rewrite the equations in terms of new dimensionless quantities.

Such dimensionless numbers are crucial for exploring fundamentals of the physics governing microfluidics. The essential fluid physics of a system is dictated by a competition between various phenomena. This competition is expressed via a series of dimensionless numbers capturing their relative importance. These dimensionless numbers (Tables 1.4 and 1.5) form a sort of 'parameter space' for microfluidic physics.

2.3 *Hydrostatics: Physics of the Stagnant*

Fluids, liquids and gases, are defined as a material which will continue to deform with the application of a shear force. These are governed by certain basic rules of physics. Fluids have a special property to mention, they flow but only under the influence of external forces; these are mainly governed by **pressure, field gradients, surface tension, and gravity**. Since we will be mainly dealing with liquids therefore, our main focus is on the concepts of hydrostatic and hydrodynamic fluidics. As the name suggests hydrostatics and hydrodynamics are processes related to static and flowing liquids, respectively. Both these processes are controlled by associated physical parameters that we will discuss in this section.

Table 1.4 Dimensionless numbers in fluid mechanics

Dimensionless number	Details	Formula
Reynolds Number	Inertial force/Viscous force convective momentum/viscous momentum *Forced Convection*	$Re = \rho UL/\eta = UL/\nu$
Prandtl Number (heat) Prandtl-Schmidt Number (mass)	Momentum/Species diffusivity Used to determine fluid or heat or mass transfer boundary layer thickness	$Pr_{heat} = \nu/\alpha = \eta C_P/K$ $Pr_{mass} = Sc = \nu/D = \eta/\rho D$
Péclet Number (heat) Péclet Number (mass)	Convection transport rate/Diffusion transportation rate	$Pe_{heat} = RePr = UL/\alpha$ $\alpha = k/\rho C_P$ $Pe_{Mass} = RePr = UL/D$
Nusselt Number (heat) Nusselt-Sherwood Number (mass)	Length scale/Diffusion boundary layer thickness Used to determine the heat (h) or mass (h_D) transfer coefficient	$Nu = \left[f_\varepsilon Re(Pr)^{1/3}\right]/2$ $Nu_{heat} = hL/k_{fluid}$ $Nu_{Mass} = h_D L/D_{fluid}$ $L = A_s/Pm$
Grashof Number (heat) Grashof Number (mass)	Natural convection buoyancy force/Viscous force Used to calculate Re for buoyant flow Controls the lengthscale to natural convection boundary layer thickness *Natural Convection*	$Gr_{heat} = g\beta(T_s - T_b)L^3/\nu^2$ $Gr_{Mass} = g\beta_C(C_{as} - C_{aa})L^3/\nu^2$ $\beta = -\left[(\partial\rho/\partial C_\alpha)_{T,P}\right]/\rho$
Rayleigh Number (heat) Rayleigh Number (mass)	Natural convection/Diffusive heat or mass transport Used to determine the transition to turbulence	$Ra_{heat} = GrPr = g\beta(\Delta T)L^3/\nu\alpha$ $Ra_{mass} = GrPr = g\beta_C(\Delta C)L^3/\nu D$
Knudsen Number (to analyze extent of continuum)	Slip length/Macroscopic length	$Kn = \beta/L$
Richardson Number	Buoyancy/Flow gradient	$Ri = g(\Delta\rho)/\rho U^2$
Eötvös (Eo) or Bond Number (Bo)	Body forces/Surface tension Used together with Morton Number to determine shape of drops or bubbles in surrounding fluid or continuous phase	$Eo = Bo = \left[(\Delta\rho)gL^3\right]/\sigma$
Capillary Number	Viscous forces/Interfacial forces	$Ca = \eta U/\sigma$
Elasticity Number	Elastic effects/Inertial effects	$El = \theta\eta/\rho R^2 = Wi/Re$
Weissenberg Number	Viscous forces/Elastic forces	$Wi = \gamma'.t_s$
Deborah Number	Stress relaxation time/Time of observation	t_s/t_o

Table 1.5 Common physical entities in fluid mechanics

Physical entity		Unit	Dimension
U	Characteristic velocity	m/s	LT^{-1}
L	Characteristic length	m	L
T	Temperature	K	Θ
T_s	Surface temperature	K	Θ
T_b	Temperature of the bulk	K	Θ
D	Mass diffusivity	m^2/s	L^2T
C_p	Specific heat	J/Kg.K	$L^2T^{-2\Theta-1}$
C_{as}	Concentration of species a at surface	Kg/m^3	ML^{-3}
C_{aa}	Concentration of species a in ambient medium	Kg/m^3	ML^{-3}
A_s	Surface area of the pipe	m^2	L^2
Pm	Perimeter	m	L
η	Dynamic viscosity	Pa.s = Ns/m^2 = Kg/ms	$ML^{-1}T^{-1}$
ν	Kinematic viscosity	m^2/s	L^2t^{-1}
σ	Surface/interfacial tension	Kg/s^2 = N/m	MT^{-2}
ρ	Density	Kg/m^3	ML^{-3}
β	Coefficient of thermal expansion	1/K	Θ^{-1}
α	Thermal diffusivity	m^2/s	L^2T^{-1}
k	Thermal conductivity	W/mK	$MLT^{-3}\Theta^{-1}$
h	Convective heat transfer coefficient	W/m^2K	$MT^{-3}\Theta^{-1}$
h_D	Convection mass transfer coefficient	m/s	LT^{-1}
λ	Mean free path	m	L
γ	Specific weight	N/m^3	$ML^{-2}T^{-2}$
R	Radius of the pipe	m	L
Θ	Stress evolution		
t_s	Stress relaxation time for the fluid	s	T
t_o	Time of observation of event	s	T
γ'	Sheer rate	1/s	T^{-1}

Pascal's Law

- Pressure applied anywhere to a fluid transmits the force equally in all directions
- Change in pressure disperses equally throughout the fluid
- Force acts at right angles to any surface in contact with the fluid
- Hydraulic press is the representative example

Hydrostatics is the physics of pressure confined within the definitions of Pascal's law and Archimedes principle constitute hydrostatics

2.3.1 Pressure and Pumping

Consider a cuboidal bottle filled with water to a height of one meter with length and width of the bottle at 5 cm each. The liquid in bottle is not continuous, instead a stack of several individual layers of water molecules, such that each layer is parallel to each other and continuously interacting with each other.

Now, **PRESSURE** is how much force is exerted on a given area and is expressed as

$$P = F/A \tag{1.1}$$

where,

P is pressure, F is force exerted, and A is the surface area on which force is exerted.

SI unit of pressure is atmosphere (atm) and is equivalent to 10^5 Pascals, another unit for pressure and have dimension Nm^{-2}.

By the virtue of the definition of pressure, the top layer of the water molecules must exert a force on the layers beneath it over the surface area of the layer. Similarly, the top layer will do so on the last layer at the bottom. It is crucial to understand that for fluids under gravity, based on (1.1), pressure exerted by an upper layer on the one underneath is directly dependent on the distance between those layers expressed as height. From Fig. 1.1a, the pressure exerted by the liquid on the bottom of the container should be calculated as

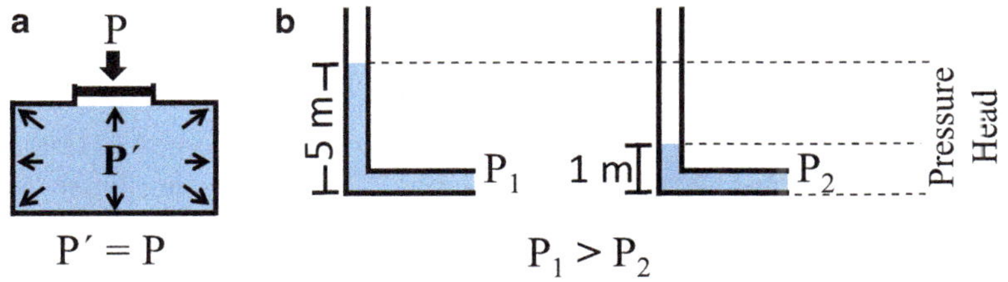

Fig. 1.1 Illustration of Pascal's law. (**a**) Pressure exerted at any point on a continuous fluid is dissipated equally in all directions on that fluid. This concept makes the basis of hydraulic press and brakes. (**b**) An extension of Pascal's law is pressure head driven flows where the height of the liquid exerts a pressure on the lower layers. This concept of height-dependent pressure is used in pumping in microfluidics. As depicted, 5 m head will exert more pressure than 1 m head

$$F = mg \tag{1.2}$$

where,
m is mass of the liquid, and g is gravity constant.
Since,

$$m = \rho V \tag{1.3}$$

where,
V is volume of container, and ρ is mass density of the liquid.
Therefore, replacing (1.3) in (1.2) will give us

$$F = V(\rho g) = hA(\rho g) \tag{1.4}$$

such that volume = height of the liquid (h) * area of the surface (A = length * width)

Similarly, replacing (1.4) in (1.1) will give us the relation of height to the pressure

$$P = hA\rho g/A = h\rho g \tag{1.5}$$

Continuing with the case that we were discussing, in Fig. 1.1b pressure exerted by a layer on the other separated by certain height within the liquid will be

$$P_2 - P_1 = \mathbf{\Delta P} = (h - h_1)\rho g = \mathbf{\Delta h}.\boldsymbol{\rho}\mathbf{g} \tag{1.6}$$

Equation (1.6) constitutes the basic of **hydrostatic pressure-based pumping** in **microfluidic systems**. '**ΔP**' is known as pressure head.

2.3.2 Buoyancy and the Problem of Microfluidic Mixing

Buoyancy is the apparent loss of weight of a body when submerged in liquid and this is mainly known as Archimedes Principle. This loss is attributed to the resistance offered by the liquid to the body. Buoyancy from Fig. 1.2 can be mathematically expressed as

$$F_{net} = F_B(\text{buoyant force}) - F_g(\text{weight}) \tag{1.7}$$

$$= (\rho_f V_f - \rho_o V_o)\, g \tag{1.8}$$

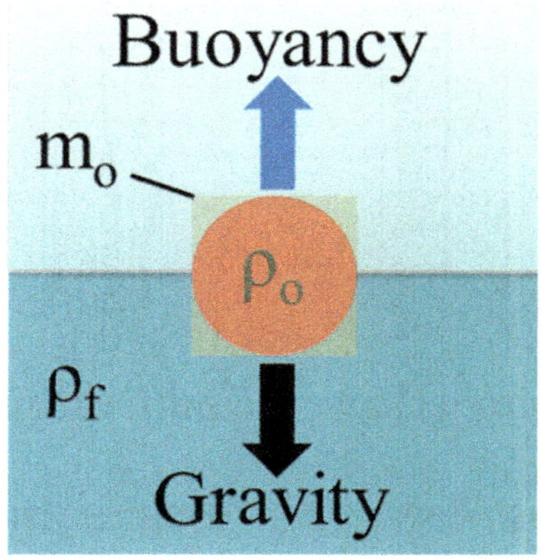

Fig. 1.2 An illustration of Archimedes principle of buoyancy on a body of density 'ρ_o' dipped in a fluid of density 'ρ_f'. The body of mass m_o experiences two opposite forces on it, gravity acting downwards and buoyancy or thrust acting upwards. Denser the body will be with respect to the fluid, the greater the gravity force will be on it. Thus the body will drown. On the contrary, if it is lighter than fluid, then it will float on the surface

For static liquid, $F_{net} = 0$; therefore, it can be deduced as

$$F_g \rho_o = F_B \rho_f \tag{1.9}$$

Where, 'o' and 'f' denote 'object' and 'fluid', respectively.

Now, based on (1.9) If,

$\rho o > \rho f$, the object will submerge and settle down to the bottom of the fluid. However, an object will float but submerged with $\rho o = \rho f$, and will float on the surface with a $\rho o < \rho f$.

This knowledge becomes the basis of buoyancy-dependent mixing in certain microfluidic set-ups. The best example is introduction of air bubbles from underneath of the static layer of liquid. The air bubbles have lower density than liquid and will move towards the top of the channel thus causing disruption in the solvent layers and introducing mixing Fig. 1.3. We will discuss other details later in this chapter.

2.4 *Hydrodynamics: Physics of the Flows*

Fluids at motion are governed by a set of variables and these are crucial in understanding the phenomena taking place within confined boundaries in microfluidics. There are few properties we will first acquaint with before looking into other aspects.

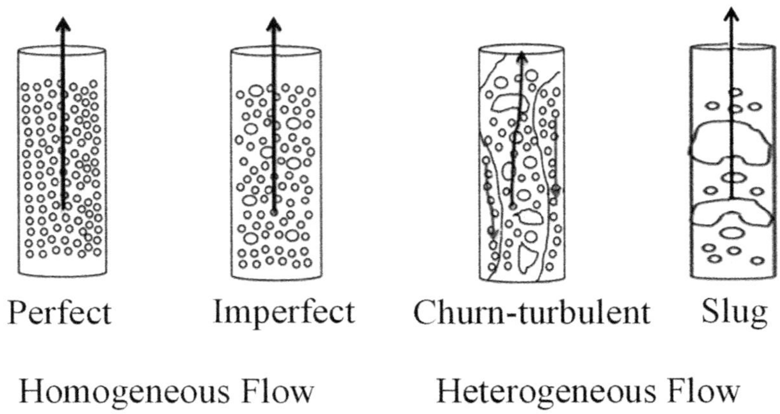

Fig. 1.3 Buoyancy-driven mixing in liquids with various types of bubble flows. Each type of bubble flow introduces mixing which could be diffusive or turbulent

2.4.1 Concept of Continuum

Typically matter is made up of atomic and sub-atomic particles. Thus, when analyzing those at micron level, matter becomes discontinuous in space with inter-atomic separation. However, when considering fluids as a continuum material, assumptions have to be made where we have to neglect that atoms are the smallest unit. And, the matter/fluid must be defined in terms of continuous fields, such as density and force density (defined as per unit volume), rather than discrete physical quantities, such as mass and force. The continuum can be confirmed for fluids by first reducing the sample volume to a very small magnitude and then measuring intrinsic properties, like density, at several points in liquid space. The density should be equal to approve the consideration of fluid continuum.

2.4.2 Important Intrinsic Properties

- Mass density (ρ; Kg/m^3): It is the mass distribution over a unit volume,
- Specific volume (V; m^3/Kg): It is the volume occupied by a unit mass,
- Weight density or Specific weight ($\boldsymbol{\gamma}$; N/m^3): It is the force due to gravity on the mass in a unit volume and is expressed as

$$\gamma = g\rho \tag{1.10}$$

- Specific gravity or Relative density ($\boldsymbol{\delta}$): it is the ratio of density of the fluid to the density of water,
- Viscosity: It is the resistance offered by the fluid to gradual deformation by neighboring fluid layers under an external force, namely shear stress or tensile stress. It is also known as thickness of the fluids. This parameter also represents

the interaction of parallel moving fluid plates with each other and with surroundings. Inter-plate collisions in a moving fluid create friction which opposes the motion of the fluid. Therefore, to move a fluid certain external stimulus, such as pressure gradient, is required. A fluid that doesn't offer any intrinsic resistance to shear force is known as an ideal or inviscid fluid while those offering resistance are called viscous or viscid.

Momentum of molecules in each respective layer is considered to be homogeneous. Additionally, due to mixing the molecules from one layers move to the other. In this case, a molecule diffusing to a fast moving layer needs to be accelerated and deaccelerated when travelling to a slow moving layer. During this these molecules carry their respective momentum with them. This is the main reason for introduction of the shear into the layers.

Dynamic/Shear viscosity (η; Poiseuille (Pl); Pa.s; N. s/m^2; Kg/ms):
It is the resistance offered by a fluid layer to adjacent layers where all the layers are moving parallel to each other but at different speeds. Thus, is also called shear viscosity. The simplest understanding can be developed with the explanation of illustration in Fig. 1.4. In panel a, suppose there are three parallel layers moving in same direction, with lowest layer being at rest and top most layer moving at a constant speed 'U', while layers should have no other gradient fields, such as concentration or temperature. For simplifying the condition, we must also assume that the plates (interface of fluid and surface) to be large enough; and the reason is that we want to omit boundary or edge effects where fluids are in contact with the surface. ***However, boundary or edge effects will have to be incorporated in theory in microfluidics***.

Therefore, when top layer is moving slow, then ideally all the layer will be parallel to each other and speed of layers will be 'zero' in the bottom layer and maximum in the top layer. Here, each layer will oppose the forward motion of the layer above it and make the layer beneath it to move forward. In such conditions, an

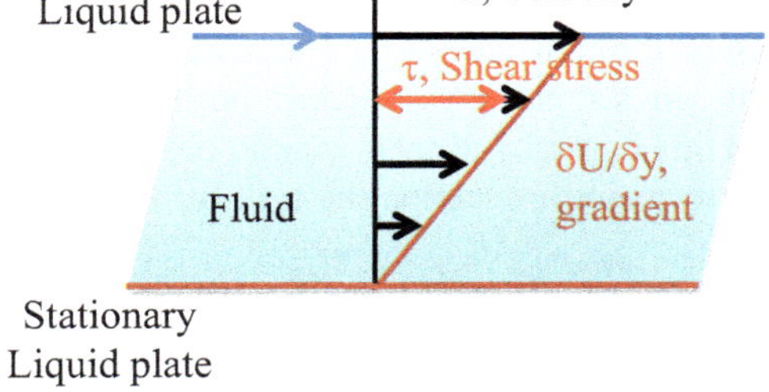

Fig. 1.4 Viscous force opposes the motion of the fluid layer moving faster than it and pushes the layer moving slower than it in its direction of motion. This introduces a resistance in fluids, which tends to exert a net opposite force resisting the fluid motion. This resistance is known as shear stress and is depicted as a function of change in velocity profile of the fluid layers moving from a static layer towards the fastest moving layer

external force will be required to overcome the fluid viscosity and keep it in motion. This force is Newton's viscous/shear force and is given by

$$F = \eta.A.U/y \tag{1.11}$$

where,

F is external force required to overcome viscosity, A is the area of each fluid layer plate, U is the velocity of each layer plate, and y is the separation between them.

U/y is rate of shear deformation of layers or shear velocity along the perpendicular to the fluid motion.

Shear stress (τ) can be written from (1.11) as

$$\tau = F/A = \eta.\partial U/\partial y = \eta.\gamma' \tag{1.12}$$

$$\text{or,}\ \eta = \tau/\gamma' \tag{1.13}$$

Kinematic viscosity (ν; m^2/s): It is the ratio of dynamic viscosity to the density of fluid

$$\nu = \eta/\rho \tag{1.14}$$

2.4.3 Types of Fluids Based on Intrinsic Properties

An in-depth knowledge of physical properties of fluids is crucial and it is the foremost thing one must know to design an efficient fluidics. Pumping is an integrated part in microfluidics. And, to effectively design the integrated pumping mechanism, knowledge of viscosity, fluid type, and fluidity becomes important. For example, a viscous fluid like honey will not flow easily through micron wide channels, but if the temperature inside channels is high then its viscosity will change making it less viscous and easy to flow. Else, an external pump will be required to force honey through channels. I first case temperature changed the fluid's viscosity while in second pressure has pushed it without affecting its viscosity. If we know this beforehand then we can design the tool to compensate these effects. We will focus in this section the type of fluids and their properties.

Newtonian vs. Non-newtonian: Case of Whole Blood Analysis in Microfluidics

According to Newton's law of viscosity, the fluid viscosity has proportionality with shear stress and shear rate, as depicted in (1.11) and (1.12). Based on this relation of viscosity (1.13), fluids can be categorized into two broad groups. The first group that has a constant viscosity for relation (1.13), and second that has changing viscosity with either of the two variables.

Fluids that has a constant η, in other words has a constant ratio of shear stress and shear rate. Such fluids are called ***Newtonian*** fluids. Water, honey, organic solvents are few examples of this type of fluids. Their viscosity only changes with temperature.

However, majority of the fluids in nature do not follow Newtonian fluid concept, thus called ***non-Newtonian***. Their viscosity changes with the change in shearing stress and shearing rate. This is why the viscosity plays crucial role in fluid properties. Now we qualify to classify fluids as Newtonian and non-Newtonian.

Let us consider the case of whole blood. Prior proceeding we must ponder to decide what type of fluid whole blood would be. Considering the composition of whole blood with ~40 % cellular material it can be classified as non-Newtonian. The reason is that if we shear the whole blood by increasing pressure the cell-fraction will not aggregate; thus, will change the blood viscosity making it less viscous. This is what we observe in systolic and diastolic blood where systolic blood is under high pressure flowing at high speed which makes it thin while diastolic blood under decreased pressure is thick where cells tend to come closer to each other and increase viscosity of the blood. On the other hand serum and plasma are Newtonian fluids and their viscosity in independent of the shear, where serum are plasma are essentially cell-free but serum is also free of clotting factors.

This knowledge of whole blood being non-Newtonian can be employed to make several kinds of microfluidic devices ranging from separating plasma to clustering cells.

Compressible and Incompressible Fluids

As the name indicates, the fluids that can be compressed into a smaller volume under an external pressure are called ***compressible fluids***. Typically, all the fluids are compressible where gases are highly compressible while liquids are slightly compressible. The fluid compressibility (β_C) is a measure of the relative change in volume due to a pressure change, and is expressed as

$$\beta_C = -1/\mathrm{V}\ (\partial \mathrm{V}/\partial \mathrm{P}) = -1/\rho\ (\partial \rho/\partial \mathrm{P}) \tag{1.15}$$

where, $\partial \mathrm{V}/\partial \mathrm{P}$ and $\partial \rho/\partial \mathrm{P}$ are change in the volume and density, respectively. V is the initial volume and ρ is the initial density.

On the contrary, if the fluid volume does not change under an external pressure, then it is considered to be ***incompressible***. There are literally no such examples of incompressibility. Incompressibility is used for the convenience of calculation purposes in fluid dynamics where an assumption is made that fluids with small or negligible compressibility are incompressible. It is important in microfluidics to assume so because then the density can be considered constant which significantly simplifies the calculation (see text box).

If the microfluidic channel is too long, there might be a huge pressure drop along the length, say 20 %. Then the density of the fluid at the inlet and outlet will be very different. This difference in density will affect the experimental composition and mathematical modelling of the experiments.

By now, we are able to understand several common terminologies that have been used while describing microfluidic systems. We will now discuss the primary laws that bind these concepts and terms together to create tools for understanding fluidics.

2.4.4 Other Important Properties

- Surface tension (σ; N/m; Kg/s^2): It is primarily a property of an interface, either liquid-air, or liquid–solid. It is the elasticity of a fluid surface to acquire minimum possible surface area. The plausible reason for surface tension is attributed to the unequal distribution of cohesion on the surface molecules due to which they continuously feels an inward pull toward the center of mass (Fig. 1.5).

The amount of surface tension (σ) is given by the force 'F' required to oppose the net inward cohesive force experienced by the top layer of length 'L', such that the top layer stop to sink toward bottom (Fig. 1.5a)

$$\sigma = F/2L \tag{1.16}$$

where, ½ is introduced in the eq. to equate the force that is acting only on one side of the surface.

Capillarity is an effect of surface tension at a solid–liquid interface, such that the liquid tends to rise in a **tube with small radius (small Bond number)** due to interplay between cohesion and adhesion between liquid molecules at liquid-air and liquid–solid interface.

The height to which a liquid will go up is expressed as

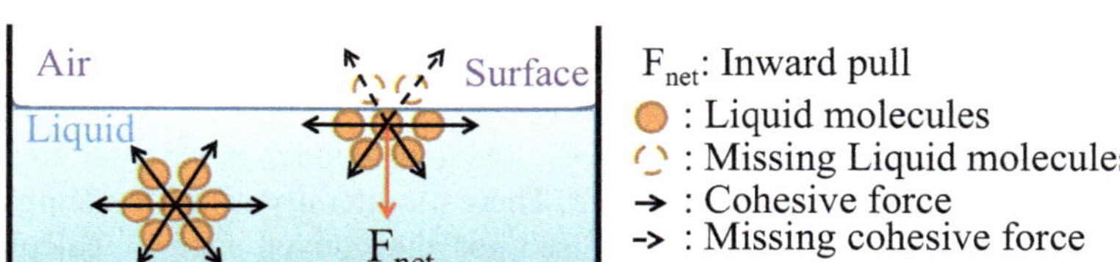

Fig. 1.5 Illustration of the tension on the surface of a liquid. On the surface layer, cohesion of liquid molecules causes a net inward pull due to which the surface behaves as a continuous film. Contrary to the surface, cohesion is cancelled by the neighboring molecules

$$h = 2\sigma Cos\Theta/\rho gr \quad (1.17)$$

where, Θ is the contact angle, σ is surface tension, ρ is density of the liquid, and r is the radius of the tube. This forms the basis of capillary pumping.

Importance of Surface Tension in Microfluidics

- Capillary pumping
- Droplet formation from a stream
- Contact angle determination
- Bubble generation for mixing

Capillary pumping: Capillary effect is employed regularly in microfluidics for removing physical pumps to minimize the bulky features. Capillary-driven pumps operate under ***Young-Laplace law*** defining the relation of difference in pressure at the interface of two fluids due to surface tension to the curvature in the surface of the liquid. This partial differential equation of Young-Laplace is expressed as

$$\Delta P = \sigma[1/r_1 + 1/r_2] = 2\sigma/r \text{ (if } r_1 = r_2) \quad (1.18)$$

where, ΔP is capillary pressure in a tube, σ is surface tension, r_1 and r_2 are the principle radii of curvature for internal and external surfaces at the interface/meniscus, and r is the radius of curvature. If r_1 and r_2 are equal then the equation reads as on extreme right.

Now, the actual radius of the tube is related to the meniscus radius by a cosine relation, such that $r = RCos\Theta$ then the (1.18) will read as

$$\Delta P = 2\sigma/RCos\Theta \quad (1.19)$$

Critical Thinking

Ignore the surface wettability for an instance. A single microfluidic channel opened at both the ends. Two drops of water were placed on both ends, such that one drop is smaller than the other drop. What should be the direction of flow?

: σ (tension) for both the liquids given the interface is same. Since r is smaller for small drop therefore, from (1.19) the capillary pressure will be more. Thus, water will move from small drop towards big drop.

In order to compensate for this pressure difference the liquid will move a distance thus giving rise to capillary pumping. The capillary pressure is crucial in designing an efficient pump because it requires a precise knowledge of the surface wettability, wetting phase and non-wetting phase. This relation is

$$P_c = P_{\text{non-wetting phase}} - P_{\text{wetting phase}} \tag{1.20}$$

Such that non-wetting phase in a typical experiment is air while water or buffer serves as a wetting phase. This is only true when surface is water-wettable, viz. hydrophilic. For hydrophobic surfaces, hydrophobic solvents serve as the wetting phase.

More regarding capillary pumping in paper microfluidics will be discussed in Chap. 2.

2.4.5 Laws Governing Dynamics

Basic Law

To understand the concept, we must first understand it intuitively what the governing principle to this branch of science is. In a general sense, laws governing fluid mechanics can be stated as ***the absence of relativistic effects for the conservation of mass, energy, and momentum***. In this process we must first (1) **identify** a system, (2) **identify** boundary of that system, (3) **identify** surroundings of the system, and (4) **identify** how it interacts with the surroundings. As described in Fig. 1.6, if '*A*' depicts mass, momentum or energy then the influx of any of these entities in the system should be equated with the efflux of equal amount of that respective entity.

This indicates that the total of any of these entities for the system will be a constant and can be written as

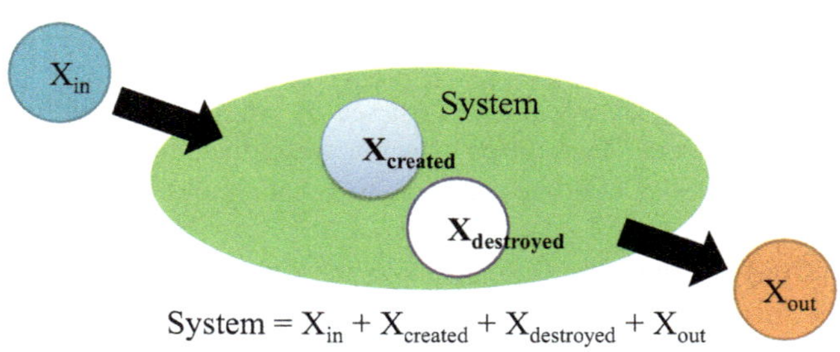

Fig. 1.6 Fundamental to understand fluid mechanics is conservation of fundamental physical components that are mass, momentum, and energy. As a general rule, system has its own mass, energy, and momentum and it has a tendency to conserve that. Any external force exerted to that body then the body should bring change in some form to conserve its mass, energy, and momentum. This is illustrated as system is equal to the total sum of the quantity of all the physical entities applied to it, applied by it, created by it and destroyed by it

$$A_{\text{in}} - A_{\text{out}} = A_{\text{system}} \tag{1.21}$$

Expression (1.21) stands true only if the extensive properties participating in conservation laws are considered, such as mass, volume, length etc. but not for intensive properties, which include pressure or temperature. The extensive properties are those which changes with the change in the amount of fluid; while intensive properties are those that do not change when an amount of fluid changes in the system.

Equations of Motion

A fluid in motion experience following forces

- Fg—gravity force
- Fp—pressure force
- Fv—Force due to viscosity
- Ft—Force due to turbulence
- Fc—Force due to compressibility

Therefore, according to the Newton's 2nd law, the net force (Fx) on a fluid with mass 'm' and acceleration 'a' in x-direction can be expressed as

$$\text{Fx} = \text{Fg} + \text{Fp} + \text{Fv} + \text{Ft} + \text{Fc} \tag{1.22}$$

Recalling the previous assumptions we discussed in fluid compressibility, the liquids with limited compressibility are considered incompressible and Fc becomes negligible.

Rewriting (1.22) with Fc = 0

$$\text{Fx} = \text{Fg} + \text{Fp} + \text{Fv} + \text{Ft} \tag{1.23}$$

Equation (1.23) is called ***Reynold's equation of motion***.

If the system is not turbulent then Ft is negligible and the resultant eq is known as ***Navier–Stokes equation of motion*** and is expressed as

$$\text{Fx} = \text{Fg} + \text{Fp} + \text{Fv} + \text{Fc} \tag{1.24}$$

And can be written as,

$$\text{P}(\partial\text{U}/\partial\text{t} + \text{U}. = \Delta\text{U}) = \Delta.\sigma = -\Delta\text{P} + \eta\Delta^2\text{U} + \text{f} \tag{1.25}$$

For real fluids with negligible viscosity, Fv is 0, the eq. is known as ***Euler's equation of motion*** and is expressed as

$$Fx = Fg + Fp + Ft + Fc \tag{1.26}$$

Conservation of Mass

It can be summarized as a time-dependent mass change over a defined fluid boundary such that mass within that boundary is constant

$$\text{Final mass} = \text{Original mass} + \text{Mass added} - \text{Mass removed} \tag{1.27}$$

or

$$\text{Final mass} - \text{Original mass} = \text{Mass added} - \text{Mass removed} \tag{1.28}$$

Equation (1.28) forms the basis of mass conservation of fluids in microfluidic systems, and can mathematically be written as

$$\text{Rate of change of mass} = \text{Net mass influx} \tag{1.29}$$

or

$$\Delta M/\Delta t = -\ \Delta I_m(\text{mass flux}) \tag{1.30}$$

Left part of (1.30) can be written in terms of ***extensive intrinsic properties, such as density and volume*** (refer back to the types of fluid section to know why intrinsic properties are used and basic conservation law to know why extensive properties are employed)

$$\Delta M/\Delta t = \partial(M)/\partial t = \partial(\rho \partial V)/\partial t = \partial/\partial t\left[\int_V \rho.\partial V\right] \tag{1.31}$$

where, ΔM is the change in mass, Δt is time interval of the mass change, differential $\partial(M)/\partial t$ is rate of change of mass, $\rho\partial V$ is the mass change in terms of changing volume

Similarly, right part of (1.30) can be written as

$$I_m = \Delta m.\Delta A \tag{1.32}$$

where, Δm is mass flowing normal to an area ΔA.

Equation (1.32) can further be expressed in terms of extensive intrinsic properties as

$$\Delta m.\Delta A = \int_s \rho.U.\partial S \tag{1.33}$$

where, U is mass flow velocity S is the surface area of the boundary region.

Now, by replacing (1.31) and (1.33) in (1.30), we will have the mass conservation equation for fluids

$$\partial/\partial t\left[\int_V \rho.\partial V\right] = -\int_S \rho.U.\partial S \tag{1.34}$$

Conservation of Linear Momentum/Inertia

It can be defined as the net momentum in a given volume at a given time is constant.

Newton's second law describes the relation of force and momentum with the expression

$$F = m.a = m.\ \partial U/\partial t = \partial(mU)/\partial t = \partial p/\partial t\ = U.\ \partial m/\partial t \tag{1.35}$$

where, m is mass of the fluid in a given area, p is momentum, U is the flow velocity, a is acceleration, $\partial U/\partial t$ is velocity rate, $\partial m/\partial t$ is mass flow rate. U. $\partial m/\partial t$ is known as ***momentum flow***.

Momentum flow can be written in terms of extensive properties

$$\partial p/\partial t\ = U.\ \partial m/\partial t = U.I_m = U.(mA) = U.\ (\rho VA) \tag{1.36}$$

Now, for momentum on this given mass of fluid to be constant,

$$\text{External forces}(F) = \text{Momentum flow rate} + \text{Momentum out} - \text{Momentum in} \tag{1.37}$$

The external forces acting on the fluid in a defined boundary are **body force** (force due to gravity) and **surface forces** (pressure, viscosity)

Thus conservation (1.37) will become

$$Fg + Fv + Fp = \text{Momentum flow rate} + \text{Momentum out} - \text{Momentum in} \tag{1.38}$$

Substituting respective values will give us the conservation of momentum equation

$$\rho g + Fv + \partial P/\partial L = \rho U.\partial V/\partial t + \partial(\rho U)/\partial t \tag{1.39}$$

Conservation of Energy

It is stated as energy within a system remains constant such that energy acting upon the body is continuously changed to other form, such as work. For fluids it is a very complicated equation that considers several forms of energy acting and dissipating out of a defined body. In its simplest form the law can be written as

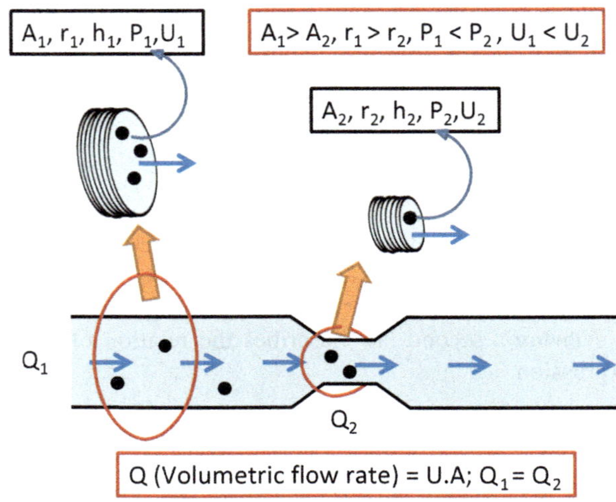

Fig. 1.7 According to conservation law, volumetric flow rate in the pipe sections depicted as *red circles* must remain same. However, channel diameters, cross-section areas, and flow velocities of the highlighted sections different. Thus, to satisfy $Q_1 = Q_2$, flow velocity of the smaller diameter > the flow velocity of higher diameter

$$\Delta E = \text{constant} \tag{1.40}$$

The simplest example of energy conservation is Bernoulli's equation where fluid flowing through a pipe having different radius as depicted in Fig. 1.7.

Bernoulli's equation for per unit volume for a given area is written as

$$\text{Pressure energy} + \text{kinetic energy} + \text{potential energy} = \text{constant} \tag{1.41}$$

$$\mathrm{P} + \left(\mathrm{mU}^2\right)/2 + \mathrm{mgh} = \mathrm{P} + \left(\rho\mathrm{U}^2\right)/2 + \rho\mathrm{gh} = \text{constant} \tag{1.42}$$

Now, from Fig. 1.7, at two different points with different cross-section areas, Equation (1.42) can be written as

$$\mathrm{P}_1 + \left(\rho\mathrm{U}_1^2\right)/2 + \rho\mathrm{gh}_1 = \mathrm{P}_2 + \left(\rho\mathrm{U}_2^2\right)/2 + \rho\mathrm{gh}_2 \tag{1.43}$$

Important Concepts

- **Stokes law: Friction and Drag on spherical particles**

Frictional force due to viscosity, which is also known as ***Stokes drag*** is given by

$$\mathrm{F}_\mathrm{d} = 6\Pi\eta\mathrm{rU} \tag{1.44}$$

where,

F_d is stokes drag, η is dynamic viscosity, r is hydrodynamic radius of the spherical particle, U is the flow velocity around the particle. 6Πηr is together is called **drag coefficient ζ**.

Applications of Stokes Law

- Hydrodynamic separation of cells
- Viscous force calculation
- Calculating shear on cells

This is important because mostly biomolecules and cells are approximately spherical. Thus Stokes law can be employed with approximation.

The **viscous force** experienced by each spherical particle is given by

$$Fv_Z = 3\eta U/2r \tag{1.45}$$

The **gravitational force** experienced by a spherical particle falling in a liquid

$$Fg = 4[(\rho_s - \rho_f)g\Pi r^3]/3 \tag{1.46}$$

where, ρ_s is particle density, ρ_f is fluid density, g is gravity constant, r is radius of particle

Terminal velocity of a spherical particle falling in a liquid under gravity

$$U_{ter} = 2[(\rho_s - \rho_f)gr^2]/9\eta \tag{1.47}$$

Stokes-Einstein law relates kinetics to Stokes law for understanding **Diffusion**

$$D = k_B T/6\Pi\eta r \tag{1.48}$$

where, D is diffusion constant, k_B is Boltzmann constant, T is temperature of the system

Drag force on a particle completely enclosed in fluid is expressed as

$$F_D = \rho U^2 C_D A/2 \tag{1.49}$$

where, F_D is drag force, C_D is drag coefficient, A is the area of reference, U is flow velocity, and ρ is fluid density.

Equation (1.49) can be rewritten as

$$F_D = \rho U^2 CAf(Re)/2 \tag{1.50}$$

where, f(Re) is function operator for Reynolds number

- **Poiseuille principle: Volumetric flow rate and pressure drop**

It describes the relation of pressure drop in a moving fluid enclosed within a tube with the flow resistance and flow rate (Table 1.6). It is also known as **Hagen-Poiseuille law** (Fig. 1.8).

Table 1.6 Volumetric flow rates for common geometries

Shape of cross-section	Volumetric flow rate (Q)	Fabrication approach
Cylindrical	$\Pi r^4 \Delta P/8\eta L$	Isotropic wet etching, Ball-end milling
Rectangular	$\Delta P w/[8(A+1)^2/A]\ \eta$	Photolithogrgaphy
Triangular	$\Delta P(3a^4)^{1/2}/320\eta l$	Anisotropic wet etching

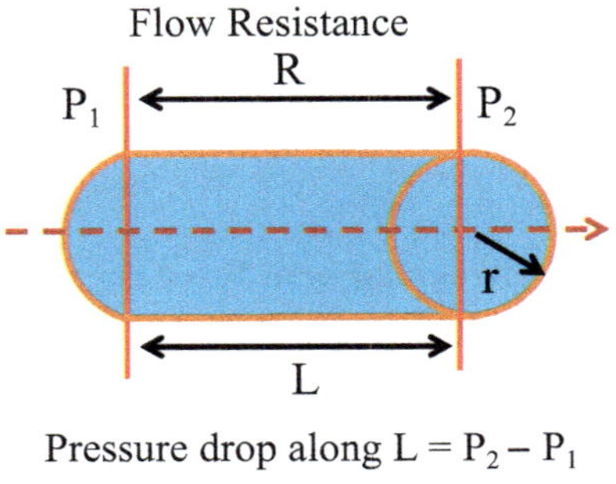

Fig. 1.8 Illustration of the pressure drop 'ΔP' along the channel length 'L' due to the flow resistance offered by the fluid to its motion from point 1 to point 2

For circular channels the relation is expressed as

$$R = (P_1 - P_2)/Q = \Delta P/Q = 8\eta l/\Pi r^4 \tag{1.51}$$

The (1.51) can be written as

$$\Delta P = RQ = 8\eta l Q/\Pi r^4 = 32\eta l U/r^2 \tag{1.52}$$

where, ΔP is pressure drop, R is the fluidic resistance, and Q is the volumetric flow rate through a cross-section area 'A' (Q = U.A)

For rectangular channels the relation in (1.51) is modified to

$$R_{hd} = C_{nc}\eta l/A^2 \tag{1.53}$$

where, C_{nc} is numerical coefficient and is given as

$$C_{nc} = 8(A+1)^2/A \tag{1.54}$$

where, *A* is aspect ratio = height of the channel (h)/width of the channel (w)

Replacing (1.53) and (1.54) in (1.52)

$$\Delta P = R_{hd}Q = C_{nc}\eta l Q/A^2 = \left[8(A+1)^2/A\right]\eta l Q/A^2 \tag{1.55}$$

Volumetric flow rate can be expressed in terms of mass flow rate with the relation

$$Q = I_m/\rho \tag{1.56}$$

Replacing (1.56) in (1.55) gives

$$\Delta P = R_{hd}I_m/\rho = C_{nc}\eta lI_m/A^2.\rho = (\eta/\rho).\left[C_{nc}lI_m/A^2\right] = \nu.\, C_{nc}l\ I_m/A^2 \tag{1.57}$$

where, I_m is mass flow rate, ν is kinematic viscosity

For circular pipes, (1.57) can be written as

$$\Delta P = RQ = 8\nu l/\Pi r^4 \tag{1.58}$$

An extension to Hagen-Poiseuille law is **Darcy–Weisbach equation**

Darcy–Weisbach equation relates head loss or pressure loss due to friction along a given circular channel and is expressed as

Pressure loss form:

$$\Delta P \text{ (pressure loss)} = f_D l\rho V^2/2D \tag{1.59}$$

Head loss form:

Replacing (1.6) in (1.59)

$$\rho g\Delta h = f_D l\rho V^2/2D \tag{1.60}$$

$$\Delta h \text{ (head loss)} = f_D lV^2/2gD \tag{1.61}$$

where, f_D is Darcy friction factor from channel wall which is

$$f_D = 64/Re \tag{1.62}$$

Fanning equation relates the ratio of local shear stress to the local fluid kinetic energy and is expressed as

$$f = \tau/\text{Kinetic Energy} = 2\tau/\rho U^2 = 16/Re \tag{1.63}$$

where,

f is fanning friction factor, τ is shear stress, Re is Reynolds number.

Pressure loss form:

$$\Delta P = \tau A/A' \tag{1.64}$$

where, A is wall area ($= 2\Pi\, rl$) and A' is cross-sectional flow area ($= \Pi r^2$), r is radius of the pipe, l is flow length.

Replacing (1.63) in (1.64)

$$\Delta P = f\rho U^2 A/2A' = f\rho U^2.\left(2\Pi\, rl/2\Pi r^2\right) = f\rho U^2.(l/r) \tag{1.65}$$

Head loss form:

Replacing (1.6) in (1.63)

$$\Delta h = f\rho U^2.\,(l/rg\rho) = f\,U^2 l/rg \tag{1.66}$$

- **Coriolis effect: Inertial frame and particle motion**

It is inertial force acting upon bodies relative to a rotating reference frame. For example, if a particle rolls on a static disc as illustrated in Fig. 1.9, by the virtue of inertia, it appears to move in a straight line to the observer in the same frame of reference. When the disc starts to rotate then the particle is still moving in the straight line if observed by someone standing in an inertial frame of reference outside of the rotating disc. However, if the observer is standing on the rotating disc in the non-inertial reference frame, then the particle will look like following a curved path, such that the particle is resisting in the change of its final destination by the virtue of inertia. Thus we can say that ***the Coriolis effect is in contrast to the normal inertia which resists the change in body's motion, whereas in this effect body resists the change in displacement***. It is crucial in inertial microfluidics where plasma can be separated from whole blood and cells of different sizes can be separated from each other. The direction of fluids in specific channels in centrifugal microfluidics as a function of inertial forces and Coriolis effect can also be achieved.

The Coriolis effect can be expressed

$$F_c = m.a_c \tag{1.67}$$

where,

Fc is Coriolis force, m is mass of the fluidic plug or particle, a_c is angular acceleration.

Since,

$$a_c = -2U\omega \tag{1.68}$$

where,

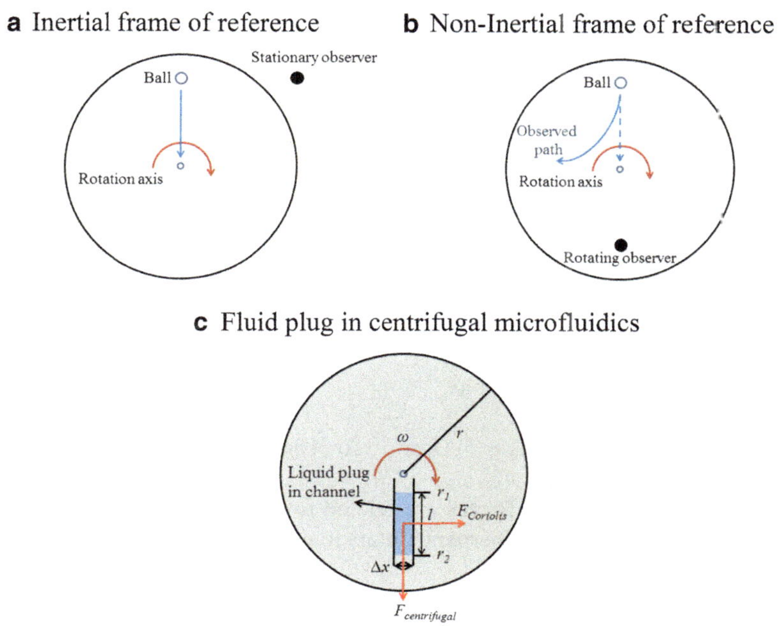

Fig. 1.9 Explanation of the Coriolis effect. (**a**) In an inertial frame of reference the observer is out of the rotating disk. The stationary observer will see the rolled ball following a straight path. (**b**) However, when observer stands on the same rotating disk on which the ball was rolled then to this rotating observer ball will seem to follow a curved path outwards. This perspective of curving of the path of ball is Coriolis effect. (**c**) This effect is used in rotating microfluidics for separating particles in a fluid plug of length 'l'. In this plug, the particle will experience an outward centrifugal force normal to the rotation axis and an outward force normal but opposite to the direction of the rotation, as depicted in '(**c**)'. Due to this effect particles will move to the wall of the channel continuously pulled the disk boundary

U is linear velocity of the liquid plug in the channel/particle (velocity relative to the rotation speed), and ω is angular velocity,

Hence,

$$F_c = -2mU\omega = -2(\rho V)U\omega \tag{1.69}$$

where,

ρ is density and V is volume of the particle.

The force is also expressed as force density, as expressed below

$$f_c = F_c/V = -2\rho U\omega \tag{1.70}$$

In centrifugal microfluidics the velocity of the liquid plug or particle (U) in the channel depends on angular velocity (ω), radial location of the fluid reservoir, channel geometry, and fluid properties, such as dynamic viscosity (η) and density (ρ). This linear velocity component is given as

$$U = D_h^2\omega^2\rho r\Delta r/32\eta l_c \tag{1.71}$$

where, D_h is hydraulic diameter of the microchannel, r is average distance of the liquid from the center of rotation, Δr is radial extent of the liquid plug (how much it has moved from its initial position), and l_c is the plug length in the channel.

Replacing (1.71) in (1.69),

$$F_c = -2(\rho V)\omega(D_h^2\omega^2\rho r\Delta r/32\eta l_c) = -D_h^2\omega^3\rho^2 r\Delta rV/16\eta l_c \tag{1.72}$$

The (1.70) can now be written as,

$$f_c = -D_h^2\omega^3\rho^2 r\Delta r/16\eta l_c \tag{1.73}$$

Centrifugal and Coriolis forces are related to each other in a sense that they operate together but normally (perpendicular) to each other as depicted in Fig. 1.9.

Centrifugal force is given by

$$F_\omega = -m\omega^2 r = -(\rho V)\omega^2 r \tag{1.74}$$

Where, r is radius of rotation.

The centrifugal force density is given by

$$f_\omega = F_\omega/V = -\rho\omega^2 r \tag{1.75}$$

Now, finding ration of (1.69) and (1.71) will give us relative effect of both the forces acting upon the particle in rotatory frame.

$$F_c/F_\omega = 2U/\omega r \tag{1.76}$$

Inertia circle is the path that moving body in a rotating reference frame will follow. The radius of this circle (r_c) and the time required to travel the edge of the frame (t_c) is given by

$$r_c = U/2\omega, \text{ and } t_c = \Pi/\omega \tag{1.77}$$

Rossby Number—length scales and Coriolis effect: It is the ratio of inertial and Coriolis forces. We can determine the effect of length scale on the efficiency of rotation in achieving Coriolis effects. The relation is expressed as

$$Ro = U/fcL = D_h^2\omega^2\rho r\Delta r/32\eta\, fcLl_c \tag{1.78}$$

where,

Ro is Rossby number, U is the relative velocity of the particle, *fc* is Coriolis factor, and L is the length scale of the motion. Coriolis factor, *fc* is expressed as

$$fc = 2\omega \mathrm{Sin}\Theta \tag{1.79}$$

where, Θ is the angle of the body to the plane of the reference surface. In case of particles in centrifugal microfluidics, Θ will be 90° thus changing the Coriolis factor to 2ω.

- **Dean number: Flows in curved pipes**

It is defined as the product of Reynolds number and the square root of the curvature ratio.

$$\mathrm{De} = \mathrm{Re}.(\mathrm{d}/2\mathrm{r})^{1/2} = (\rho \mathrm{Vd}/\mu).(\mathrm{d}/2\mathrm{r})^{1/2} \tag{1.80}$$

where,

De is Deans number, Re is Reynolds number, r is curvature radius of the channel/ tube, d is travelled length of the liquid, and V is axial velocity.

2.5 *Key Dimensionless Numbers Explained*

2.5.1 Reynolds Number: Inertial Focusing to Separate Plasma from Whole Blood

The Reynolds number is one of the most crucial dimensionless numbers in fluid mechanics. However, when we discuss it with reference to microfluidics, its relevance is practically limited. The reason is that the fluids employed in microfluidics-related applications have small values for their respective Reynolds numbers that make the inertial effects irrelevant.

Still, importance of the Reynolds number can't be undermined. One best example to explain the importance of inertia in microfluidics is separation of plasma from whole blood. A straight channel, as illustrated in Fig. 1.10, is curved at one end. The liquid flowing through this channel will feel a sudden curve on its path. At the corner, liquid still tends to go straight due to which in the process of changing path, it loses momentum at the corner. In this case,

Time taken for this liquid to turn around the corner is expressed as

$$\mathrm{t_i} \sim \mathrm{w}/\mathrm{U_0} \tag{1.81}$$

where, t_i is turn time, w is width of the curve, U_0 is velocity of the fluid before turning.

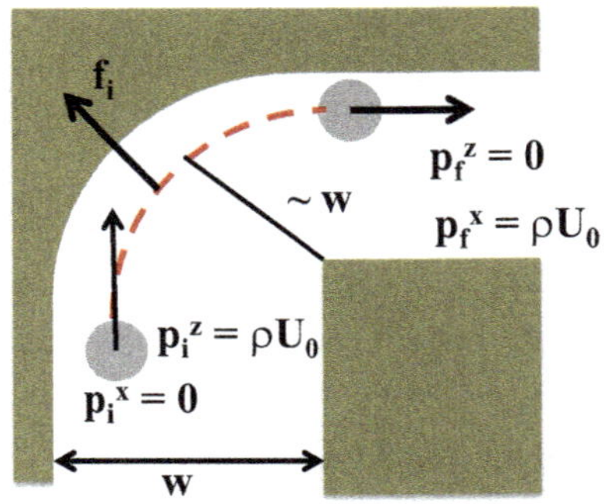

Fig. 1.10 Inertial separation of the particles moving with a velocity U_0 and following a curved path approximately equal to the width of the channel. Due to the conservation of momentum, particle will experience an outward push known as inertial centrifugal force with a density f_i as described in (1.84). P_i is inertial flow pressure where x represents in x direction and z represents in z direction. Initially, prior to the curve, along x axis, the momentum is 0 while all the momentum is focused along z axis. At the curve particle will lose z momentum that will translate to the x-momentum. During this transition, the particle will experience f_i normally outwards towards the putter wall of the microchannel

Now, the liquid during this time from (1.81) will lose a fraction of its linear momentum density, which can be given as

$$\mathrm{p_i} \sim \rho \mathrm{U_0} \tag{1.82}$$

The lost fraction of momentum ($\mathrm{p_d}$) in (1.82) will be transferred as a force, named inertial/centrifugal force, which is directed outwards in the same direction the liquid was initially flowing prior entering the curve. This inertial force density can be calculated as

$$\mathrm{f_i} \sim \mathrm{p_i}/\mathrm{t_i} \tag{1.83}$$

By replacing (1.81) and (1.82) in (1.83) for $\mathrm{p_d}$ and $\mathrm{T_i}$ we will get

$$\mathrm{f_i} \sim \rho \mathrm{U_0}/\mathrm{ti} \sim \rho \mathrm{U_0^2}/\mathrm{w} \tag{1.84}$$

These three equations form the basis of particle separation in non-circulating fluidic chips.

2.5.2 Pėclet Number: Diffusivities Across Channel Width and No-Membrane Dynamic Filtering

In day-to-day life turbulent fluid mixing is crucial. To elaborate the time scale vs length scale in the absence of this mixing, let us consider that we are holding a cup

Table 1.7 Diffusivities of common biological elements

Biological element	Solute	Proteins	Virus	Bacterium	Mammalian Cell
Size	0.1 nm	5 nm	100 nm	1 μm	10 μm
Diffusivity (**μm²/s**)	2000	40	2	0.2	0.02

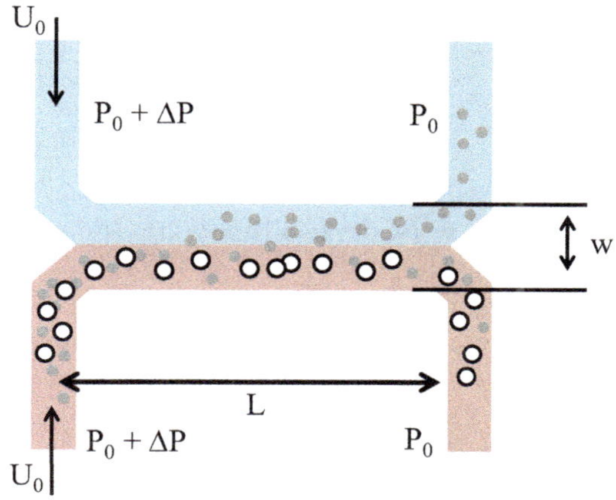

Fig. 1.11 Illustration of the working mechanism of a membrane-less dynamic particle separator. The working principle depends on the different diffusivities of different particles and the time and length scale required for diffusion in transverse direction according to their respective diffusivities. Two liquids enter through a T-junction and are left to diffuse over a length L in the arm of the H-filter. However, the liquids will never mix due to laminar regimen. A solution with different particles is injected through one channel. During the flow in the arm section, where both the fluids stay in contact, smaller particles with higher diffusivity and lower Pèclet number will move to the other layer and thus separate from the rest of the initial solution

of coffee; in this case we will not be able to smell the aroma of coffee held in our hands for next several days. If real world fluids were low Reynolds number then due to laminarity in flow, diffusion will be the only means of mixing, as we have already seen the relation between time and length scales. On the contrary, in microfluidics, due to low-Reynolds number regimen, mixing is predominantly by diffusion; this could be lengthy on a time scale. The common diffusivities for few biological moieties are enlisted in Table 1.7. Now, there are several applications requiring rapid mixing, and this is where Pèclet number helps us understanding the extent of mixing in our micro-devices.

Let us discuss the case of the 'H' filter. It is named so due to its geometrical appearance (Fig. 1.11). Left T-junction is used as inlets keeping those at pressure $P_0 + \Delta P$; while T-junction on the right is used as outlet keeping it at pressure P_0. The arm of the 'H' is the central channel where diffusion takes place. The H-filter works on the basis of diffusion, which is the only mean of transverse movement of particles between two parallel moving fluid layers.

According to the mass transport and Fick's law,

Table 1.8 Conditions for particle separation in no-membrane H-filter

	t_{conv} (time of particle convection)	t_{diff} (time of particle diffusion)	Pe_{Mass} (Pėclet number)
	L/U_0	$w^2/4D$	$RePr = U_0L/D$
For diffusion	$t_{conv} \leq t_{diff}$		Low Pe,
For no diffusion	$t_{conv} >> t_{diff}$		High Pe
			Either longer channel length or slower flow velocity or both are required to achieve diffusion

$$D(\text{diffusion coefficient}) = w^2(\text{width to diffuse})/2t_{diff}(\text{time to diffuse}) \quad (1.85)$$

The time to diffuse is dependent on the particle velocity in the fluid (U_0) and the length (L) it will travel before diffusing completely to the adjacent stream. Thus, eq. (1.85) can be written as

$$D = w^2U_0/2L \quad (1.86)$$

However, the particle has to diffuse through only half of the total width of the H-connecting arm because the other half is filled with different liquid. Due to this w becomes w/2 and eq. (1.86) will be

$$D = w^2U_0/4L \quad (1.87)$$

Also, particles in liquid flow experience convection along with the diffusion. The time of particle convection become important as referred in Table 1.8. Thus the particle diffusivity becomes crucial as well. Typical diffusivities for common biological moieties are mentioned in Table 1.7.

As mentioned in Table 1.8, for particle to diffuse into the other half across the channel width filled with different liquid, then the prerequisite must be the diffusivity of the particle should be high; the aspect ratio of channel length to its width should be smaller along with flow velocity. This all will result in low Pėclet number and smaller diffusion time. The opposite will be true for a particle not to diffuse into the adjacent fluid across its width.

3 Conclusion

This chapter provides basic information, such as basic principles and related theory, for developing a fundamental background to understand the so-looking complex physics of microfluidics. On the contrary, the foundation of fluid mechanics starts dating back to high school physics. In this chapter, we have brushed all those concepts we studied back then and added some advanced theoretical knowledge built upon that. In principle, this chapter covers everything that one should get acquainted to for understanding microfluidics. This will allow non-physicists and engineers to interface with microfluidic engineers and thus, can improve communication, which is the biggest challenge when non-physical scientists and doctors interact with engineers.

Further Reading

Recommended reviews and articles

1. Squires TM, Quake SR (2005) Microfluidics: fluid physics at nanoliter scale. Rev Mod Phys 77 (3):977. doi:10.1103/RevModPhys.77.977
2. Sackmann EK, Fulton AL, Beebe DJ (2014) The present and future role of microfluidics in biomedical research. Nature 507:181. doi:10.1038/nature13118
3. Beebe DJ, Mensing GA, Walker GM (2002) Physics and applications of microfluidics in biology. Annu Rev Biomed Eng 4:261. doi:10.1146/annurev.bioeng.4.112601.125916
4. Duncombe TA, Tentori AM, Herr AE (2015) Microfluidics: Reframing biological enquiry. Nat Rev Mol Cell Biol 16:554–567. doi:10.1038/nrm4041
5. Burger R, Kirby D, Glynn M, Nwankire C, O'Sullivan M, Siegrist J, Kinahan D, Aguirre G, Kijanka G, Gorkin RA, Ducree J (2012) Centrifugal microfluidics for cell analysis. Curr Opin Chem Biol 16(3–4):409. doi:10.1016/j.cbpa.2012.06.002
6. Di Carlo D, Irimia D, Tompkins RG, Toner M (2007) Continuous inertial focusing, ordering, and separation of particles in microchannels. PNAS 104(48):18892. doi:10.1073/pnas.0704958104
7. Gorkin R, Park J, Siegrist J, Amasia M, Lee BS, Park JM, Kim J, Madou M, Cho YK (2010) Centrifugal microfluidics for biomedical applications. Lab Chip 10(14):1758. doi:10.1039/b924109d
8. Zhang J, Yan S, Yuan D, Alici G, Nguyen N-T, Warkiani ME, Li W (2016) Fundamentals and applications of inertial microfluidics: a review. Lab Chip 16:10. doi:10.1039/C5LC01159K
9. Martel JM, Toner M (2014) Inertial focusing in microfluidics. Annu Rev Biomed Eng 16:371. doi:10.1146/annurev-bioeng-121813-120704
10. Temiz Y, Lovchik RD, Kaigala G, Delamarche E (2015) Lab-on-a-Chip devices: how to close and plug lab? Microelectron Eng 132:156. doi:10.1016/j.mee.2014.10.013
11. Hou HW, Bhagat AAS, Lee WC, Huang S, Han J, Lim CT (2011) Microfluidic devices for blood fractionation. Micromachines 2(3):319. doi:10.3390/mi2030319
12. Ren Y, Leung WWF (2016) Numerical investigation of cell encapsulation for multiplexing diagnostics assays using novel centrifugal microfluidic emulsification and separation platform. Micromachines 7(2):17. doi:10.3390/mi7020017
13. Bhagat AAS, Kuntaegowdanahalli SS, Dionysiou DD, Papautsky I (2011) Spiral microfluidic nanoparticle separator. In: Proc SPIE 6886, Microfluidics, BioMEMS, and Medical Microsystems VI, 68860 M doi:10.1117/12.767350

Recommended Books and Lecture Notes

14. Bansal RK (2010) A textbook of fluid mechanics and hydraulic machines. Laxmi Publications, New Delhi, India [ISBN: 8131808157]
15. Kleinstreur C (2014) Microfluidics and nanofluidics: theory and selected applications. Wiley, Hoboken [ISBN: 978-0-470-61903-2]
16. Wilkis JO (2012) Fluid mechanics for chemical engineers. Prentice Hall International Series in Physical and Chemical Engineering Sciences. New Jersey [ISBN: 0–13–148212–2]
17. Bruus H (2014) Governing equations in microfluidics. In: Microscale acoustofluidics, pp 1–28. eISBN:978-1-84973-706-7. doi:10.1039/9781849737067-00001
18. Achim W (2015) A guided tour through buoyancy driven flows and mixing. Master. buoyancy driven flows and mixing, France, pp 66.<cel-01134112v2> https://tel.archives-ouvertes.fr/LEGI/cel-01134112v2
19. Bruus H (2006) Theoretical microfluidics. Lecture notes third edition. http://homes.nano.aau.dk/lg/Lab-on-Chip2008_files/HenrikBruus_Microfluidics%20lectures.pdf

Chapter 2
Microfluidics Overview

Geeta Bhatt, Sanjay Kumar, Poonam Sundriyal, Pulak Bhushan, Aviru Basu, Jitendra Singh, and Shantanu Bhattacharya

1 Introduction

Microelectromechanical systems and micro-fluidics are two fast emerging domains in diagnostics research. The Microsystems technology emerged as a fall out of the microelectronics industry mostly due to the obsoleteness of some of the microelectronic processes owing to integration density issues. The area was first widely explored in the mechanical and physical sensing domains and found wide interests primarily because of low overall size, high yields of production and ability to integrate with a variety of processes. The technology saw a turnaround towards chemical/biochemical sensing starting from the end of 80s as prompted by the fast molecular diagnostic requirements imposed by the gene sequencing industry fuelled by the Human Genome project. Microfluidics is mostly concerned with handling of miniscule samples of fluids of volume 10^{-9}–10^{-18} L which is well suited to the handling of different expensive analytes important for diagnostics

G. Bhatt • S. Kumar • P. Sundriyal • P. Bhushan • J. Singh
Microsystems Fabrication Laboratory, Kanpur, Uttar Pradesh, India

Department of Mechanical Engineering, Indian Institute of Technology, Kanpur, Uttar Pradesh, India

A. Basu
Microsystems Fabrication Laboratory, Kanpur, Uttar Pradesh, India

Design Programme, Indian Institute of Technology, Kanpur, Uttar Pradesh, India

S. Bhattacharya (✉)
Microsystems Fabrication Laboratory, Kanpur, Uttar Pradesh, India

Department of Mechanical Engineering, Indian Institute of Technology, Kanpur, Uttar Pradesh, India

Design Programme, Indian Institute of Technology, Kanpur, Uttar Pradesh, India
e-mail: bhattacs@iitk.ac.in

C.K. Dixit, A. Kaushik (eds.), *Microfluidics for Biologists*,
DOI 10.1007/978-3-319-40036-5_2

work. This technology has very prominent advantage with respect to low overall chip area and high integration density. For handling small volumes of fluid of the range indicated above various micro-channels and micro-confinements are devised using a variety of techniques in which the mixing, reacting, handling and transporting etc. take place. The main motivations of this field are powerful analytical and diagnostic techniques which have been parallel devised by chemists, biochemists and material scientists over the last couple of decades to understand the life processes for sustenance of life itself. These may include modern methods as used in chemical diagnostics like [1]:

1. Micro-analytical methods (Chemical analysis methods for higher sensitivity and higher resolution)
   (a) High-pressure liquid chromatography (HPLC)
   (b) Matrix assisted laser desorption/ionization time of flight methods (MALDI-TOF)
   (c) Capillary electro-osmosis and electrophoresis methods (CE)
2. Sensitive detection of for chemical and biological hazards which may have military connotations (being used as bio-warfare tools)
3. Molecular biology driven methods to recognize basic structures of biological entities deterministically which may include:
   (a) High throughput DNA sequencing
   (b) Causative genomics
   (c) Protein crystallography and folding
   (d) Immunological mechanisms through the study of binding chemistries of various pathogenic and non-pathogenic biological entities so on so forth

The fluid which is analysed in the microfluidics domain can be handled in various ways. Hence depending on this, microfluidics is classified in three different types, continuous flow microfluidics (in which there is continuous flow of fluid through the micro-channels), droplet based microfluidics (in which discrete manipulating volumes are formed in the immiscible phase) and digital microfluidics (in which discrete, independently controlled droplets are manipulated in the open environment i.e., on the substrate) [2]. Depending upon the requirements, various materials like silicon, glass and various elastomers (polydimethysiloxane (PDMS), SU8 (negative photoresist)) are used in microfluidics for making various micro-channels, micro-valves etc. Hence keeping in view the various requirements of the field of microfluidics, a large domain of researchers are involved in the field pertaining to the advanced applications of the field of Microfluidics.

In recent years, microfluidics has been an extensively explored domain owing to its high applicability to develop low-cost diagnostic devices. Clinical diagnostics is one of the promising application areas for deployment of such lab-on-chip systems also better known as point-of-care (POC) systems [3]. Lab-on-chip technology is preferred over conventional laboratory lab oriented techniques due to their faster performance and overall miniaturized size which leads to reduced use of analytes and thus promotes low cost clinical diagnostics which can be of immense utility in resource poor settings. Various advantages that such devices offer are reduced detection time, increased sensitivity, greater control of molecular interactions, cost efficiency, reduced chemical wastage, lesser human intervention etc. With a purpose of exploring various dimensions of microfluidics in clinical diagnostics, this chapter summarizes various aspects of the field of point of care diagnostic devices including their fabrication technologies like Laser micro-machining, lithography and MRDI process for making micro-channels/micro-valves; micro-fluidic systems and its various fluid handling modules like micro-mixers, micro-pump, micro-valve and micro-cantilevers etc. various applications of micro-fluidics like electrophoresis (gel electrophoresis, capillary electrophoresis and surface elec-trophoresis), dielectrophoresis, polymerase chain reaction (PCR) and gene delivery and further various sensing and detection techniques like electrochemical sensing, optical sensing, mass based sensing and surface plasmon resonance (SPR) sensing etc.

2 Basic Fabrication Techniques

In this section we would like to discuss the various fabrication techniques that are normally used for micro-fabrication of high aspect ratio micro-channels.

2.1 LASER

The term LASER is an acronym for Light Amplification by Stimulated Emission of Radiation. Laser devices produce intense light beams which are monochromatic, coherent, and highly collimated. The wavelength of laser light is monochromatic and all photons are coherent. Laser beams show very low divergence and can travel over great distances, can be focused to a very small spot with high intensity and find a variety of applications in different fields.

Atoms possess energies only in particular discreet energy levels although when in bulk there may be a bulk behaviour of the orbital energies. The electrons within these atoms are naturally present in their ground state and they go to higher energy levels when excited through light beams of an external source. This process is known as absorption. After a short duration of time is lapsed the electrons returns back to their initial ground states and in the process the atom emits a photon. This process is known as spontaneous emission. In a traditional light source both absorption and emission occur together. If an outside photon having precisely the amount of energy needed for spontaneous emission is struck on the excited atoms, this external photon is increased to two photons one provided by the excited atom. Both released photons have the exact same phase. This process is known as stimulated emission and it is a fundamental process for the operation of a laser

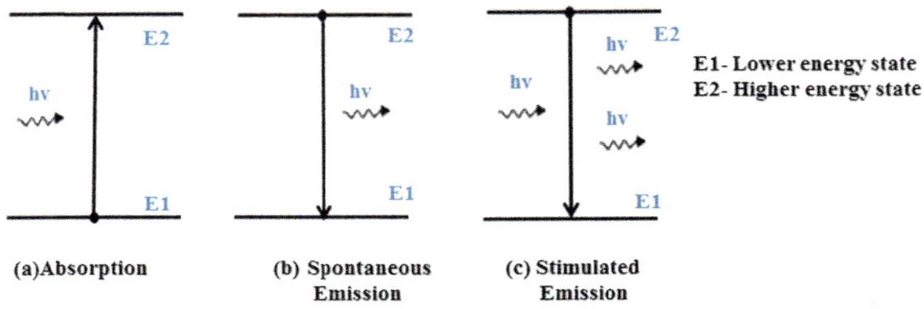

Fig. 2.1 Schematic of Laser process

source (Fig. 2.1). In this process, the key is the photon having exactly the same wavelength as that of the emitted light [4]. Application of laser machining for micro-channel fabrication and mask making is discussed in the following sections.

2.1.1 Application of Laser Machining for micro-channel making

Laser Photo-ablation was first introduced in 1997 by Roberts's group for making polymer microfluidic channels [5]. The integration of laser micro-beams with micro-fluidic devices is beneficial for rapid throughput fabrication strategies particularly for Lab on chip applications to achieve the manipulations of biological entities and biological fluids within micro-fluidic platforms [6, 7]. One specific advantage of pulsed laser micro-beam irradiation is that it does not require any specialized instrumentation except a XYZ stage which would follow commends from a drawing software. This reduces the design complexity and cost of the individual micro-devices tremendously while increasing the speed of their fabrication by several folds. So it is economical to dispose the devices so made after a single use. Secondly, the laser micro-beam can be positioned to any optically accessible location within the micro-device thus provides high flexibility in designs, enables potential parallelization of cellular analysis at multiple device locations etc. [7].

In laser machining processes, a high power laser is used to break the bonds of polymer molecules thereby removing the decomposed polymer parts from the region being ablated by the laser. Excimer laser has been successfully used for making micro-channels with 193 or 248 nm pulses with a pulsing frequency range of 10 to several kilo-hertzs. Micro-channels can be made by a maskless direct laser etching process or through lithographic patterning processes [8]. Laser etching is suitable for machining a wide range of polymeric materials including polymethyl methacrylate (PMMA) [9], polystyrene (PS), polycarbonate (PC), polyethylene terephthalate (PET), polyethylene terephthalate glycol (PETG), polyvinylchloride (PVC) and polyimide [10]. Surface chemistry can be modified due to the formation of reactive species during the laser ablation process which makes this process very attractive to biological assays.

Laser fabricated channels have high surface roughness than injection molded, imprinted or hot embossed channels although a newly intended hybrid machining strategies particularly on PMMA claims to have an average surface roughness of a few hundred nanometers [11]. Surface roughness depends on the absorption of the lasing frequency of the polymer. For example, PMMA channels made at 248 nm have high roughness and porosity [12]. Parameters to govern quality of the fabricated channels are laser power, scanning speed, polymer absorptivity, laser pulse rate and number of passes made to realize a complete channel.

2.1.2 Application of LASER Machining for Mask Making

Photo-masks are very important for miniaturization of devices. However, due to finite scanning speeds of laser pattern generators this process has an overall low-throughput. MEMS fabrication is typically a multilayer fabrication process when we consider devices and is also highly iterative thus needing multiple changes based on device performance so that the design can achieve finality. Hence precise mask fabrication is a critical step towards the MEMS grade precision and accuracy needs. For making such masks a MEMS laboratory needs a laser pattern generator for easy control on mask- fabrication process [13, 14]. Nonconventional machining has been widely used for micro-machining purposes but their use for mask making is not much explored [15]. Kumar et al. has shown how non-conventional manufacturing processes can be utilized towards the fabrication of small MEMS grade structure [16].

Laser machining is a highly localized and non-contact process to ablate micro-features and structures has three simple steps, viz., (a) interaction between the matter and beam, (b) absorption/heat conduction and an associated temperature rise, and (c) melting and vaporizing of the material. The various advantages as offered by laser micro-machining are the easy and precision control and rapid machining.

Figure 2.2 shows a one-step demagnification and laser ablation technique as applied to mask making as reported by Kumar et al. [16].The aluminium mask which is made with large sized features through electro-discharge machining processes is mounted on the excimer laser system and the shadow the mask is subsequently demagnified on a thin chrome film after proper alignment and focusing. An optimum solution can be extracted from the different machining parameters, including energy, pulse frequency, pulse duration, and pulse numbers, etc., so that edge roughness of fabricated features can be minimized. An energy optimization can also be performed by energy value calculation used for metal film ablation without affecting substrate. The minimum resolvable feature-size using this is roughly 10 μm [16]. The mask-making strategy with a combination of advanced machining technologies, easily available within an advanced machining laboratory, can be very helpful for iterative micro-systems designing.

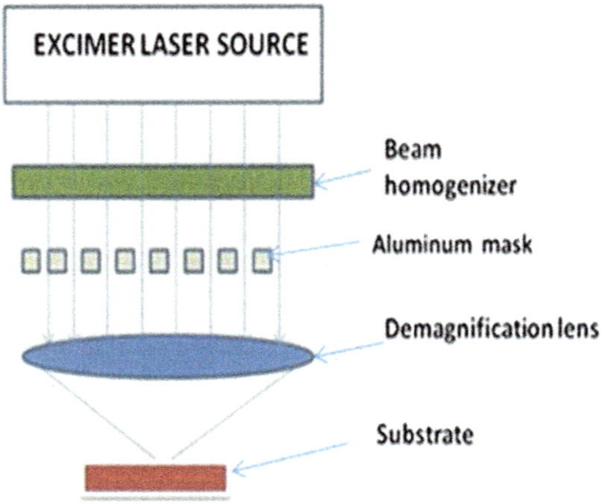

Fig. 2.2 Schematic of the demagnification lens placement after shadow mask [16]

2.2 *Photo-Lithography*

Photo lithography is a non-contact process which deploys the power of light exposure to print extremely small features (up to sub micron levels) into photo-chemicals and resists. Major steps in optical lithography are pattern transfer, alignment and exposure; which are explained in the forthcoming sections in details.

2.2.1 Pattern Transfer

In lithography processes a pattern is transferred to a photosensitive material by selective exposure to a radiation source (UV source in photolithography, X-Ray source in X-Ray lithography, electron beam in e-beam lithography etc.) (Fig. 2.3). Physical properties of the photosensitive chemical change when it is exposed to such radiation source. The changed properties of the photo-chemical render the exposed regions to be constitutionally different from the unexposed regions and this different created by light is utilized to print features and structures on the surface of the photochemical [17]. The changed properties are different with positive and negative tone photochemicals. In a positive tone resist the exposed regions are debonded and in the negative tone resist these regions are cross-bonded (Fig. 2.4a, b).

If the resist is exposed to a specific wavelength of light the chemical resistance of the resist to developer solution differs. If the resist is placed in a developer solution after selective exposure to a light source, one of the two regions will be etched (exposed or unexposed). If the exposed material is etched away by the developer, the material is positive resist as shown in Fig. 2.4a. If and the unexposed region is etched away, it is considered to be a negative resist as shown in Fig. 2.4b.

Fig. 2.3 Pattern transfer process

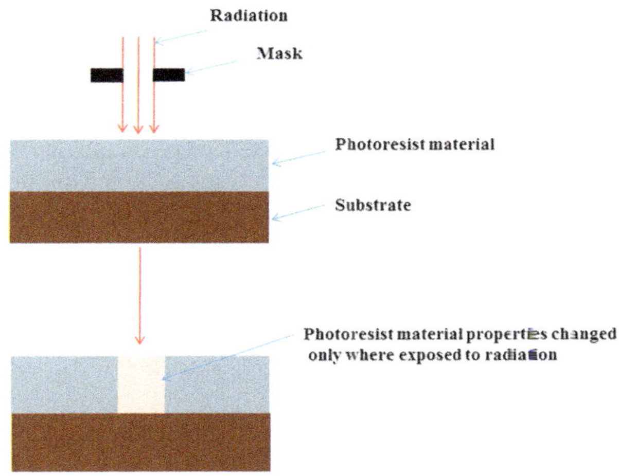

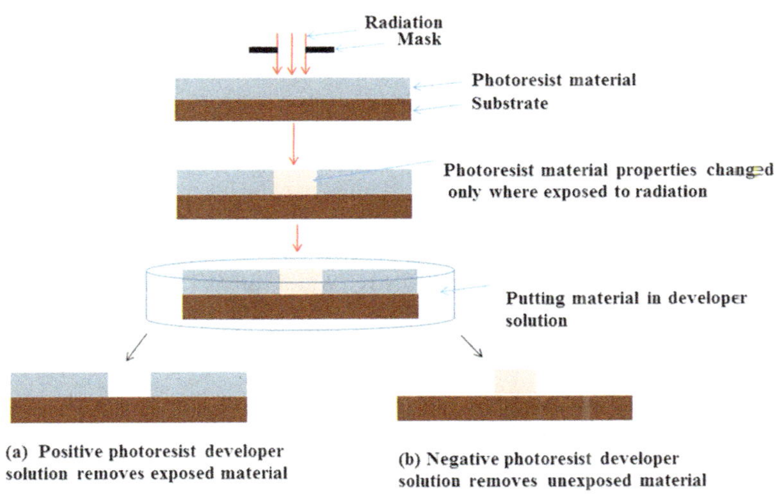

Fig. 2.4 (**a**) Pattern in positive photoresist, (**b**) Pattern in negative photoresist

2.2.2 Alignment

Pattern alignment plays a major role in photo-lithography for making controlled feature sizes and shapes. The pattern transferred to a wafer has a set of alignment marks which have highly precise features. These marks are used as reference for positioning other patterns at different layers with respect to a pattern of one layer each alignment mark should be labelled for easy identification, positioning and time saving [17]. Complex MEMS features are mostly multi-level and use multiple masks for different operations related to micromachining on a single chip platform.

2.2.3 Exposure

The main exposure parameters to achieve accurate pattern transfer from the mask to the photosensitive layer depend on radiation source wavelength and the exposure dose required achieving the desired property change of the photo-resist. Typical sources of UV light are mercury vapour lamps and excimer lasers. A chemical reaction takes place between the light and resist when UV light hits the resist. The mask unprotected areas undergo a chemical reaction [17]. Application of lithography is discussed in the next section.

2.3 *Soft Lithography*

Soft lithography covers a domain of processes which are based on non-photolithographic techniques principally based on replica moulding and self-assembly processes for micro and nanoscale fabrication. It is a convenient, effective and inexpensive method for making micro and nanostructures. Under the replica moulding process an elastomeric patterned stamp with relief structures on its surface is used to generate patterns with very small features (30–100 μm size) by a casting process using a soft polymeric material called PDMS. Replica Further the patterned polydimethylsiloxane (PDMS) stamps may be used to print molecules on surfaces with great precision and accuracy a process better known as micro-contact printing [18]. In the replica moulding process a silicon master is fabricated with patterned features composed of photoresists which are used for PDMS micro-channel fabrication. Silicon wafers are firstly coated with photoresist and undergo a photolithography step to generate a set of pattern on the surface of the wafer [19, 20].

Replica moulding with PDMS is carried out by pouring PDMS prepolymer (PDMS and curing agent) onto a pretreated (with mold release agent) and patterned Silicon wafer which acts as a mold. To obtain an enclosed micro-fluidic device, the PDMS replica is reversibly or irreversibly sealed to a variety of planar substrates like silicon and glass. The irreversible bonding of such multi wafer stacks can be carried out using a brief exposure of both bonding surfaces prior to bonding to Oxygen plasma [21]. Soft lithography using PDMS offers several advantages like low cost of fabrication, rapid processing, reusability of the SU-8 master for multiple runs, facile sealing and bonding to a number of different substrates particularly in case of reversible bonding, and multilayer fabrication to create complex three-dimensional systems etc.. PDMS has good optical transparency from 230 nm onwards and covers the whole visible spectrum. Its elastomeric properties allow for easy interconnections between macro to microfluidic platforms and the properties are particularly helpful in realizing on-chip fluid handling components like valves, pumps, mixers, reactors, sensors etc. [22].

2.4 Micro-Scale Replication by Double Inversion (MRDI) Process

Biological and chemical applications require micro-fluidic devices with micro/nanoscale mechanical structures. Present fabrication techniques suffer from a low pattern transfer quality particularly if the feature sizes are in the nanoscale during the simultaneous embossing of the microscale and nanoscale patterns into a thermoplastic polymeric substrate since the polymer flow becomes insufficient. Fabrication of 3-D micro and nano-structures require expensive and time-consuming lithography assisted techniques and in some cases very high cost is involved in realizing these mechanical structures through the use of processes like electron or ion beam writing or nanolithography etc. [23]. In order to address the problems of lithography driven processes we have developed a low cost and high throughput replication based process called micro replication by double inversion (MRDI).

Figure 2.5 shows schematic of a standard replication process. The various steps involved in such a process are (a) pouring the liquid monomer onto the master (b) master and replica separation after photo-polymerisation by exposure to UV-light (c) The use of the replica (with some surface pre-treatment) as a new master for a further duplication of the features into another polymeric substrate. After polymerisation and separation the second replica contains the same structures as the initial one. (Represented in Fig. 2.5a–d). The master in MRDI process can also be made using laser micromachining as shown in Fig. 2.6.

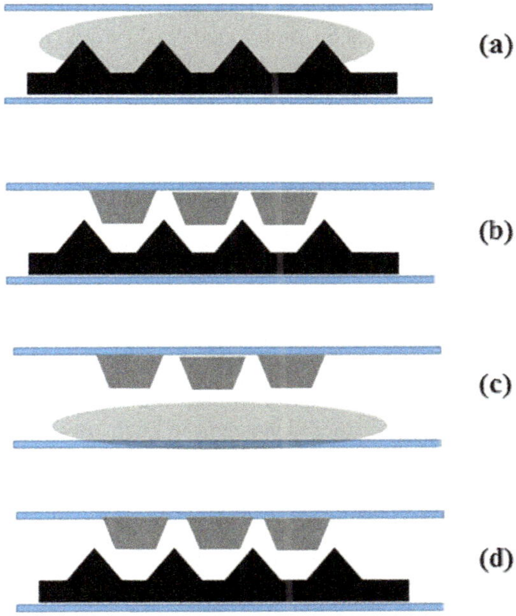

Fig. 2.5 Schematic representation of replication process

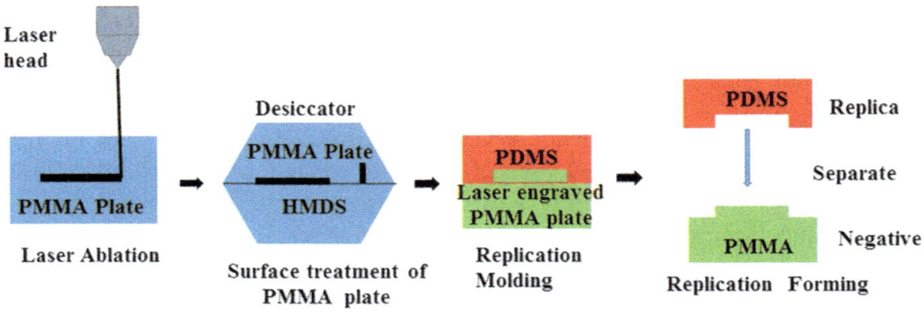

Fig. 2.6 Schematic representation of MRDI process

The principle advantages offered by the MRDI process are: (1) a sturdy mold, (2) repeated use of the mold for replication of polymer, (3) an overall inexpensive process as compared to other photolithography-driven processes,(4) A reusable mold, and (5) the use of laser ablation or other micromachining techniques for mold making.

PDMS replica mold is widely used for channel making in microfluidic applications although when we talk of embedded channels and features there are some very serious limitations of replication and moulding processes [24]. For example if the structure that is to be realized is a long micro-channel which is embedded within a chunk of the polymer PDMS both replication and moulding process done in the way as explained and also the MRDI may prove out to be failures and if done as a sandwiching between two replicated surface may impose limitations in terms of accurate alignment. In order to address these issues a replication and moulding technique has been developed with wires where the features and structures are replicated as embedded features within polymeric domains [25].

2.5 *Embedded Structures with Replication and Moulding Processes*

An easy fabrication procedure has been developed for three dimensional structures with soft materials like PDMS [26]. This has been used to develop micro-channel arrays within PDMS matrices which in a separate module as described below been tested for micro vibration control. An array of upto six rows of micro-channels with 20 numbers of micro-channels in each row have been fabricated using this procedure and in a very innovative manner The process was initiated with a micro drilling exercise wherein 200 μm diameter holes (20 in numbers) were drilled using MIKROTOOLS (DT 100, Singapore) machine tool in a simple plastic mold box using a CAD package on the side walls of this mold box. The centre to centre distance between these holes in x and y direction were taken as 2 mm. In order to

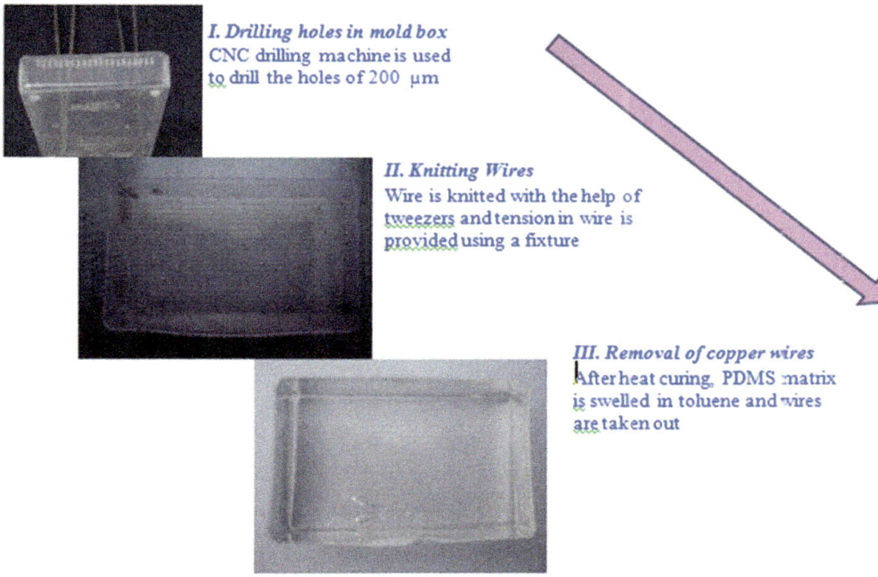

Fig. 2.7 Dimensional micro-channel arrays within PDMS blocks using thin circular copper wires. (Reproduced from Singh et al. [27] with permission from the Institute of Electrical and Electronics Engineers)

realize high aspect ratio micro-channels, thin wires of copper etched to a diameter of 50–80 μm were inserted between opposing holes in a mold box with a special tweezers and provided with finite tension using a special jig developed for this purpose. PDMS commercially available in a 10:1 ratio of a silicone rubber to curing agent was poured into the plastic box over the wires and replicated the wire array. This was followed by a release mechanism which comprised of a swelling step wherein the matrix with embedded wire array was taken out curing the PDMS and dipped in toluene for a finite amount of time. This swelling of PDMS matrix was performed mainly to release the grip over the embedded wires. This was followed by another shrinking step wherein the assembly was heat cured and the whole PDMS shrinks back to a smaller size. Figure 2.5 shows the process flow chart of this technique (Detailed in Fig. 2.7).

There has also been utilization of this wire based replication technique for the development of embedded fluid handling structures like microvalves etc. [28].

Figure 2.8 shows an embedded assembly of micro-valving as reported by Singh et al. In Fig. 2.8 there are two embedded channel in a piece of PDMS where the central channel shows a 80 μm channel and the solenoidal track around this central channel shows an embedded valve structure which can be inflated/deflated using compressed air so that it can squeeze the inner channel and act as a fluid valving device.

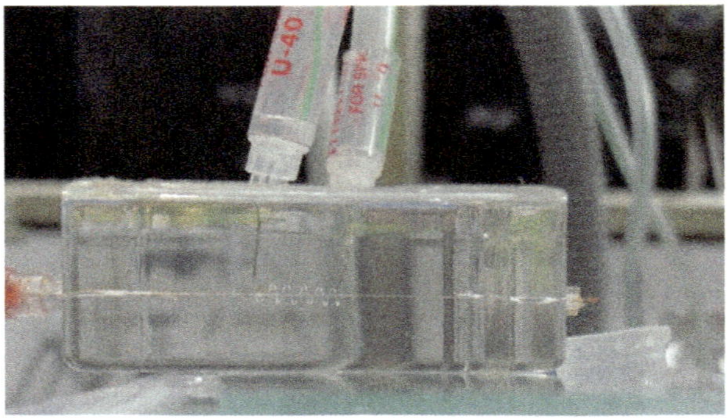

Fig. 2.8 Solenoid microvalve (Reproduced from Singh et al. [29] with permission from the Springer)

There are many other soft lithography approaches which have been reported in literature like solvent assisted micromoulding, nanoimprint lithography and dip pen lithography etc. Where the patterning is carried out at various scales with soft materials [30].

3 Microfluidics for Flow Control and Some Novel Effects

The early applications of microfluidic technologies have been mostly in chemical/ biochemical analysis. The field of microfluidics offers numerous capabilities like rapid sensing and detection of analytes corresponding to limited concentration, rapidity of performance, easy usage with minimal human intervention, good resolution using very small quantity of sample volume and thus offering solutions to low cost diagnostics etc. Micro-flows are mostly laminar with values of Reynold's no. in the range of less than 1 and this laminarity is an advantage in some of the typical requirements of sensitive diagnostics like single cell isolation, single cell manipulation and detection, drug delivery etc. Microfluidics also offers possibilities to control concentration of sample in both space and time [1].

The laminar nature of Micro-flows makes molecular diffusion the only means to facilitate mixing in micro-scaled devices in the absence of convective transverse fluid motion. For a typical microscale device the diffusion length scale being large and the diffusion constant being very small for fluids ensures a very high diffusion time. This coupled with the laminar nature of microflows gives rise to a very interesting domain where various spatial and temporal strategies are utilized by various micro-chip designs to micro-scale actuation for enabling a reduced inter-diffusion length causing the flows to mix vigorously and sometimes in a controlled manner.

There has been a lot of research on micro-mixing methodologies, which focus primarily on increasing the contact surface between two streams for diffusion through multi-lamination strategies. The multi-lamination in flows can be achieved easily by stacking different streams in parallel, perpendicular, radial, as well as angular orientations in respect of the flow direction. Various strategies have been developed for micro-mixing which has been broadly classified as Active and Passive mixing.

3.1 Micro-mixer Design and Characterization

A micro-mixer is a device which can passively (without using energy) or actively (with external energy) is enabled to mix multiple fluids. This device is associated heavily with key technological advancements in many fields like chemical engineering, pharmaceuticals, bio-chemistry, analytical chemistry and high-throughput synthesis and drug screening etc. The micro pathways or parts which are heavily deployed in promoting active or passive mixing may vary from a long serpentine channel to a piezoelectrically vibrating membrane.

The development of micro-mixers has progressed rapidly in the last decade. Initially, the devices manufactured used to be housed within silicon or glass wafers [31] and from then a number of micro-mixers with polymeric parts have been fabricated and successfully developed [32]. As a result of their overall simple design, passive micromixers have found a lot of applications as analytical chemistry tools. While surveying the various the polymeric microfluidic systems, a simply designed microfluidic system with efficient passive micromixing is a natural choice for many applications in chemical and biochemical analyses [33]. Researchers have proposed various designs of micro-mixers (Fig. 2.9) and their use including simple

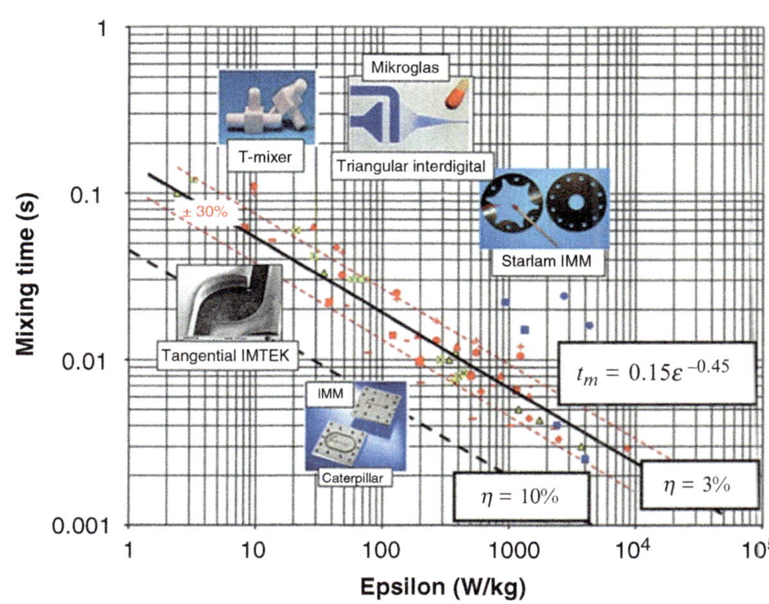

Fig. 2.9 Comparison of various micromixers with respect to time and specific power dissipation (Reproduced from Falk and Commenge [34] with permission from the Elsevier)

T-type or Y-type micromixers and also parallel and sequential Lamination mixer [34].

Nanoparticle synthesis using microscale controlled mixing is a new essence that is being carried out in micromixers and microreactors. The nanoparticles so synthesized have a much skewed distribution and can be custom-made in regards to the average size of the distribution and also the standard deviation on the particle diameters [35].

Many materials which are either amenable to microfabrication such as Silicon and glass or to laser micromachining such as PDMS and PMMA as discussed in previous sections can be employed for the fabrication of the flow channels which constitute the path in a micromixer. Sometimes when the requirement of fabrication is extremely complex precision orientated techniques like micro-stereolithography [36], two photon processes [37] etc. are deployed to realized complex 3-D flow paths.

3.2 *Bilayer Staggered Herringbone Micromixers (BSHM)*

A novel modification of passive micro-mixers has been attempted by Choudhary et al. [38] wherein a method is developed to fabricate micro fluidic devices using MRDI technique as described above onto soft polymeric material PDMS [38]. It has been observed that the novel fabrication method—MRDI was able to produce excellent replication results. The limitation that the process faced however came from the lack of control on speed and power in the laser ablation step which had to be iterated with design of experiments (DOE) strategy. The dimensions used for fabrication were simulated and arrived at in this work (channel width = 100 μm; channel depth = 100 μm; herringbone = 30 μm deep and 50 μm wide). Well shaped herringbone structures were created on the path of flow and mixing was thoroughly evaluated as a function of the overlapping herringbones arm. It was found that as the overall ratio of the herringbone arms on the top and bottom sides of the flow path was reduced the flow length reduced. The image of fabricated herringbone micromixer has been shown in Fig. 2.10a, b and a snap shot of the experimental mixing data evaluated through dilution of a fluorescence marker dye is provide in Fig. 2.10c.

3.3 *Long Microchannel Arrays as Vibration Pads*

Long microchannels arrays have been also found to be suitable for custom made vibration damping characteristics and can passively reduce mechanical vibrations at a range of low frequencies and as such can be used for passive micro-vibration

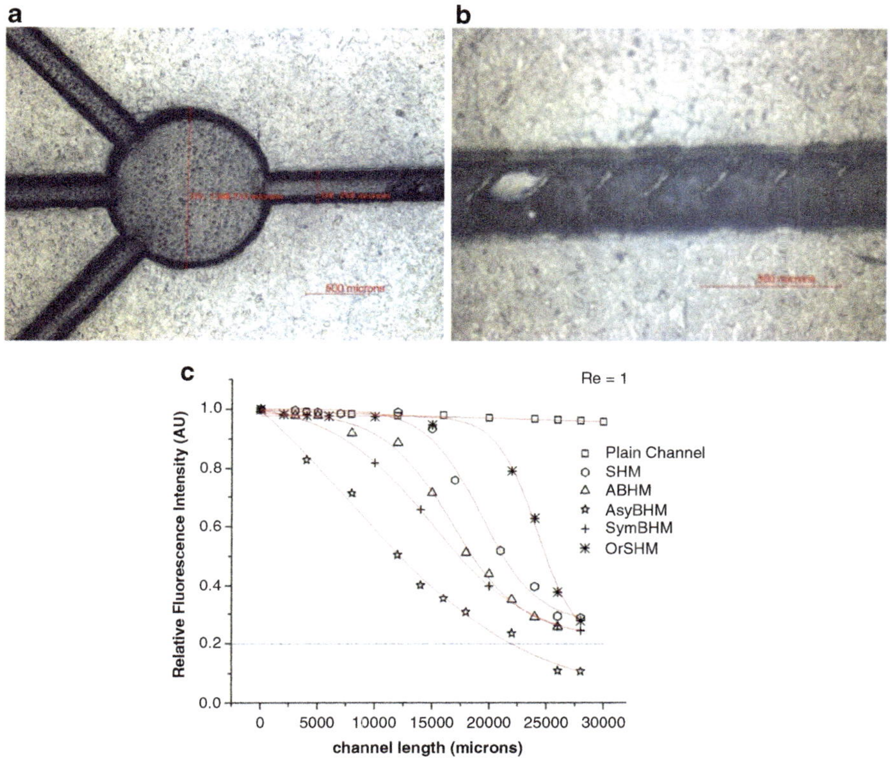

Fig. 2.10 PDMS Replica obtained after soft lithography and herringbone structures ((**a**), (**b**)); (**c**) Mixing performance by dilution of a fluorescence dye in a plain channel, symmetric Herringbone structure, alternate bilayer Herringbone mixer, Asymmetric bilayer herringbone mixer, symmetric bilayer herringbone mixer and oblique ridges and staggered herringbone mixer (Reproduced from Choudhary et al. [27] with permission from the Springer)

damping. The channel array is embedded in blocks of viscoelastic materials in an oil filled and hermetically sealed manner [27]. In this work, the passive response of a replicated array of oil-filled micro-channels, structured within a block of PDMS is reported. Constrained and unconstrained vibration-damping experiments are performed on this block, by applying an excitation signal transversely at the geometric centre of the lower face of the block; its vibration suppression ability is detected. Figure 2.11 shows the constrained and unconstrained configurations schematically and the respective loss factor data with no. of embedded channel layers.

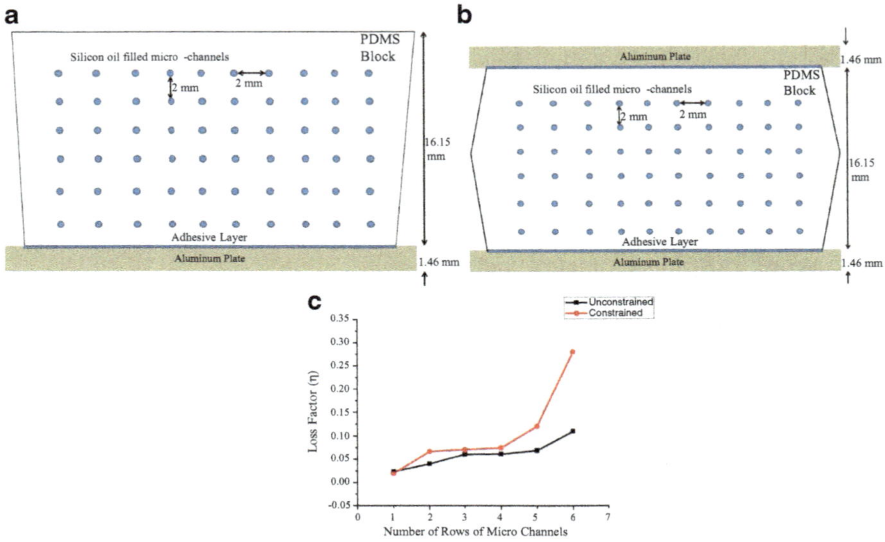

Fig. 2.11 (**a**) Schematic of unconstrained mode damping. (**b**) Schematic of constrained mode damping. (**c**) Loss factor plot with respect to number of rows of embedded channels in both constrained and unconstrained damping analysis (Reproduced from Singh et al. [27] with permission from the Institute of Electrical and Electronics Engineers)

3.4 Micro-pump

Micro-pump are MEMS based fluid handling devices which again work passively or actively to initiate flow through micro-channels. One of their critical purposes is to initiate micro-flows of samples of fluids (either chemical or biochemical) for reaction assays for diagnostics applications. The pump volume is generally of the order of volume of the sample that is being transported. Micro-pumps can be of mechanical or non-mechanical type depending on the actuation principle that they may possess. In mechanical micro-pumps the actuation methods can be electrostatic, magnetic, piezoelectric, pneumatic and thermo-pneumatic etc. while in non-mechanical systems actuation scheme like electro-osmotic, electro-hydrodynamic, electrochemical or ultrasonic etc. can be deployed. It is generally observed that the flow rates in case of mechanical micro-pumps are orders of magnitude more than the non-mechanical micro-pumps.

The first ever peristaltic micro-pump was patented by Smits in 1989. This was a silicon based device fabricated using already established MEMS fabrication technology, and was provided with a set of piezoelectric valves which would deflect and squeeze out on chambers in a sequential manner. This pump had a maximum flow rate of 3 μL/min at zero back pressure and a Maximum Pressure head of 0.6 m H_2O. Further, it was observed that the discharge rate varied linearly with the back pressure till a threshold frequency of 15 Hz was met beyond which it started to fall

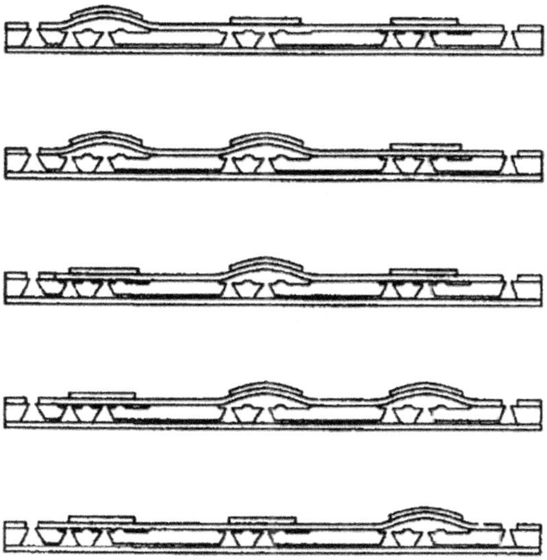

Fig. 2.12 Pumping sequence of the peristaltic pump (Reproduced from Smits [38] with permission from the Elsevier)

off. At frequencies more or equal to 50 Hz, no pumping action was observed which was attributed to viscosity of the pumped fluid. In another work reported later [39], the same group reported an improvised version of micro-pump capable of delivering 100 μL/min at zero back pressure as shown in Fig. 2.12.

Peristalsis with soft polymeric membranes have served the purpose of achieving micro flows although literature has only thoroughly looked at their performance metrics like flow rate [40], ability to withstand back pressure [41], actuation mechanism [42, 43] and fabrication strategies. The most widely used discreet peristaltic strategy that is worth mentioning in this context [44] uses two replicated and molded layers of Poly-dimethylsiloxane (PDMS) bonded over a hard substrate with the top layer being used as the actuator layer wherein controlled compressed air is used to pinch circular chambers inter-connected by thin replicated channels in the second layer. Some earlier reports on micro-pumps as reported by Kant et al. [45] involves such architectures as described in the next section.

3.4.1 Micro-pumping System with Peristaltic Motion

Kant et al. have provided an innovative design of peristalsis based micro pump with an optimized fluid chambers possessing enhanced discharge efficiency per unit volume of the pumping architecture and reduced reverse flow, very often important from the standpoint of blood cell sorting assays, within the pumping chamber where full delivery of fluid containment is critical. Researchers have given a simulation on COMSOL to optimize the chamber design to evaluate the effect of actuator membrane interaction. Optimized geometrical profile formulated above was seen to allow the maximum contact area between fluid containment and actuating

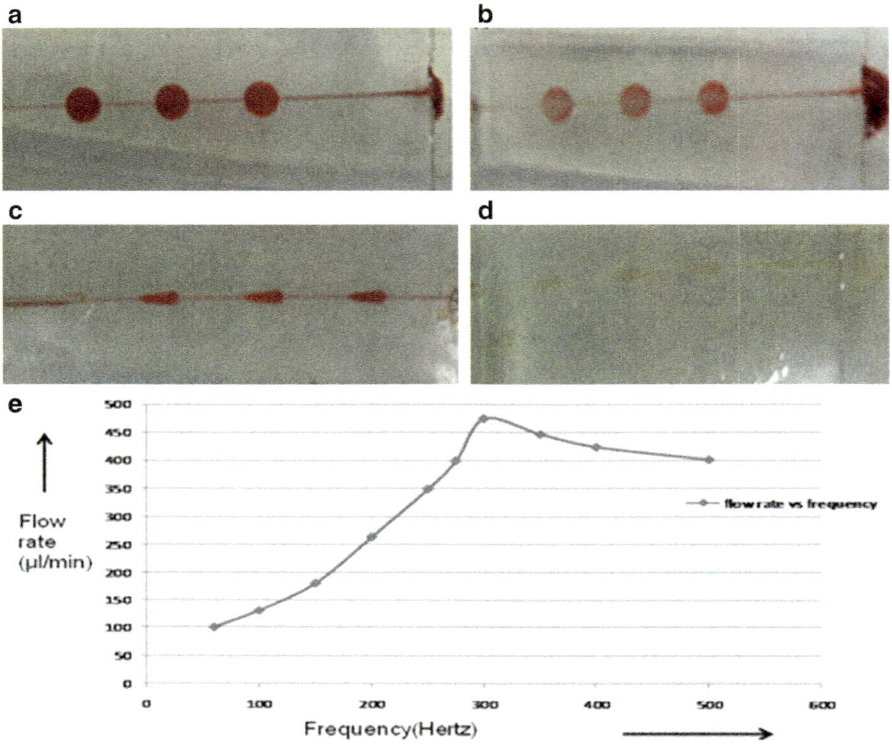

Fig. 2.13 (**a**) Optical micrograph of pumping chambers in the circular design case before start of wash cycle (**b**) at the end of wash cycle (**c**) optical micrograph of pumping chambers in the optimized design case before start of wash cycle (**d**) at the end of wash cycle (**e**) Discharge with respect to operating frequency of the modified design as in case (**c**) (Reproduced from Kant et al. [45] with permission from the Springer)

membrane thus reducing the problem of fluid retainability inside the chambers. Figure 2.13 shows optical micrograph of pumping chamber for both type of design before and after the wash cycle and the discharge performance of the modified chamber design as in part (c) and (d). Additionally Fig. 2.13e shows the discharge related to (c) configuration.

New design of pumping system (experimentally) decreases the percentage retainability of biological and other fluids contained within the chambers which make it a comparatively high efficiency micro-pumping system as compared to conventional design with circular membrane and chambers.

3.5 Micro-valve

Micro-valves are designed to control the switching action of flows in microchannels. This helps in controlling the flow rates so that proper manipulation of the biological entities being carried by biological fluids can occur and hence the overall efficiency of the microfluidic system to handle and deliver increases. Micro-valves can also be characterized as active and passive microvales. Atwe et al. has developed a novel pH-sensitive hydrogel based micro-valve for intelligent valving system for metered flow [46]. Figure 2.14a shows the design of micro-valve using hydrogel developed through precursors chitosan and polyvinyl alcohol (PVA)

The hydrogel solution has been shown to be prepared through Chitosan and polyvinyl alcohol (PVA) in acetic acid (CH_3COOH) and crystallized using glutaraldehyde, crosslinking agent in thin wafers form and this has been found to be very sensitive to pH changes. The pore structure of hydrogel had been investigated

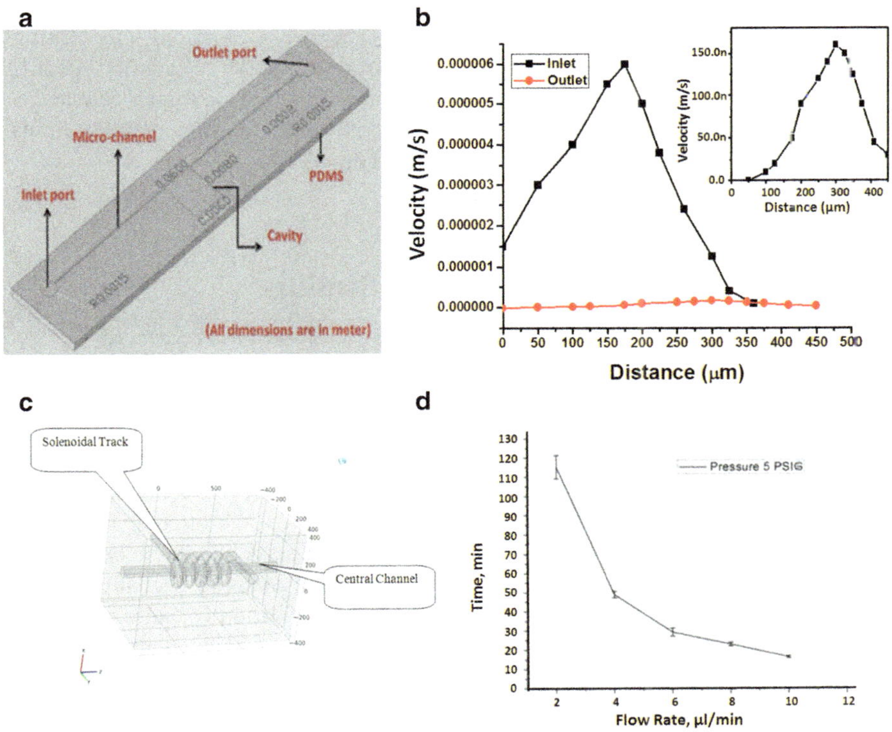

Fig. 2.14 (**a**) Design of the micro-valve; (**b**) Valve performance through PIV measurements as a function of diametral distance from wall of the channel; (**c**) architecture of a solenoidal microvalve; (**d**) valve performance plot of flow rate versus time corresponding to a compressed air pressure of 5 psi. (Reproduced from Atwe et al. [46] and Singh et al. [47] with permission from the Springer)

through Field Emission Scanning Electron Microscopy (FESEM) and thin wafers of the gel were physically placed inside PDMS microchannels which blocked the main flow as the pH in the actuating channels changed suitably. Flow metering in channels was observed till complete realized of valving. This device was further precisely characterized with micro PIV using a solution containing fluorescent polymeric micro beads. The principle advantage device is the small range of pH (3–7) over which the valving response was observed.

Singh et al. [29] has reported a pneumatically actuated solenoid micro-valve design with performance evaluation in terms of flow constriction. Figure 2.8 reported earlier has already shown a schematic of this valve. The valve performance has been evaluated with particle image velocimetry (PIV) (see Fig. 2.14b–d) [29]. These valves are definitely an advantage over quake valves which were done using two layers of microstructured PDMS with the bottom layer containing the micro-channels as very thin and the top layer containing the valve structured at the bottom of the top layer. These valves have so miniscule response time that they almost close immediately qualifying them to be digital in nature. Design modification has been performed in these digital valves to solve the leakage/leaching problem, with a completely different synthesis process, wherein the closing arrangement of a Quake valve has been varied from top down to radially inwards across the whole cross-section of the micro-channel. The valve's design enables its wide applicability to microfluidics in drug delivery; study the flow of body fluids across vasculature like flow within embedded channels etc.

4 Micro-cantilevers for Mass Based Sensing and Diagnostics

Micro-cantilever is a beam fixed or anchored at only one end. The beam carries load to the support where it is forced against by a moment and shear load. A micro-cantilever is a device that can act as a sensor may be biosensor or gas sensor else for various other applications by detecting changes in cantilever bending through optical deflection method or piezoresistive method or vibrational frequency method. The cantilever bending occurs when a specific mass of analyte or biomolecule is adsorbed on the surface of the cantilever and leads to a change in the surface energy of this thin cantilever structure. Miniaturisation of cantilever helps sensing the presence of trace elements in chemical and biological analytes as well as small mechanical displacement etc. Fabrication of micro-cantilevers is a multiple step lithography driven process where the cantilever structure is first realized using surface micromachining and this is succeeded by bulk micromachining to release the cantilever structure additively built in the first step. The fabrication of cantilever micro-structures can be done using different materials like silicon, silicon dioxide, aluminium, gold and of different polymers.

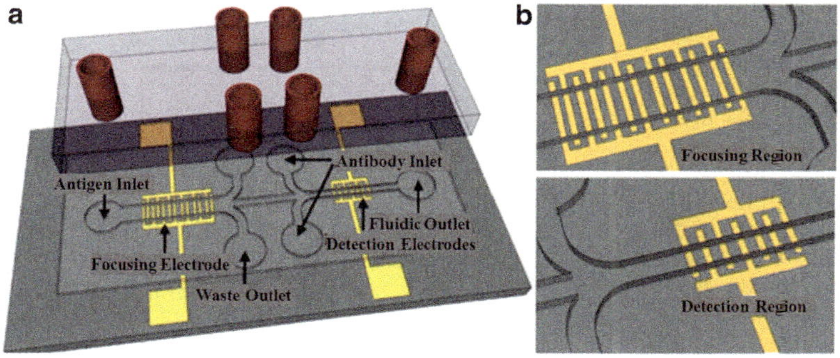

Fig. 2.15 Three dimensional schematic of (**a**) impedance based biosensor for pathogen detection. (**b**) Magnified view of the focusing region and detection region (Reproduced from Ghoshdastidar et al. [48] with permission from the Royal Society of Chemistry)

Both micro-cantilevers as well as other three-dimensional micro-structures have been exclusively used for sensing and diagnostics. Ghoshdastidar et al. [48] have made micro-machined impedance bio-sensor for accurate and rapid detection of E. coli through interdigited electrode structures. They have fabricated two sets of interdigitated electrodes made of gold, embedded in a SU-8/PDMS micro-channels. Using dielectrophoretic trapping the microorganism E. coli was focused on the centre of the microchannel where a second set of electrodes were used to sense through impedance spectroscopy the type and the concentration of the micro-organism. Figure 2.15 shows the three dimensional schematic of biosensor platform.

Boehm et al. [49] introduced a methodology for fast detection of bacteria through a micro-fluidic lab-on-chip device to detect cells using antibody immobilization. Composite self-excited PZT-Cantilever was fabricated and their frequency of resonance was measured both in water and air. Resonant frequency of the second mode was observed to decrease due to pathogen attachment by Campbell et al. [50]. Weeks et al. [51] have reported detection of Salmonella strains using silicon nitride based micro-cantilever. V shaped cantilevers were used which were 180 mm long and 18 mm wide. The cantilever was functionalized using antibody S. Heidb. The cantilever deflection with respect to the Salmonella concentration cfu/ml was obtained and it was directly observed that with the increase in concentration of the sample, deflection was increased. Initially S. Typh strains were injected at 100 cfu/mL but no deflections were noticed but as the S. Heidb strains were injected deflections were obtained within 20 s.

Ilic et al. [52] have done virus detection using nanoelectromechanical devices using a non-pathogenic insect baculovirus to test the ability to specifically bind and detect small number of virus particles. Arrays of surface, antibody-coated poly-crystalline silicon nanomechanical free standing cantilever beams were used to detect the binding from varying amounts of concentrations of baculo viruses in a buffer solution. Figure 2.16 shows the fabrication technique and SEM image of the fabricated micro-cantilever. Because of their small mass, the 0.5 mm $\times$ 36 mm

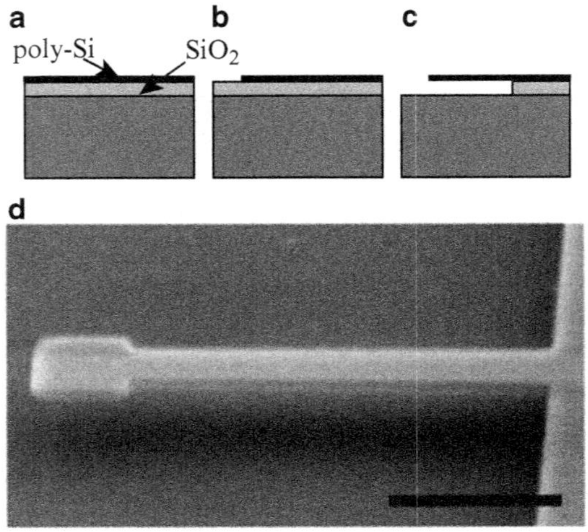

Fig. 2.16 Fabrication sequence of the nanomechanical oscillators; (**a**) Thermal oxidation and LPCVD deposition of the polycrystalline silicon device layer; (**b**) Lithographic definition of the oscillator; (**c**) Sacrificial silicon dioxide removal using HF; (d) Free standing cantilever (L = 6 μm, w = 0.5 μm, t = 150 nm) SEM image (Reproduced from Ilic et al. [52] with permission from the American Institute of Physics)

cantilevers have good amount of mass sensitivities of the order of 10 – 19 g/Hz, enabling the detection of an immobilized AcV1 antibody monolayer corresponding to a mass of about 3×10^{-15} g. Reviewing some of the papers it can be observed that cantilevers deflect in the range of 10^5–10^6 cfu/mL of micro-organisms. Thus although these numbers are not good with respect to PCR methods which can go up to sensitivities of few tens of cells but still from a rapidity point of view these test may serve to be better.

Secondly there may be problems during frequency shifts which can be found by the work reported by Ilic et al. [53] while detecting the E. Coli due to low quality factor of the cantilever structure in liquid and also the relative position of the immobilized microbe along the cantilever from its tip end. Figure 2.17 shows the frequency shift while capturing E. coli cells on cantilever surfaces.

Our research group has also started fabrication of micro-cantilever structures with thin films which are of significant importance in biosensing applications. Fabrication of thin film micro-cantilever possesses major difficulties like releasing the structures and etchants selectivity. One of the most important problem is stiction problem which occurs during wet etching. Releasing the cantilever without using any sacrificial layer has been demonstrated by our work [54]. Other than that Shipley photoresist S-1813 when hard baked, can act as a good protective layer from etchants. Several films of aluminium (Al) of thicknesses ranging between 200 and 800 nm have been deposited on cleaned Si-wafers through sputtering process. On the top surface of this Al thin film cantilever structures were patterned through positive photoresist S-1813 (M/s Shipley). The positive photoresist structure acted as a mask or protective layer for the TMAH etching process that was used subsequently for releasing of the cantilever structures. The Al thin film was first etched off through the vias opened up for the remaining portion of the film for the top elevation of the mask drawing. The etching of the aluminium from the opened positions were further carried out using Transene solution at a temperature of 50 °C.

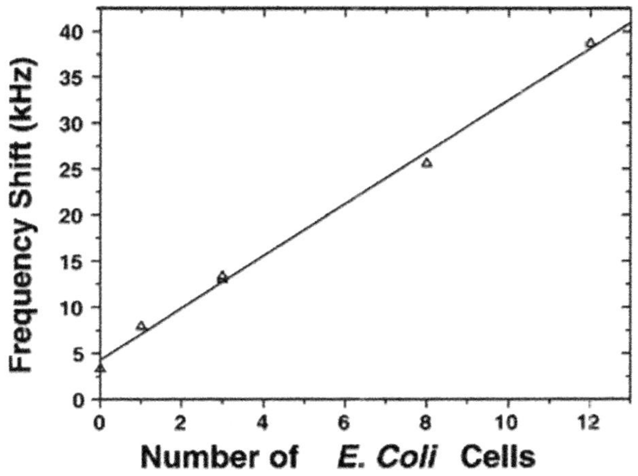

Fig. 2.17 Frequency shift observed with increase in number of E. coli cells (Reproduced from Ilic et al. [53] with permission from the American Institute of Physics)

For releasing the Al micro-cantilever a second step of masking and anisotropic wet etching was done with TMAH etchant [55]. Figure 2.18a shows a detailed fabrication flowchart of the process used for etching of silicon cantilevers. Figure 2.18b shows the FESEM image of a thin film and Fig. 2.18c shows a force deflection characterization using a nano-indenter. Our studies have revealed a very high resilience (almost equal to that of natural rubber) of these metallic cantilever structures [56].

Our group has also been heavily involved in fabrication of polymeric cantilevers using the photosensitive epoxy based polymer SU-8. These SU8 cantilevers have been developed using a one-step lithography based process using maskless gray-scale lithography (MGL). Generally in photolithographic process there are only two states '0' or '1' i.e., the photoresist either stays or gets dissolved after exposure or during development but in gray-scale lithography a selective exposure process is possible by changing the gray-scale values. Also variation in exposure dose varies according to the penetration depth as offered by the photoresist to the laser beam.

In this work, we have tried to develop a methodology for fabricating three dimensional interdigitated micro cantilever structures of SU-8 through grayscale lithography. The difference in our cantilever is in terms of sectional thickness which has been realized at smallest level of 2 μm. Earlier researchers have shown a total thickness of 45 μm using different grades of SU-8. Our fabrication technique enables to fabricate at 1/20th the dimension achieved by earlier researchers. Thin cantilevers find a lot of prominence in the area of sensitive detection of biological entities. Figure 2.19 shows the FESEM image of fully dimensioned interdigitated micro cantilever structures. The hanging SU-8 structures are metallized and serve as interdigitated electrodes which may be able to capture and position single cells using dielectrophoresis process. The miniaturized nature of our architecture enables us to perform our studies on bacterial cells providing an opportunity to carry out stiffness based segregation cells.

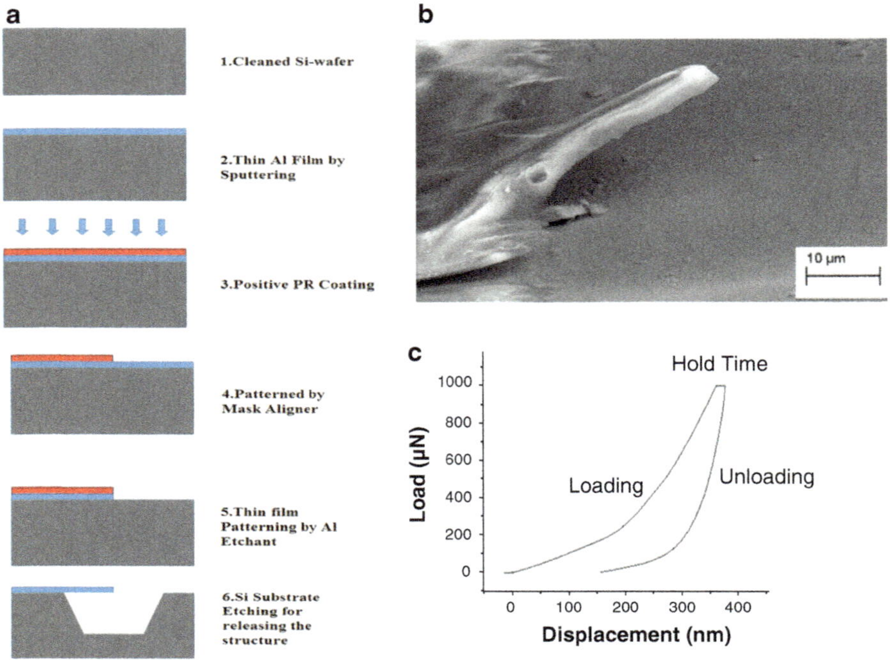

Fig. 2.18 (**a**) Fabrication flow chart for the process used (**b**) FESEM image of Al microcantilever 78 μm long and 28 μm width (**c**) Force displacement characterization on a nanoindenter

Fig. 2.19 Interdigitated cantilevers placed in microfluidic channel

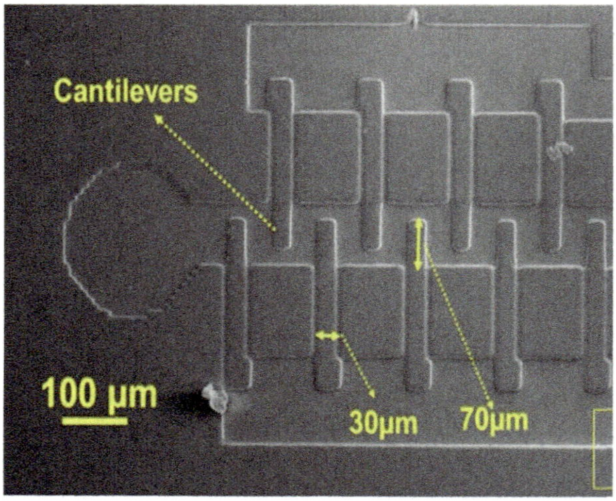

5 Applications in Clinical Diagnostics

Using the basic principles and tools of microfluidics, various diagnostic applications are realized. These applications are focused on separation, modification, combination, replication of biological entities that would result in a signal transduction which may further be able to get detected with instrumentation or normal eyes. Applications range in many domains like paper microfluidics, PCR microchips, electrophoresis microchips, Gene delivery modules etc.

5.1 *Paper Microfluidics: Applications in Clinical Diagnostics*

Lateral flow technology is well suited for point-of-care disease diagnostics because it allows complex bio/chemical processing to be performed without the need for external instrumentation. It fulfils the criterion of world health organization (WHO) i.e., ASSURED (affordable, sensitive, specific, user-friendly, rapid and robust, equipment-free, and deliverable to users). Since the fabrication cost of paper based device is very less as compared to the equipment used towards bio-diagnostics like PCR and ELISA readers thus this technique is more affordable to a wide section of people who cannot afford expensive healthcare particularly in developing countries. There are several other advantages of using paper based diagnostic devices such as thin and easy to transport, lightweight, disposability, chemical modifiability and bio-compatibility [57]. Paper based devices can be fabricated using various methods such as photolithography [58], plotting with an analogue plotter [59], ink jet etching, plasma treatment [60], paper cutting, wax printing [61], ink jet printing [62], flexography [63], screen printing, laser treatment etc. [64].

The lateral flow tests often rely on antigen–antibody interactions to detect targets of interest in bodily fluids, such as serum, blood, or urine. Depending on the assay format, either the antigen or antibody is immobilized on the paper substrate as the capture agent. The targets of interest bind to the immobilized capture agent, resulting in visually distinguishable lines or spots generated by colorimetric, fluorescent, or enzymatic conjugates [65]. When conjugated gold/silver nanoparticles-antibodies bind to specific biomarkers thus changing the overall size of the nanoparticle assembly and thus their absorption properties which results in a colorimetric assay. The change in colour can be even detectable through a commercial smart phone camera [66].

The basic design of a lateral-flow test strip, shown in Fig. 2.20, comprises four porous pads. The sample pad, conjugate pad and absorbent pad are usually made by filter paper. While the test line and control line is made on nitrocellulose membrane. The nitrocellulose membrane is a microporous structure which is made from the nitrocellulose and nonwoven materials (glass, fiber or cellulose etc.). The nitrocellulose membrane is preferred as a substrate for the formation of biochemical

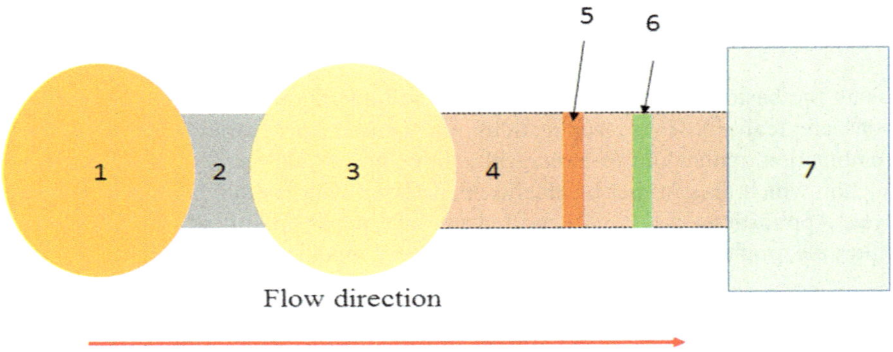

Fig. 2.20 The schematic diagram of paper based colorimetric detection assay. (1) Sample pad: blood sample with viral proteins pipetted onto it; (2) filter paper: through which blood is migrated towards conjugated pad; (3) conjugated pad: conjugated antibody-nano nanoparticles pipetted onto it; (4) nitrocellulose membrane (5) test line: visualization of color change if there is any positive test result. (6) Control area (7) absorbent pad: to wick the extra fluid

complexes because of the various reasons; First, high rate of adsorption of the protein on it. Second, chemistries that make the membrane wettable with aqueous solution which helps in protein adsorption. Third, the pore size of membrane can be controlled according to requirement. The general detection steps of paper assay is depicted in Fig. 2.20.

In the past decade, lateral flow or paper based assay have been widely used in various types of biologically relevant applications including paper-based molecular assays, paper-based ELISA (P-ELISA), paper-based nucleic acid assays, and paper-based cell assays for rapid diagnostic of protozoan and viral diseases like malaria, dengue etc. [57]. Compare to other existing detection techniques, the Paper based rapid diagnostic kits (RDT) require little or no pre sample preparation and give results in few minutes. Guidelines for the evaluation of malaria diagnostic assays have been published and provide a standardized approach to diagnostic assessment [67]. In recent years various researchers has performed testing of Plasmodium falciparum on paper based device. Pereira et al. [68] fabricated the 3-D paper based device to detect malaria biomarker in a single step. The concentration and detection steps were integrated into single step that occurs entirely within a portable paper-based diagnostic strip. Figure 2.21 shows the schematic device for malaria detection on a single chip.

Weaver et al. [69] performed chemical colour tests embedded on a paper card which can significantly identify formulations corresponding to very low quality anti-malarial drugs. The presence or absence of chloroquine (CQ), doxycycline (DOX), quinine, sulfadoxine, pyrimethamine, and primaquine antimalarial medications, were examined. The sensitivity of the developed test card varies from 90 to 100 % with no false positives in the absence of pharmaceutical. For detection of extremely low concentrations of parasite target, a combined system has been developed which combines the isothermal amplification (Recombinase Polymerase

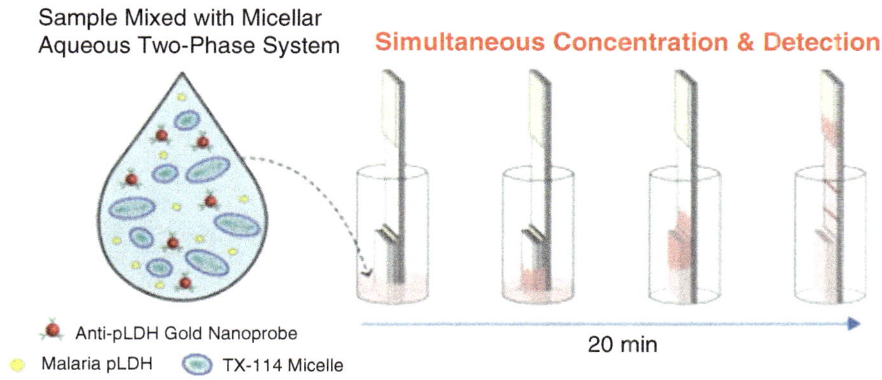

Fig. 2.21 Schematic diagram of paper based device for malaria detection in a single step (Reproduced from Pereira et al. [68] with permission from the Elsevier)

Amplification) and colorimetric detection of a malaria DNA sequence [70]. Paper based micro-fluidic device has also been used in detection of uric acid, glucose using gold nanoparticles [71].

Dengue viral has been diagnosed using the conventional techniques like PCR (polymerase chain reaction) and ELISA (enzyme-linked immunosorbent assay) [72], even these methods are highly accurate but they need a highly clean lab and expert people to perform the test. In the developing countries like, India where the lab facilities are not appropriate in all the locations especially in rural areas, the paper based detection can play a complementary role. Yen et al. [73] developed a paper based testing device which can give a patient a diagnosis within few minutes on whether they have dengue, Ebola or yellow fevers. The red, green and orange silver nanoparticles were immobilized to the respective antibodies that bind the spiked proteins. These antibodies were then attached to a small piece of paper. Instead of whole blood samples, salivary fluid has also been used for dengue detection because it is an important source of biomarkers and is useful for rapid point-of-care diagnostics. Salivary fluids carries the immunoglobulins (e.g., IgGs and IgMs), lymphocytes and plasma cells, which may serve as biomarker in point of care diagnostics. But major limitation of salivary fluid as a sample is that it cannot be applied directly to commercially available lateral flow test strips because it causes the non-specifically binding of conjugated particles to the nitrocellulose membrane. To overcome this problem, Yi et al. [65] recently developed a rapid test for the detection of anti-DENV IgG in saliva. They introduced samples and reagent in separate flow paths. The sample flowed through a matrix of fiber glass which reduces the non-specific adhesion caused by the salivary substances. Figure 2.22 shows the design of device for dengue detection.

The device gave good results in saliva samples spiked with IgG but requires further improvement to detect IgG extracted directly from the blood of dengue-infected patients. To improve the sensitivity (It is the ability of a paper based assay test to correctly recognize patients who have a given disease or disorder.) and

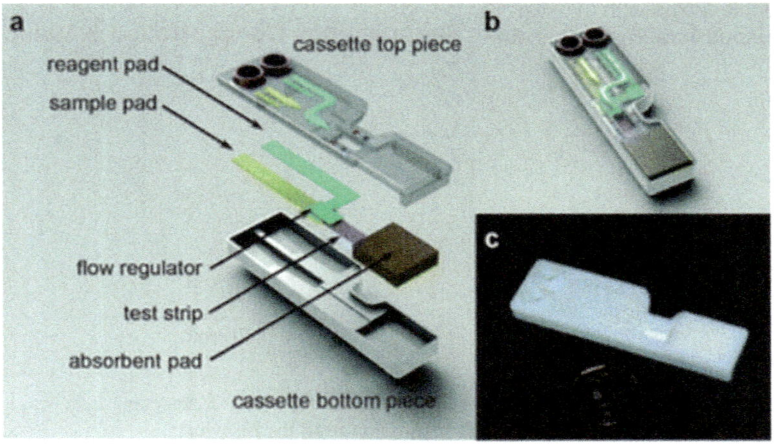

Fig. 2.22 Device for the stacking lateral flow immunoassay; (**a**) The exploded view of the device. The test assembly consists of a sample pad, a reagent pad, a flow regulator, a test strip and an absorbent pad. The test assembly is housed in the cassette; (**b**) The assembled stacking flow device; (**c**) Photograph of the stacking flow device prototyped with a 3D printer (Reproduced from Yi et al. [74] with permission from the Royal Society of Chemistry)

specificity (It is the ability of a paper based assay test to correctly exclude the healthy individuals) of the dengue detection, a methodology has been developed which combines reverse transcription loop-mediated isothermal amplification (RT-LAMP), paper based device and fluorescence based colorimetric detection [74].

5.1.1 Basic Principles of Fluid Flow in Paper Micro-fluidics

The flow through the paper occurs due to capillary action. The capillary action is governed by three main variables i.e., cohesive force, surface tension and adhesive force (Fig. 2.23). Capillary action happens only when the adhesive force is greater than cohesive force. When liquid is brought into contact with a dry paper, it will start absorbing liquid because there is an unbalanced pressure difference within the bulk phases [75]. The fluid can rise against gravity and it stops when the hydrostatic pressure balances the interfacial pressure difference.

Let us consider capillary rise in a cylindrical tube of inner radius r.

The equilibrium height h of the liquid within a capillary is obtained by

$$\Delta p = \frac{2\gamma_{LV}\cos\theta}{r} \tag{2.1}$$

The hydrostatic pressure is given as:

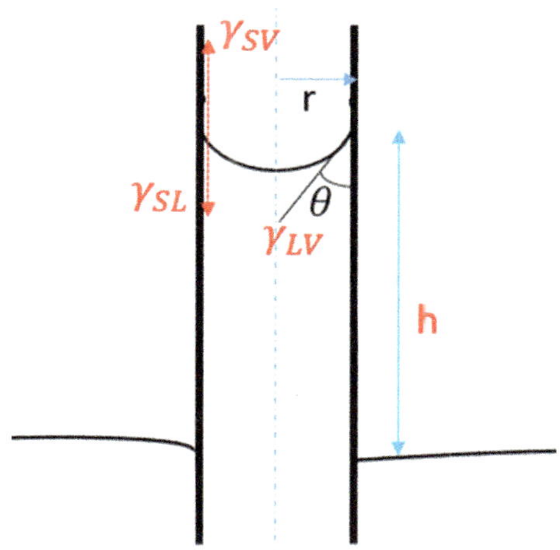

Fig. 2.23 Capillary rise in the tube

$$\Delta p = (\rho - \rho_v)gh \approx \rho gh \tag{2.2}$$

From (**2.1**) and (**2.2**):

$$\Rightarrow h = \frac{2\gamma_{LV}\cos\theta}{r\rho g} \tag{2.3}$$

Where γ_{LV} = interfacial surface tension of the liquid–vapor, γ_{SV} = interfacial surface tension between solid-vapor, γ_{SL} = interfacial surface tension between solid–liquid, ρ & ρ_v are the densities of liquid and the vapor and θ is the contact angle.

The transport of fluid on paper channel in 1-D can be estimated using the Lucas–Washburn equation [76]:

$$L = \sqrt{\frac{\gamma_{LV} Dt \ \cos\theta}{4\mu}} \tag{2.4}$$

Where L is the distance moved by the fluid front, t is time, D is the average pore diameter, γ_{LV} is the surface tension, and μ is viscosity.

The above formula is useful in designing of paper channel. The length of paper channel is directly proportional to the square root of time. The volumetric flow rate (Q) of the fluid in a paper channel of constant width can be described by the Darcy's law:

$$Q = \frac{-kA}{\mu L}\Delta P \tag{2.5}$$

Where k is the permeability of the paper to the fluid, μ is the viscosity of the fluid, A is the cross sectional area of the channel perpendicular to flow, and ΔP is the pressure difference along the direction of flow over the length L [77]. Apart from the above governing equations the delivery of fluid depends on various other parameters such as properties of the porous materials, including pore size, pore structure and surface treatments etc.

Although these paper micro-fluidic point-of-care devices are capable of rapid detection at a lower price they are easy to use and eco-friendly. These devices have some limitations like the sample virus gets embedded in the pores of the paper device due to which only a small portion of the conjugated antibody-antigen reaches the test line resulting in an overall reduction of sensitivity etc. [78].

5.2 Electrophoresis

Electrophoresis is the process in which particles dispersed in the fluid move under the effect of uniform electric field. It was observed by Ferdinand Frederic Reuss in 1807. This effect comes under play due to the presence of charge interface between the particle surface and fluid around it. Electrophoresis helps in separating the molecules according to their size (smaller molecules travel with higher speed). Electrophoresis basically deals with applying a electric field on a particle and with respect to the charge present in the molecule, the molecule starts moving towards the oppositely charged electrode. Hence on this basis, electrophoresis can be divided in two types, one is cataphoresis (electrophoresis of positively charge ions) and other is anaphoresis (electrophoresis of negatively charged ions).

Figure 2.24 shows the effect of uniform electric field on the positively charge ion as well as on the neutral body. The positively charged ion starts moving towards the negative pole while the neutral body experiences no external force.

Electrophoresis can be done using three techniques namely:

1. Gel Electrophoresis
2. Capillary Electrophoresis
3. Surface Electrophoresis

(a) **Gel Electrophoresis**

Gel Electrophoresis is the technique used in electrophoresis to separate the molecules depending on their size using a porous matrix. While using this porous matrix, which is generally made up of agarose gel, an electric field is passed making one side of the matrix positive and opposite size of the matrix as negative. In this manner when we keep our negative charges on the side of the negative pole, and start the field, it will quickly start moving to the positive terminal through the gel matrix. Agarose is the polysaccharide polymer material. When the solution of this is made, it makes a porous kind of structure with very small size pores, which are just enough to allow molecules like DNA/RNA to pass. When the field is applied, the smaller molecules start

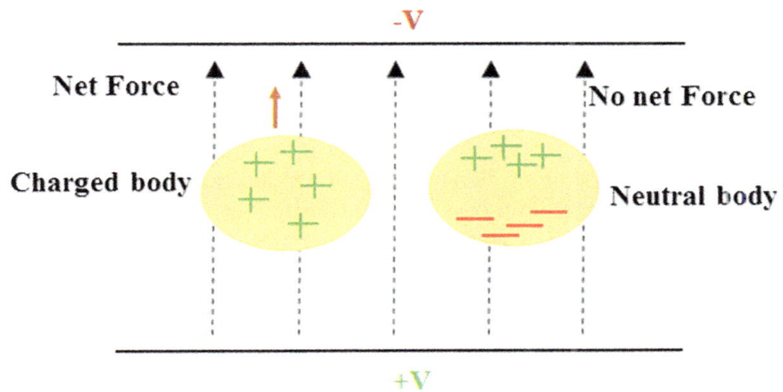

Fig. 2.24 Effect of uniform field on the various type of particles (positively charged particles, neutral body)

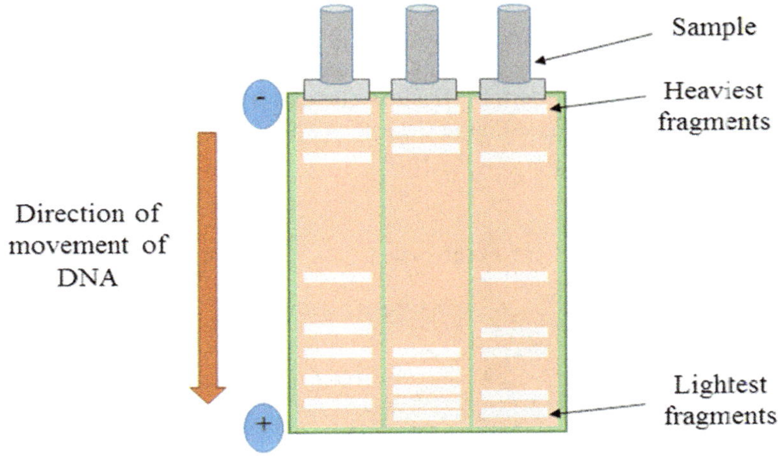

Fig. 2.25 Gel Electrophoresis

moving with the higher speed while longer molecules take more time. In this way separation can be done. If we attach a fluorescence molecule to the DNA being separated, we can exactly get the count of number of molecules with a particular size by counting the fluorescence. Figure 2.25 shows the schematic of gel electrophoresis.

(b) **Capillary Electrophoresis**

Capillary electrophoresis method also shares the similar principle as that of the gel electrophoresis. The only difference is that in case of gel electrophoresis, there are kind of multiple capillaries (pores through gel) being used for causing electrophoresis, while in case of capillary electrophoresis, there is only one small opening through which DNA travels, i.e., single capillary and the separation takes place in the similar way.

(c) **Surface electrophoresis**

This technique of electrophoresis is mainly used for handling longer strands of DNA. If the DNA strands are long, there is the problem of making them to travel along agarose or so. Hence to handle the heavier molecules, mainly of the size above 10 kbp (kilo base pair), this technique is used. In this case the friction between surface and DNA strand plays a very major role, while making DNA to move on the surface. Li et al. has reported that the mobility of dsDNA, while moving along the surface is influenced by intensity of the electric field, ionic strength and the migration distance [79]. Going with the modifications, gel free microchannel electrophoresis has also been fabricated by Lee and Kuo [80] for fractionating larger DNA. It was observed that the channel's bottom surface was modified in this process. This process helped in fractionating molecules of size 3.5–21.2 kbp. Ghosh et al. [81] fabricated PDMS micro-channels to make dsDNA to travel along. It was observed that the dsDNA travels along the corner rather than the channel base, with the fast speed. The physics behind the movement in such case was also optimized using molecular dynamics simulation. Hence it was remarked that the orthogonally placed pair of surface channel corners is a more favourable option rather than the flat base. Figure 2.26 shows the schematic of the fabricated device.

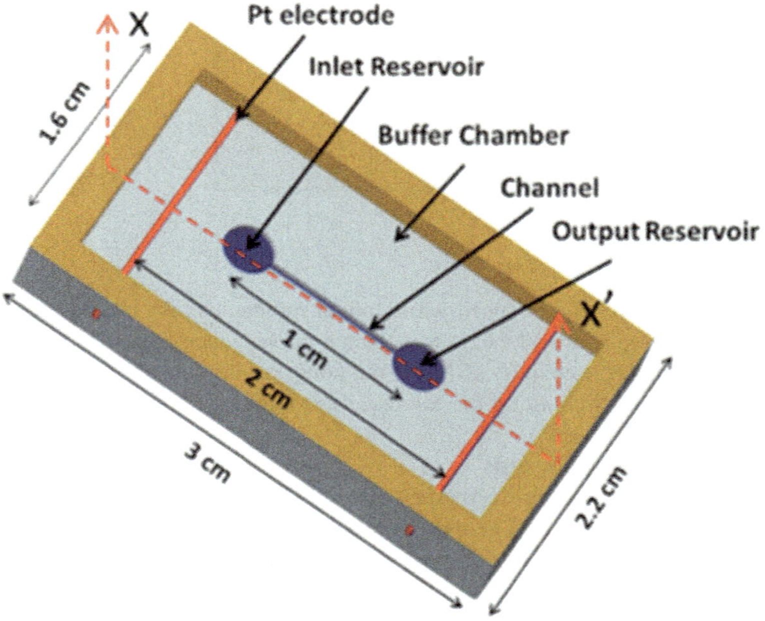

Fig. 2.26 Color online Schematic of the SE device, (dimensions: reservoirs diameter = 3 mm, width of channel = 400 μm, depth of features = 250 μm) (Reproduced from Ghosh et al. [81] with permission from the American Institute of Physics)

5.3 Dielectrophoresis

Dielectrophoresis (DEP) is a phenomenon in which movement of dielectric particles occurs under the effect of force exerted due to non-uniform field. It was demonstrated in 1950s by Herbert Pohl. DEP force depends on various parameters like electrical properties of particle and its surrounding medium, shape and size of dielectric particles and frequency and magnitude of applied field. DEP is helpful in cell separation, cell concentration and nanoparticle/nanowire manipulation. DEP process can be classified in two different types, positive DEP in which dielectrophoresis takes place in the direction of increasing electric field strength and negative DEP in which dielectrophoresis takes place in the direction of reducing field strength. Figure 2.27 shows the effect of non-uniform electric field on the dielectric particles.

DEP force [82] is given by:

$$\mathbf{F_{DEP}} = \mathbf{2\pi r^3 \varepsilon_0 \varepsilon_m Re[K(\omega)]} \nabla |\mathbf{E_{rms}}|^2 \tag{2.6}$$

Where, r is the radius of the particle, ε_o is permittivity of free space, ε_m is real part of the permittivity of suspending medium, E_{rms} is root mean-square electric field and K(ω) is Clausius-Mossotti factor (measure of effective polarizability of the particle). Clausius-Mossotti factor is given by:

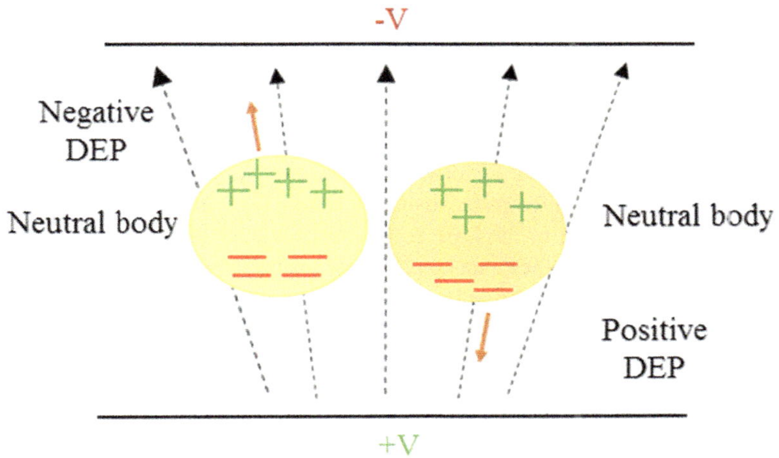

Fig. 2.27 Effect of non-uniform electric field on dielectric particle, positive and negative DEP

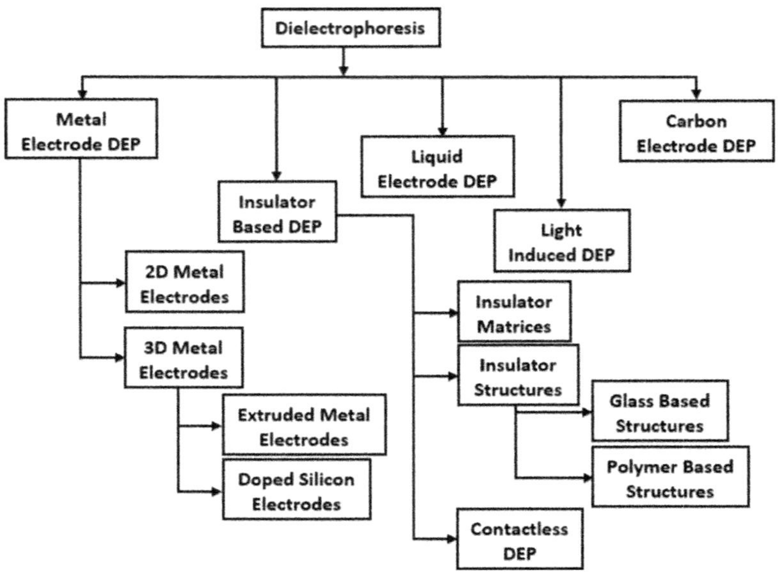

Fig. 2.28 Classification of DEP techniques

$$\mathbf{K}(\omega)=\frac{\left(\varepsilon_p^*-\varepsilon_m^*\right)}{\left(\varepsilon_p^*+2\varepsilon_m^*\right)};\ \underset{(\mathbf{i}=\mathbf{p},\mathbf{m})}{\varepsilon_p^*}=\varepsilon_i-j\frac{\sigma_1}{\varepsilon_0\omega} \tag{2.7}$$

Where, p and m refer to the particle and medium, ε is permittivity, σ is conductivity and ω is the angular frequency of applied field ($\omega = 2\pi f$). Dielectrophoresis can be carried out in various circumstances, i.e., using metal electrodes, liquid electrodes, carbon electrode, insulator based DEP and light induced DEP. The detailed classification of the DEP techniques is given in Fig. 2.28.

As Fig. 2.28 represents, DEP can be performed using various types of electrodes, even electrodeless DEP is possible using techniques. Initially metal electrode DEP were used in which metal electrodes used to be printed on the substrate surface. Printed electrodes further possess various characteristics and various designs of electrodes (interdigitated electrodes (IDEs), other discrete shapes etc.) It has advantage of performing the process at low voltage value as the electrodes are placed very near to each other. Very low volume of solution (volume corresponding to ≈ 30 μm height from the base) can effectively undergo DEP process. Beyond this height, the efficiency reduces steeply. In metal electrode category, 2D and 3D types of electrodes are there. 2D electrodes are also called as the planar electrodes. To increase the volume respective to further height, 3D metal electrodes (Extruded metal electrodes, thick layer of doped silicon electrodes) were evolved. With time, several limitations of the metal electrode DEP like constraint on the volume that can undergo DEP and erosion of metal electrodes with time has led to evolution of

Table 2.1 Fabrication techniques for accomplishing DEP

Dielectrophoresis technique		Fabrication techniques	Characteristic features	Problem
Metal electrode DEP [83]	2D metal electrodes	• Metal deposition (Sputtering) • Polymer patterning • Etching	• Voltage requirement: 10 s of volts • Volume ≈ 30 μm from bottom	Electrode fouling
	3D metal electrodes [84]	• Electrodeposition (Gold electrodes) • Thick film photolithography (Doped Silicon Electrodes [85])		
Insulator based DEP	Insulator matrices [86]		Channel filled with dielectric beads or shapes	Higher voltage requirement
	Insulator structures [87]	• Glass based: ✓ Wet etching • Polymer based [88]: ✓ Injection moulding ✓ Micromilling ✓ Photolithography	Symmetric repeated insulator structures of various shapes	
	Contactless DEP [89]		Electrodes not solid but conducting fluid	
Liquid electrode DEP [90]		Metal patterning and polymer photolithography	Vertical equipotential surface creation at side walls	
Light induced DEP [91]			Light excites photoconductive layer and create an electric field gradient	
Carbon electrode DEP [92]			• Low voltage requirement • Lower electrolysis of sample • More chemical stability than metal electrodes	

various new techniques. Table 2.1 represents various fabrication techniques corresponding to the stated techniques along with their characteristics features and problems involved with these modifications.

The characterised DEP modules have been used for various applications, including separation, concentration etc. DEP of various nanoparticles like polystyrene particles and polystyrenes particles mixed with Saccharomyces cerevisiae cells [93] is carried out in asymmetric insulator traps. This kind of DEP has resulted in separation of particles on the basis of size. It was observed that large size particles leave the system earlier than those of the smaller size particles. This separation scheme has been observed valid for individual particles as well as for mixture of cells and polystyrene particles. DEP of DNAs is also carried out by various authors.

Gold IDEs have been used for detecting DNA molecules [94]. A very important remark about the DEP capture of DNA that was observed is DEP capture of DNA takes place in a very specific frequency range. Initially DEP capture increases and at a particular frequency capture stabilizes showing fluorescence plateauing. After this state, DNA capture reduces with increase in frequency. Castle-walled microelectrodes were also used for DEP of DNA [95]. Very interesting observations were made in this study. Below crossover frequency, positive DEP was observed and the DNA capture was seen at the gaps between the printed electrodes. And above crossover frequency, negative DEP was seen and the capture of DNA molecules was seen at the electrodes. It was also emphasized that the crossover frequency changes with change in the length of the DNA molecule. Further applications of DEP are explored in the consequent discussion on PCR in molecular identification.

5.4 Polymerase Chain Reaction Microchips

PCR [96] is the method by which replication of DNA takes place, it is done for some pre-selected sample by conducting a polynucleotide amplification reaction. This is a very useful technique which helps in molecular diagnostics by making several copies of one DNA in very small amount of time. It was discovered in 1985.

For carrying out PCR reaction, a solution has to be prepared which has several component in it like DNA template, primers, Taq-polymerase, Deoxynucleic triphosphates (dNTP's like adenosine (A), cytosine (C), guanine (G), thymine (T)), buffer solution having divalent ions (Mg^{2+} ions) (approximate 1.5 mM concentration in solution, 1:10 dilution). Further a thermal cycler is required to maintain the temperature of the solution in three steps (denaturation, annealing and extension).

Three basic steps of PCR cycle are denaturation (at 94 °C) of dsDNA to two ssDNA, annealing (at 54 °C) of forward and reverse primers to the denatured ssDNA from 5′ to 3′ direction and finally extension (at 7294 °C), attachment of dNTP's and backbone formation by Taq polymerase enzyme. Numbers of cycles that take place in PCR process depend on the quantity of DNA, primers and dNTP's. In general 25–35 cycles are the standard for PCR process. The standard process comprises of denaturing step of 5 min, then 25–35 cycles of 30 s at 9494 °C, 45 s at 5494 °C, 2 min at 7294 °C and final extension of 7 min at 7294 °C.

The basic principle of PCR and thermal cycle steps is elaborated in Fig. 2.29. In PCR, there is exponential increase in number of DNA copies which are generated.

PCR is mainly two types, liquid PCR and solid PCR. Liquid PCR process is the normal PCR process in which all the DNA, primers and dNTP's are in solution. While in solid PCR, primers are immobilized on the surface and DNA and dNTP's remain in the solution above primers. Further there is two amplification mechanism, interfacial amplification and surface amplification that take place in the solid PCR. Optimization of design and fabrication process for inexpensive reusable DNA amplification chip with polydimethylsiloxane chamber (3 μL) on silicon base with spin-on glass layer (140 nm thick) in between is done by Bhattacharya

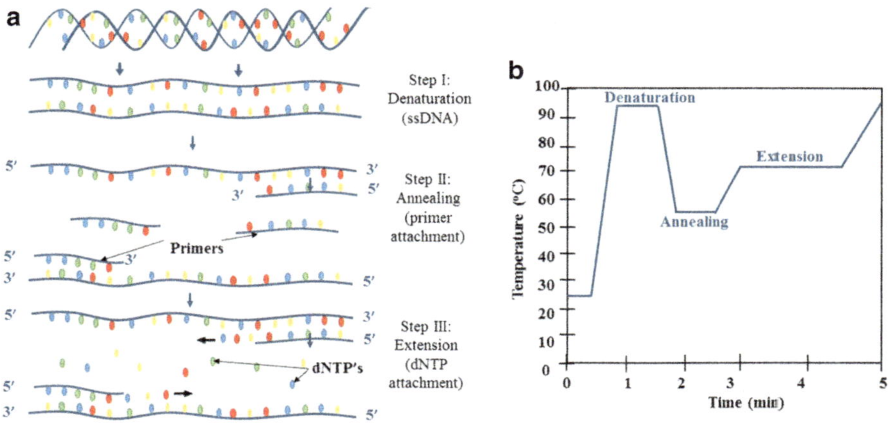

Fig. 2.29 PCR process (**a**) Process description; (**b**) Temperature requirement during PCR process

et al. [97]. Amplification chip contains microfabricated platinum heater structures on the bottom of silicon base. Feed channels and inlet–outlet ports are bonded to SOG film (bond strength: maximum 83 psi) with PCR cycle control by thermal cycle functionalized using sensing thermocouple. 527 fragments of IBR viral genome with 0.07 pg/μL concentration have been amplified in 51 PCR cycles. Figure 2.30 shows the schematic of the fabricated device.

Negligible nonspecific binding of template DNA to the hydrophobic interiors has been shown by fluorescence measurement design. Fabrication of a microchip platform for sensitive detection of microorganisms using integrated sorting; concentration and real time PCR based detection system is done by Nayak et al. [98]. Microchip is designed in such a way that to it pre-concentrates the specific microorganisms from highly dilute sample and real time molecular identification is performed using q-PCR process in a pico-litre micro-channel by use of optimized interdigitated electrodes and micro-scale thermal cycling mechanism. Gold nanoparticles (coated with secondary antibody (Goat anti-mouse IgG)) are attached to targeted microorganisms (E-coli DH5α) to mediate immune-conjugation. Sorting and concentration is done through dielectrophoresis (DEP) technique and finally detection is done by quantitative-PCR (q-PCR) using flucrescence measurement. A primary anti E. coli antibody captured E. coli DH5α cells with binding through nanoparticle bridges. Figure 2.31 shows schematic of detection of molecules.

Fluorescence detection has shown that the DEP frequency is different for the micro-organisms with conjugation and without conjugation. Additionally the fluorescence pattern observed was seen different for in case of conjugated and non-conjugated bacterial cells (Fig. 2.32).

An on-chip system for the electrokinetic capture of bacterial cells and their identification using PCR is designed by Bhattacharya et al. [99] using a PCB

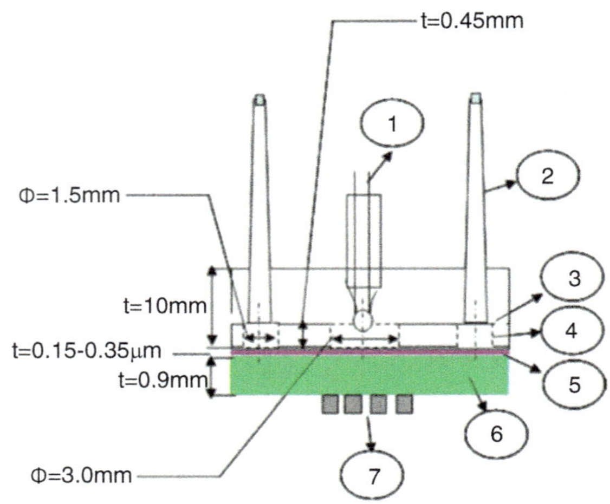

Fig. 2.30 Schematic of the silicon PDMS cassette. (1) Glass housed thermocouple, (2) epoxied inlet/outlet ports, (3) polydimethyl siloxane channels, (4) inlet/outlet reservoirs, (5) SOG layer, (6) thermally oxidized silicon wafer, (7) heaters (Reproduced from Bhattacharya et al. [97] with permission from the Institute of Electrical and Electronics Engineers)

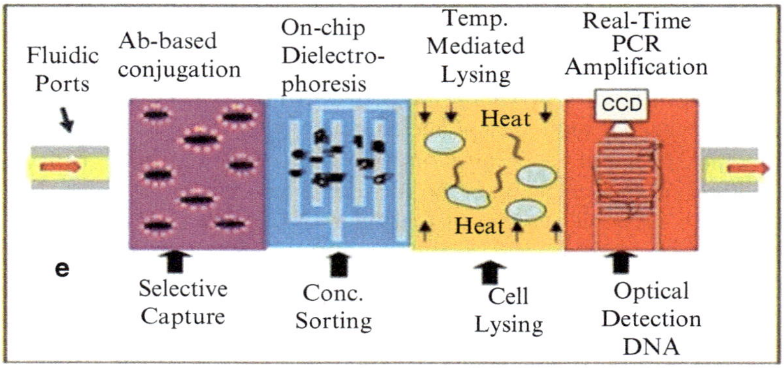

Fig. 2.31 Schematic of detection microorganisms (Reproduced from Nayak et al. [98] with permission from the Nature Publishing Group)

bonded silicon-glass platform with electrodes, micro-channels and chambers built on it is used for identification purpose (Fig. 2.33).

DEP forces are used to divert cells using one set of interdigitated electrodes in big chamber while the diverted cells are collected in smaller chamber with second set of interdigitated electrodes to trap and concentrate the cells. In the trapped cells, under DEP force, PCR mix is injected for PCR amplification with SYBR green detection for identification of Listeria monocytogenes V7 cells. Using DEP forces, higher sensitivity was achieved from 10^6 to 10^4 cfu/mL cells. Very specific identification technique for identifying as low as 60 cells in 600 nL volume. Further optimization of fabrication process for PCR micro-chip through system identification technique is performed [100]. Design and optimal placement of a thin film resistance based temperature detector (RTD) in the PCR microchip is studied and the most optimal design is suggested. RTD integration is done to eliminate the need

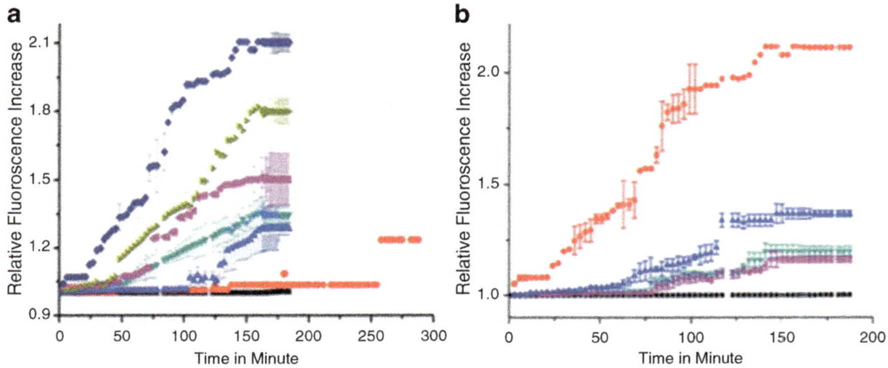

Fig. 2.32 (**a**) Plot showing the trend in increase in fluorescence intensity during DEP of different bacterial cell (normal) concentrations with respect to time, -●-: 10^2 cfu/mL; -▲-: 10^3 cfu/mL; -►-: 10^4 cfu/mL; -▼-: 10^6 cfu/mL; -◄-:10^7 cfu/mL; -◆-: 10^9 cfu/mL; -■-: control, (**b**) Plot showing the trend in increase in fluorescence intensity during DEP of different bacterial cell (conjugated) concentrations with respect to time, -●-: 109 cfu/mL; -▲-:106 cfu/mL; -►-:104 cfu/mL; -▼-: 103 cfu/mL; -■-: control (Reproduced from Nayak et al. [98] with permission from the Nature Publishing Group)

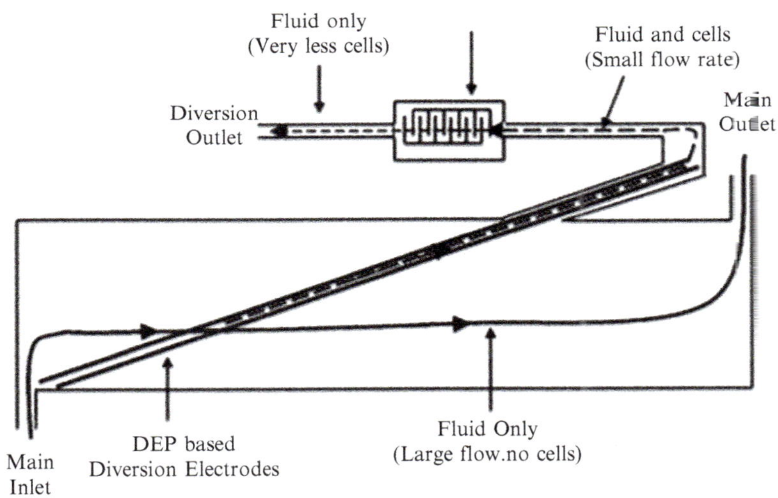

Fig. 2.33 Principle of operation of the dielectrophoresis-based diversion and diversion of cells (Reproduced from Bhattacharya et al. [99] with permission from the Royal Society of Chemistry)

for the thermocouple. In this research the optimal placement test run for placement of RTD is performed. The configuration chosen for RTD placement is heater is placed at the bottom side of the chip while RTD is placed on the upper side of the chip. RTD is used in four wire measurement mode which minimizes measurement errors due to lead resistance. Various placement methods for RTD are studied to obtain a linear relation in input and output parameters. Parametric identification using Auto Regressive with Exogenous Inputs (ARX) method is performed and 2nd and 4th order temperature models are given which helped in deciding on the final configuration. Detection of DNA is also performed using gold nanoparticle functionalised with primers [101, 102]. Using this type of mechanism, colorimetric change from red to purple is observed, if complementary sequence is present i.e., if hybridization takes place. If hybridization takes place, formation of gold nanoparticle aggregate is observed which is responsible for colour change. This kind of work has also been done by Shen et al. [103] by making some manipulation in the process by immobilising primers on gold nanoparticle surface by thiol bond. Nanostructured aggregate formation on hybridization is observed.

5.5 Gene Delivery Using Nanoscale Material and Electrophoretic Transport of DNA

Gene delivery is the process in which cell is modified by introducing foreign DNA into the cell. This process is very important with respect to gene therapy and genetic modification of crops. Various methods that come into picture while considering gene therapy are viral method (virus has capability to inject its DNA into cell) and non-viral methods (electroporation, microinjection, gene gun, hydrostatic pressure and sonication). Viral method is easy to arrange but it has drawback of having various random insertion sites, while non-viral can be controlled with respect to insertion sites and number of insertions. Various processes have several advantages and respective limitations. Here we are summarizing a bit about electroporation and combination of electroporation and hydrostatic pressure wave method. Electroporation is the method of gene delivery in which high voltage ($\approx$100,000–500,000 V/m) is used for transfecting DNA into cells [104]. This high voltage leads to high transfection but many a times results in high cell mortality as well. Hence if some means can be devised to have higher yield with low electric field strength combined with some other means, it may be a very successful aspect. So in this direction a prominent combination of low electric field ($\approx$20,000 V/m) electroporation and hydrostatic pressure through shock wave is analysed to attain high yield efficiency [105]. This high pressure wave was generated using nano-energetic materials.

6 Various Sensing and Detection Techniques

Interest in BioMEMS has been immensely growing for applications like biosensors for detection of protein/DNA, cells and other biomolecules, immunoisolation devices, drug delivery etc. [106]. There are various detection techniques implemented for signal transduction in biosensors which includes mechanical, electrical and optical primarily. This section presents various BioMEMS sensing modalities.

6.1 Electrochemical Sensing

Electrochemical sensing is one of the oldest techniques used dating back to 1950s. This technique uses electrodes to measure electrical properties to sense materials present in the sample without damaging the sample system. Typically there are several types of electrochemical sensor. The three major types include amperometric, measurement of current, conductometric, measurement of conductive properties and potentiometric, measurement of potential difference or charge accumulation across the electrodes. Impedimetric, which measures impedance change and field-effect, for measurement of current because of the gate potentiometric effect are also, used these days [107].

Amperometric sensors have been typically developed for determination of biochemical oxygen demand (BOD) since conventional techniques take 5–6 days for the measurement. Tonning et al. proposed an interesting approach for sensing of Vibrio fischeri bacterium, Pseudokirchneriellasubcapitata freshwater marine alga and Daphnia magna freshwater crustacean. The fabricated sensor consisted of several cells and electrodes for sensing and finally the data was processed using chemometry mathematical methods [108]. Also, these sensors are commonly used for detection of glucose. Mu et al. investigated nickel oxide modified glucose sensors that use carbon electrodes [109]. These sensors have a response time of 5 s hence demonstrating excellent sensitivity. One of the recent interesting findings include thick-film textile based amperometric sensors. Direct screen printed amperometric sensors incorporated in clothing has been reported by Yang et al. These sensors were printed on the elastic of clothing and studied for mechanical stress, bending and stretching [110]. Specific enzyme sensors can be developed using this technique which can prove useful for the purpose of monitoring alcohol consumption, stress or sweat monitoring for athletes etc.

Microfluidics is commonly used to position cells, but a novel device has been fabricated to trap the cells directly over the sensing electrodes. A microfluidic trap device to measure exocytosis from cells targeting them to electrochemical electrodes has been demonstrated by Gao et al. (Fig. 2.34) [111]. The device is fabricated in a manner such that there is no need for precise pressure control or handling of fluids on the chip.

Conductometric sensors are used in varied fields to measure the change in conductivity of a reaction solution caused by microbial activities. They are preferred over other sensors because of their inexpensive fabrication process, no requirement of reference electrode, high accuracy and cancellation of interferences. In the field of gas sensing, a novel and highly sensitive NO_2 gas sensor using caesium-doped graphene oxide has been reported by Piloto et al. Doping of caesium to graphene oxide results in the carbon atoms reducing the graphene oxide, demonstrating very low detection limits for NO_2 (90 ppb) [112]. Latif and Dickert have

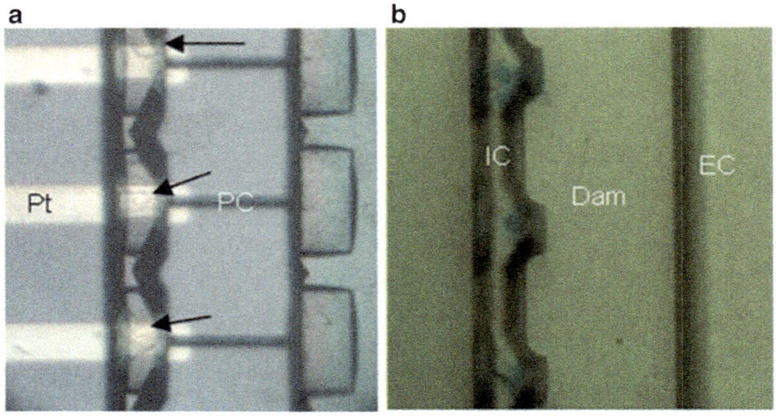

Fig. 2.34 Photomicrographs depicting cells targeted to docking sites (Reproduced from Gao et al. [111] with permission from the Royal Society of Chemistry)

developed a unique conductometric sensor which is a combination of interdigital electrode structures with imprinted polymers for engine's lubricating oil monitoring [113]. Using a sol–gel process, titania and silica are imprinted and used as sensitive coatings. The change of conductance in these layers is measured as they are exposed to varying concentrations of the oil. Conductometric biosensors are extensively used for environmental monitoring as well. They are used for detection of several pollutants, organophosphorous pesticide [114], heavy metal ions [115], formaldehyde [116], 4-chlorophenol [117] and nitrate [118]. In spite of being a novel field, arrays of such sensors can be used for various applications in environmental modeling providing the advantage of improved accuracy and low costs.

Potentiometric sensors work on a principle where an ion-selective or gas sensing electrode is monitored for changes in potential with respect to a reference electrode. When biological molecules bind to the sensing electrode the potential changes due to consumption of the electrolyte by it. Gold coated silicon electrode with alkanethiol molecules immobilized on it is used as a sensing element for detection of cancer and proteins [119]. Real-time wearable sensors are paving their way in owing to their self-tracking nature. From tracking heart rate to monitoring one's mood these sensors help the person quantify their health and make improvements accordingly. A unique tattoo based epidermal pH ion-selective sensor has been developed [120]. It is a flexible wearable sensor which can be an asset for physiological monitoring purposes. Potentiometric sensors are targeting ultra-low cost and robust sensors, which has led to development of paper based potentiometric sensors. Bendable electronics and carbon nanotubes have been combined to develop a novel low-cost sensor for diagnostic purposes which entails advantages like excellent electrical properties, better conductivity and higher surface area contact with the sample [121].

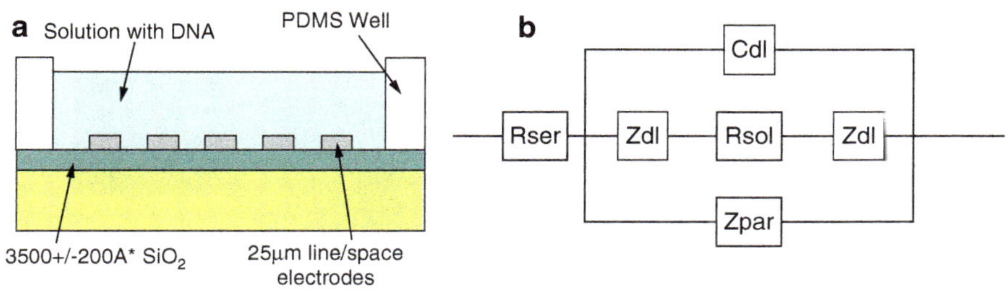

Fig. 2.35 (**a**) Schematic of the sensing device (**b**) Equivalent circuit model of solution with DNA molecules (Reproduced from Liu et al. [123] with permission from the American Institute of Physics)

Impedimetric sensors measure the changes in impedance when biomolecules bind to the electrode hindering the electron transfer. Based on this, carbon electrodes deposited with PBA-modified graphene oxide were reported for detection of glycated haemoglobin (HbA1c) for diagnosis of diabetes [122].

Liu et al. reported a device for label-free electrical detection of DNA through impedance measurement (Fig. 2.35). Impedance vs. frequency characteristic was extracted to detect the presence of DNA molecules in a nanomolar range for a 400 bp molecule [123]. Such electrochemical sensors have great potential for point-of-care diagnostic devices, providing low-cost, high precision and portability.

6.2 *Optical Sensing*

In the recent decade optical biosensors have drawn extensive attention and have rapidly developed changing the face of communication technologies. Due to its advantages like high sensitivity, low-weight, high capacity to transfer information, invulnerability to electromagnetic interference and low-cost, varied applications are integrating these sensors [124]. The sensing mechanism works on the basis of changes in optical properties such as UV–Vis absorption, bio or chemiluminescence, reflectance and fluorescence brought by the interaction of the biocatalyst with the target analyte [125].

A SU8 based optical biosensor has been reported by Yardi et al. A novel technique has been developed to tag standalone optical fibres to a substrate using laser exposed SU8 micro-droplet (Fig. 2.36). The stitched optical fibres show high transmissibility for both aligned and misaligned configurations of the fibres [126].

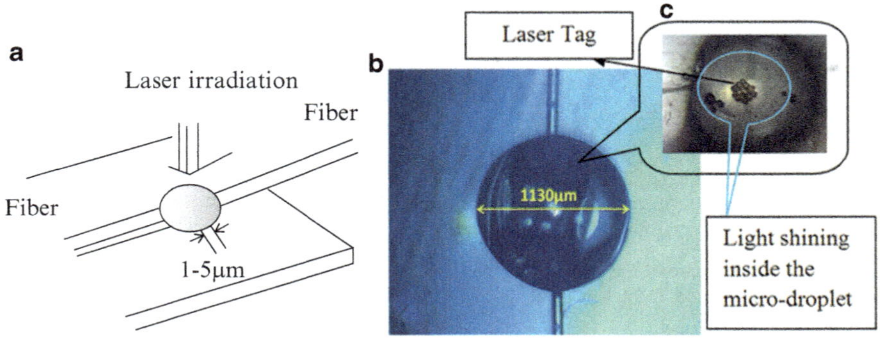

Fig. 2.36 (**a**) Schematic of fibers connected to SU8 droplet (**b**) Optical micrograph of the SU8 droplet (**c**) Image of light shining in the droplet (**c**) Image of light shining in the droplet (Reproduced from Yardi et al. [126], manuscipt accepted and under publising in Springer)

6.3 Mass Based Sensing

Mass based sensors rely on transduction of mechanical energy. Cantilever based mass sensors are the most commonly used mechanical sensors that measure changes in mass through its oscillating frequency shift. Cantilevers came to use as highly sensitive biosensors after the advent of atomic force microscopy (AFM) [127]. They were used as a tip in AFM to measure the force between the tip and the sample through tip deflection or changes in resonant frequency of a vibrating cantilever. A micro-cantilever mass based sensor works on the same principles as above. Interaction of the cantilever with biomolecules changes the resonant frequency of the cantilever [128]. There exists a linear relationship between the added mass and the shift in the resonant frequency of the cantilever. The sensitivity of these devices can be increased by increasing the quality factor, decreasing the mass of the cantilever and using highly sensitive detection instruments to measure small variations in frequency. Several fields like communication, sensing, optics, optical microscopy etc. extensively use these cantilever based devices.

Nano-cantilevers have been reported with detection limits in the range of zeptograms (10^{-21} g) for top-down technology and yoctograms (10^{-24} g) for bottom-up technologies [129]. Nano-cantilever based sensors have reduced sensitivity in liquids due to large damping. A novel sensing mechanism which uses excitation of higher vibration modes to increase the sensitivity has been reported by Braun et al. Bacterial virus T5 interactions were observed using this gold coated cantilever in parallel to measurements at the reference cantilever [130]. Using this mechanism, the quality factor has been increased from 1 (at the 1st mode) to 30 (at 16th mode). Naik et al. have demonstrated high frequency cantilever-based resonators for biomolecular spectrometry. IgM protein was detected in real-time by observing the jumps in the resonance frequency as the protein gets adsorbed onto the cantilever one by one [131]. Utilization of amplitude shift instead of a resonant

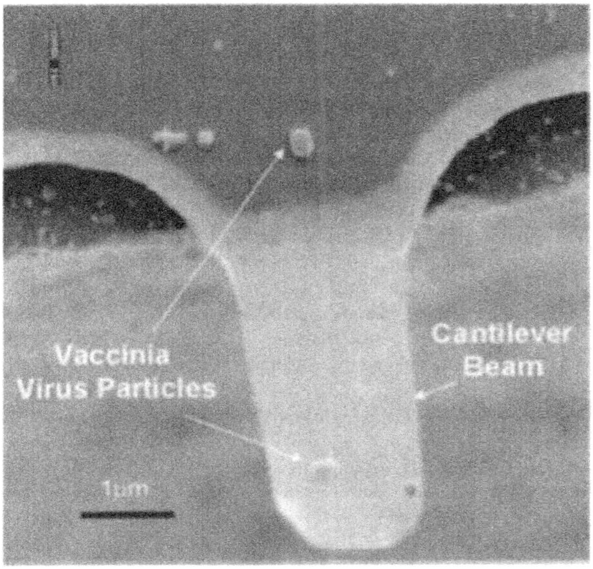

Fig. 2.37 SEM image showing the cantilever with a vaccinia virus particle (Reproduced from Gupta et al. [133] with permission from the American Institute of Physics)

frequency shift to increase the sensitivity and eliminate the frequency tracking elements from the sensor has been reported by Kumar et al. [132].

Gupta et al. used silicon cantilevers to detect bacterial cells (Listeria cells) and viruses (vaccinia virus) (Fig. 2.37). Single virus detection was carried out on 4–5 μm long, 1–2 μm wide and 20–30 nm thick cantilevers [133, 134]. Detection of Bacillus anthracis spores in air and water using a planar rectangular shaped cantilever was demonstrated by Davilia et al. [135].

BioMEMS have made a significant impact on clinical diagnostics and the affordability of these devices. They are being used for detection of protein, DNA, detection of cancer, HIV and other infectious diseases. They have enabled low-cost, rapid and point-of-care diagnostic of diseases allowing early and better diagnosis.

References

1. Whitesides GM (2006) The origins and the future of microfluidics. Nature 442 (7101):368–373
2. Fair RB (2007) Digital microfluidics: is a true lab-on-a-chip possible? Microfluid Nanofluid 3 (3):245–281
3. Srinivasan V, Pamula V, Fair R (2004) An integrated digital microfluidic lab-on-a-chip for clinical diagnostics on human physiological fluids. Lab Chip 4:30–315
4. http://www.drdo.gov.in/drdo/data/Laser%20and%20its%20Applications
5. Robert M, Rossier J, Bercier P, Girault H (1997) UV laser machined polymer substrates for the development of microdiagnostic systems. Anal Chem 69:2035
6. Hellaman A, Rau K, Yoon H, Bae S, Palmer J, Philips K, Albritton N, Venugopal V (2007) Laser-induced mixing in microfluidic channels. Anal Chem 79:4484–4492

7. Quinto-Su PA, Lai HH, Yoon HH, Sims CE, Allbritton NL, Venugopalan V (2008) Examination of laser microbeam cell lysis in a PDMS microfluidic channel using time-resolved imaging. Lab Chip 8(3):408–414
8. Cheng J, Wei C, Hsu C, Young T (2004) Direct-write laser micromachining and universal surface modification of PMMA for device development. Sens Actuators 99:186
9. Klank H, Kutter J, Geschke O (2002) CO(2)-laser micromachining and back-end processing for rapid production of PMMA-based microfluidic systems. Lab Chip 2:242
10. Lippert T, Wei J, Wokaun A, Hoogen N, Nuyken O (2000) Polymers designed for laser microstructuring. Appl Surf Sci 168:270
11. Kant R, Gupta A, Bhattacharya S (2015) Studies on CO2 laser micromachining on PMMA to fabricate microchannels for microfluidic applications. In: 5th international and 26th All India Manufacturing Technology, Design and Research Conference (AIMTDR 2014) December 12th–14th, 2014, IIT Guwahati, Assam, India
12. Johnson T, Waddell E, Kramer G, Locascio L (2001) Chemical mapping of hot-embossed and UV-laser-ablated microchannels in poly(methyl methacrylate) using carboxylate specific fluorescent probes. *Applied Surface Science* 181:149–159
13. Terasawa T (1989) 0.3 μm optical lithography using phase-shifting mask. In: Proceedings of SPIE, p 142
14. Lu Y (2006) A digital micro-mirror device-based system for the micro fabrication of complex, spatially patterned tissue engineering scaffolds. J Biomed Mater Res A 77:396–405
15. Apte P, Rizvi NH (2002) Developments in laser micro-machining techniques. J Mater Process Technol 127:206–210
16. Kumar A, Gupta A, Kant R, Akhtar SN, Tiwari N, Ramkumar J, Bhattacharya S (2013) Optimizaton of laser machining processes for the preparation of photomasks, and its application to microsystem fabrication. J Micro/Nnolith MEMS and MOEMS 13(1):1–8
17. https://www.memsnet.org/mems/processes/lithography.html
18. Whitesides G, Ostuni E, Takayama S, Jiang X, Ingber D (2001) Soft lithography in biology and biochemistry. Ann Rev Biomed Eng 33:335–373
19. Duffy D, Mcdonald J, Schueller O, Whitesides G (1998) Rapid prototyping of microfluidic system in PDMS. Anal Chem 70:4974–4984
20. McDonald J, Duffy D, Anderson J, Chiu D, Wu H, Schueller O, Whitesides G (2000) Fabrication of microfluidic systems in poly(dimethylsiloxane). Electrophoresis 21:27–40
21. Folch A, Toner M (2000) Microengineering of cellular interaction. Ann Rev Biomed Eng 2:227–256
22. Ismagilov R, Rosmarin D, Kenis P, Chiu D, Zhang W, Stone H, Whitesides G (2001) Pressure-driven laminar flow in tangential microchannels: an elastomeric microfluidic switch. Anal Chem 73:4682–4687
23. Chou S, Krauss P, Renstrom P (1996) Nanoimprint lithography. J Vac Sci Technol B 14:4129
24. Whitesides S, Whitesides G (2000) Fabrication of topologically complex three-dimensional microfluidic systems in PDMS by rapid prototyping. Anal Chem 72:3158–3164
25. Singh RK, Ghubade A, Basu B, Bhattacharya S (2009) A novel replicamoulding process for realizing three dimensional microchannels within soft materials. In: ICEMS
26. Singh RK (2014) Micro-manufacturing of 2/3 dimensional
27. Singh RK, Kant R, Pandey SS, Asfer M, Bhattacharya B, Panigrahi PK, Bhattacharya S (2013) Passive vibration damping using polymer pads with microchannel arrays. Microelectromech Syst 22(3):695–707
28. Rajeev Kumar Singh AGSB (2013) Design and fabrication of 3-dimensional helical structure in polydimethysiloxane for flow control applications. Microsyst Technol, pp 1–11
29. Singh RK, Kant R, Singh S, Suresh E, Gupta A, Bhattacharya S (2015) A novel helical microvalve for embedded micro-fluidic applications. Microfluid Nanofluid 19(1):19–29
30. Singh RK, Ghubade A, Chaudhury R, Bhattacharya S (2009) Fabrication technology forbiomedical systems using non-conventional micromachining

31. Osborn JL, Barry L, Elain L, Peter K (2010) Microfluidics without pumps: reinventing the T-sensor and H-filter in paper networks. Lab Chip 10(20):2659–2665
32. Bhattacharya S, Jordan B, Darryl J, Gangopadhyay S (2003) A flow visualization experiment for a first course in micro-fluidics. In: Proceedings of ASEE, University of Texas at Arlington, TX
33. Nguyen N-T, Wu Z (2005) Micromixers—a review. J Micromech Microeng 15 R1–R16
34. Nam-Trung N, Wu Z (2004) Micromixers—a review. J Micromech Microeng 15(2):1–16
35. Kant R, Singh H, Bhattacharya S Nano-scale particle etching using micro-mixer. Microfluid Nanofluid, Manuscript under review
36. Zhang X, Jiang XN, Sun C (1999) Micro-stereolithography of polymeric and ceramic microstructures. Sens Actuators A 77(2):149–156
37. Richard BG, Hahn EL (1975) Coherent two-photon processes: transient and steady-state cases. Phys Rev A 11(5):1641
38. Choudhary R, Bhakat T, Singh RK, Ghubade A, Mandal S, Ghosh A, Rammohan A, Sharma A, Bhattacharya S (2011) Bilayer staggered herringbone micro-mixers with symmetric and asymmetric. J Microfluid Nanofluid 10:271–286
39. Smits JG (1990) Piezoelectric micropump with three valves working peristaltically. Sens Actuators A21–A23:203–206
40. Trouchet D, Ajdari A, Tabeling P, Goulpeau J (2005) Experimental study and modeling of polydimethylsiloxane peristaltic micropumps. J Appl Phys 98:044914
41. Stemme E, Stemme G (1993) A valveless diffuser/nozzle-based fluid. Sens Actuators 39:159–167
42. Carrozza M, Croce N, Magnani B, Dario P (1995) A piezoelectricdriven stereolithography-fabricated micropump. J Micromech Microeng 5:177–179
43. Böhm S, Timmer B, Olthuis W, Bergveld P (2002) A closed-loop controlled electrochemically actuated micro-dosing system. J Micromech Microeng 10:498–504
44. Jeong O, Park S, Yang S, Pak J (2005) Fabrication of a Peristaltic PDMS micro pump. Sens Actuators 123:453–458
45. Kant R, Singh H, Nayak M, Bhattacharya S (2013) Optimization of design and characterization of a novel micro-pumping system with peristaltic motion. Microsyst Technol 19 (4):563–575
46. Atwe A, Gupta A, Kant R, Das M, Sharma I, Bhattacharya S (2014) A novel microfluidic switch for pH control using Chitosan based hydrogels. Microsyst Technol 20(7):1373–1381
47. Singh RK, Kumar A, Kant R, Gupta A, Suresh E, Bhattacharya S (2014) Design and fabrication of 3-dimensional helical structures in polydimethylsiloxane for flow control applications. Microsyst Technol 20(1):101–111
48. Ghoshdastider S, Barizuddin S, Dweik M, Almasri M (2013) A micromachined impedance biosensor for accurate and rapid detection of E. coli O157: H7. RSC Adv 3(48):26297–26306
49. Boehm A, Gottlieb P, Hua S (2007) On-chip microfluidic biosensor for bacterial detection and identification. Sens Actuators B 126(2):508–514
50. Campbell GA, Mutharasan R (2005) Detection and quantification of proteins using self-excited PZT-glass millimeter-sized cantilever. Biosens Bioelectron 21(4):597–607
51. Weeks B, Camarero J, Noy A, Miller A, Yoreo JD (2003) Development of a microcantilever-based pathogen detector. In: Nanotech
52. Ilic B, Yang Y, Craighead H (2004) Virus detection using nanoelectromechanical devices. Appl Phys Lett 85(13):2604–2606
53. Ilic B, Czaplewski D, Zalalutdinov M, Craighead H, Neuzil P, Campagnolo C, Batt C (2001) Single cell detection with micromechanical oscillators. J Vac Sci Technol 19(6):2825–2828
54. Basu AK, Bhattacharya S (2016) Fabrication and resilience measurement of thin aluminium cantilevers using scanning probe microscopy. Taylor and Francis, London
55. Yan G, Chan PC, Hsing I et al (2001) An improved TMAH Si-etching solution without attacking exposed aluminum. Sens Actuator 89:135–141

56. Basu AK, Dwivedi P, Bhattacharya S (2016) Fabrication of 3-dimensional interdigited structure in microfluidic channel by one step maskless greyscale lithography. In: Bangalore India Nano, Bangalore
57. Chen Y-H, Kuo ZK, Cheng C-M (2015) Paper—a potential platform in pharmaceutical development. Trends Biotechnol 33(1):4–9
58. Cheng CM, Martinez AW, Gong J, Mace CR, Phillips ST, Carrilho E, Mirica KA, Whitesides GM (2010) Paper-based ELISA. Angew Chem 49:4771–4774
59. Nie J, Zhang Y, Lin L, Zhou C, Li H, Zhang L, Li J (2012) Low-cost fabrication of paper-based microfluidic devices by one-step plotting. Anal Chem 84(15):6331–6335
60. Li X, Tian J, Nguyen T, Shen W (2008) Paper-based microfluidic devices by plasma treatment. Anal Chem 80:9131–9134
61. Dungchai W, Chailapakul O, Henry C (2011) A low-cost, simple, and rapid fabrication method for paper-based microfluidics using wax screen-printing. Analyst 136(1):77–82
62. Li X, Tian J, Garnier G, Shen W (2010) Fabrication of paper-based microfluidic sensors by printing. Colloid Surf B 76(2):564–570
63. Olkkonen J, Lehtinen K, Erho T (2010) Flexographically printed fluidic structures in paper. Anal Chem 82:10246–10250
64. Chitnis G, Ding Z, Chang C-L, Savran CA, Ziaie B (2011) Laser-treated hydrophobic paper: an inexpensive microfluidic platform. Lab Chip 11:1161–1165
65. Zhang Y, Bai J, Ying JY (2015) A stacking flow immunoassay for the detection of dengue-specific immunoglobulins in salivary fluid. Lab Chip 15:1465–1471
66. Koesdjojo MT, Pengpumkiat S, Wu Y, Boonloed A, Huynh D, Remcho TP, Remcho VT (2015) Cost effective paper-based colorimetric microfluidic devices and mobile phone camera readers for the classroom. J Chem Educ 92:737–741
67. Unicef (2007) Malaria diagnosis: a guide for selecting rapid diagnostic test (RDT) kits
68. Pereira DY, Chiu RY, Zhang SC, Wu BM, Kamei DT (2015) Single-step, paper-based concentration and detection of a malaria biomarker. Anal Chim Acta 882:83–89
69. Weaver AA, Lieberman M (2015) Paper test cards for presumptive testing of very low quality antimalarial medications. Am Soc Tropical Med Hygiene 92:17–23
70. Cordray MS, Kortum RRR (2015) A paper and plastic device for the combined isothermal amplification and lateral flow detection of Plasmodium DNA. Malar J 14:1
71. Kumar A, Hens A, Arun RK, Chatterjee M, Mahato K, Layek K, Chanda N (2015) A paper based microfluidic device for easy detection of uric acid using positively charged gold nanoparticles. Analyst 140:1817–1821
72. Teoh BT, Sam SS, Tan KK, Danlami MB, Shu MH, Johari J, Hooi PS, Brooks D, Piepenburg O, Nentwich O, Smith AW, Franco L, Tenorio A, AbuBakar S (2015) Early detection of dengue virus by use of reverse transcription recombinase polymerase amplification. J Clin Microbiol 53:830–837
73. Yen C-W, Puig HD, Tam JO, Márquez JG, Bosch I, Schifferli KH, Gehrke L (2015) Multicolored silver nanoparticles for multiplexed disease diagnostics: distinguishing dengue, yellow fever, and Ebola viruses. Lab Chip 15:1638–1641
74. Lo S-J, Yang S-C, Yao D-J, Chen J-H, Tu W-C, Cheng C-M (2013) Molecular-level dengue fever diagnostic devices made. Lab Chip 13:2686–2692
75. Hamraoui A, Nylander T (2002) Analytical approach for the Lucas–Washburn equation. J Colloid Interface Sci 250:415–421
76. Byrnes S, Thiessen G, Fu E (2013) Progress in the development of paper-based diagnostics for low-resource point-of-care settings. Bioanalysis 5:2821–2836
77. Fu E, Ramsey SA, Kauffman P, Lutz B, Yager P (2011) Transport in two-dimensional paper networks. Microfluid Nanofluid 10:29–35
78. Mansfield MA (2005) The use of nitrocellulose membranes in lateral-flow assays. In: Drugs of abuse. Springer, New York, pp 71–85
79. Li B, Fang X, Luo H, Peterson E, Seo Y-S, Samuilov V, Rafailovich M, Sokolov J, Gersappe D, Chu B (2006) Influence of electric field intensity, ionic strength, and migration

distance on the mobility and diffusion in DNA surface electrophoresis. Electrophoresis 27 (7):1312–1321
80. Lee HH, Kuo Y (2008) Surface modification of Gel-Free microchannel surface electrophoresis device for DNA identification. Jpn J Appl Phys 47(4R):2300
81. Ghosh A, Patra TK, Kant R, Singh RK, Singh JK, Bhattacharya S (2011) Surface electrophoresis of ds-DNA across orthogonal pair of surfaces. Appl Phys Lett 98(16):164102
82. Jones TB, Jones TB (2005) Electromechanics of particles. Cambridge University Press, Cambridge
83. Price JA, Burt JP, Pethig R (1988) Applications of a new optical technique for measuring the dielectrophoretic behaviour of micro-organisms. Biochim Biophys Acta 964(2):221–230
84. Voldman J, Gray ML, Toner M, Schmidt MA (2002) A microfabrication-based dynamic array cytometer. Anal Chem 74(16):3984–3990
85. Iliescu C, Xu GL, Samper V, Tay FE (2004) Fabrication of a dielectrophoretic chip with 3D silicon electrodes. J Micromech Microeng 15(3):494
86. Lin IJ, Benguigui L (1982) High-intensity, high-gradient electric separation and dielectric filtration of particulate and granular materials. J Electrostat 13(3):257–278
87. Chou C-F, Tegenfeldt JO, Bakajin O, Chan SS, Cox EC, Darnton N, Duke T, Austin RH (2002) Electrodeless dielectrophoresis of single-and double-stranded DNA. Biophys J 83 (4):2170–2179
88. McGraw GJ, Davalos RV, Brazzle JD, Hachman JT, Hunter MC, Chames JM, Fiechtner GJ, Cummings EB, Fintschenko Y, Simmons BA (2005) Polymeric microfluidic devices for the monitoring and separation of water-borne pathogens utilizing insulative dielectrophoresis. In: MOEMS-MEMS Micro and Nanofabrication, pp 59–68
89. Shafiee H, Caldwell JL, Sano MB, Davalos RV (2009) Contactless dielectrophoresis: a new technique for cell manipulation. Biomed Microdevices 11(5):997–1006
90. Demierre N, Braschler T, Muller R, Renaud P (2008) Focusing and continuous separation of cells in a microfluidic device using lateral dielectrophoresis. Sens Actuators B 132 (2):388–396
91. Hoeb M, Radler JO, Klein S, Stutzmann M, Brandt MS (2007) Light-induced dielectrophoretic manipulation of DNA. Biophys J 93(3):1032–1038
92. Jaramillo MC, Torrents E, Martin-Duarte R, Madou MJ, Juarez A (2010) On-line separation of bacterial cells by carbon-electrode dielectrophoresis. Electrophoresis 31(17):2921–2928
93. Saucedo-Espinosa MA, LaLonde A, Gencoglu A, Romero-Creel MF, Dolas JR, Lapizco-Encinas BH (2016) Dielectrophoretic manipulation of particle mixtures employing asymmetric insulating posts. Electrophoresis 37(2):282–290
94. Bakewell DJ, Morgan H (2006) Dielectrophoresis of DNA: time-and frequency-dependent collections on microelectrodes. IEEE Trans NanoBiosci 5(1):1–8
95. Kasahara H, Ding Z, Nakano M, Suehiro J (2015) Effect of DNA length on dielectrophoretic characteristics of DNA-labeled microbeads. In: IEEE international conference on industrial technology, pp 3341–3346
96. Wilding P, Kricka LJ (1996) Polymerase chain reaction. U.S. Patent 5,587,128, 24
97. Bhattacharya S, Gao Y, Korampally V, Othman MT, Grant S, Kleibceker SB, Gangopadhyay K, Gangopadhyay S (2007) Optimization of design and fabrication processes for realization of a PDMS-SOG-silicon DNA amplification chip. J Microelectromech Syst 16 (2):401–410
98. Nayak M, Singh D, Singh H, Kant R, Gupta A, Pandey SS, Mandal S, Ramanathan G, Bhattacharya S (2013) Integrated sorting, concentration and real time PCR based detection system for sensitive detection of microorganisms. Sci Rep 3:3266
99. Bhattacharya S, Salamat S, Morisette D, Banada P, Akin D, Liu Y-S, Bhunia AK, Ladisch M, Bashir R (2008) PCR-based detection in a micro-fabricated platform. Lab Chip 8 (7):1130–1136
100. Korampally V, Bhattacharya S, Gao Y, Grant S, Kleiboeker SB, Gangopadhyay K, Tan J, Gangopadhyay S (2006) Optimization of fabrication process for a PDMS-SOG-Silicon based

PCR Micro Chip through system identification techniques. In: IEEE international symposium on computer-based medical system, pp 329–334
101. Cai M, Li F, Zhang Y, Wang Q (2010) One-pot polymerase chain reaction with gold nanoparticles for rapid and ultrasensitive DNA detection. Nano Res 3(8):557–563
102. Deng H, Xu Y, Liu Y, Che Z, Guo H, Shan S, Sun Y, Liu X, Huang K, Ma X, Wu Y (2012) Gold nanoparticles with asymmetric polymerase chain reaction for colorimetric detection of DNA sequence. Anal Chem 84(3):1253–1258
103. Shen H, Hu M, Yang Z, Wang C, Zhu L (2005) Polymerase chain reaction of Au nanoparticle-bound primers. Chin Sci Bullet 50(18):2016–2020
104. Zimmerman U, Pilwat G, Riemann F (1974) Dielectric breakdown of cell membranes. In Membrane transport in plants, pp 146–153
105. Patel VK, Kant R, Bhatt G, Ganguli A, Singh D, Nayak M, Gupta A, Gangopadhyay K, Gangopadhyay S, Gurunath R, Bhattacharya S Synchronized electro-mechanical shock wave induced bacterial transformation, Paper under review
106. Grayson A, Shawgo R, Johnson A, Flynn N, Li Y, Cima M (2004) A BioMEMS review: MEMS technology for physiologically integrated devices. Proc IEEE 92:6–21
107. Grieshaber D, MacKenzie R, Voeroes J, Reimhult E (2008) Electrochemical biosensors-sensor principles and architectures. Sensors 8:1400–1458
108. Ponomareva O, Arlyapov V, Alferov V, Reshetilov A (2011) Microbial biosensors for detection of biological oxygen demand (a Review). Appl Biochem Microbiol 47:1–11
109. Mu Y, Jia D, He Y, Miao Y, Wu H (2011) Nano nickel oxide modified non-enzymatic glucose sensors with enhanced sensitivity through an electrochemical process strategy at high potential. Biosens Bioelectron 26:2948–2952
110. Yang Y, Chuang M, Lou S, Wang J (2010) Thick-film textile-based amperometric sensors and biosensors. Analyst 135:1230–1234
111. Gao Y, Bhattacharya S, Chen X, Barizuddin S, Gangopadhyay S, Gillis K (2009) A microfluidic cell trap device for automated measurement of quantal catecholamine release from cells. Lab Chip 9:3442–3446
112. Piloto C, Notarianni M, Shafiei M, Taran E, Galpaya D, Yan C (2014) Highly NO2 sensitive caesium doped graphene oxide conductometric sensors. Beilstein J Nanotechnol 5:1073–1081
113. Latif U, Dickert F (2011) Conductometric sensors for monitoring degradation of automotive engine oil. Sensors 11:8611–8625
114. Dzydevich S, Shul'ga A, Soldatkin A, Nyamsi Hendji A, Jaffrezic-Renault N, Martelet C (1994) Application of conductometric for sensitive detection of pesticides biosensor based on the cholinesterases. Electroanalysis 6:752–758
115. Zhylyak G, Dzyadevich S, Korpan Y, Soldatkin A, El'Skaya A (1995) Application of urease conductometric biosensor for heavy-metal ion determination. Sens Actuators B 24:145–148
116. Soldatkin A, Dzyadevich S, Korpan Y, Arkhipova V, Zhylyak G, Piletsky S (1998) Biosensors based on conductometric detection. Biopolymers Cell 14:268
117. Anh T, Dzyadevych S, Van M, Renault N, Duc C, Chovelon J (2004) Conductometric tyrosinase biosensor for the detection of diuron, atrazine and its main metabolites. Talanta 63:365–370
118. Xuejiang W, Dzyadevych S, Chovelon J, Renault N, Ling C, Siqing X (2006) Conductometric nitrate biosensor based on methyl viologen/Nafion®/nitrate reductase interdigitated electrodes. Talanta 69:450–455
119. Wang Y, Zhang Z, Jain V, Yi J, Mueller S, Sokolov J (2010) Potentiometric sensors based on surface molecular imprinting: detection of cancer biomarkers and viruses. Sens Actuators B 146:381–387
120. Bandodkar A, Hung V, Jia W, Valdés-Ramírez G, Windmiller J, Martinez A (2013) Tattoo-based potentiometric ion-selective sensors for epidermal pH monitoring. Analyst 138:123–128

121. Novell M, Parrilla M, Crespo G, Rius F, Andrade F (2012) Paper-based ion-selective potentiometric sensors. Anal Chem 84:4695–4702
122. Siva Rama Krishna V, Bhat N, Amrutur B, Chakrapani K, Sampath S (2011) Detection of glycated hemoglobin using 3-aminophenylboronic acid modified graphene oxide. Life Science Systems and Applications Workshop (LiSSA), 2011 IEEE/NIH, pp 1–4
123. Liu Y, Banada P, Bhattacharya S, Bhunia A, Bashir R (2008) Electrical characterization of DNA molecules in solution using impedance measurements. Appl Phys Lett 92:143902
124. Yang M, Li S, Jiang D (2014) Review on optical fiber sensing technologies for industrical applications at the NEL-FOST. In: EWSHM-7th European workshop on structural health monitoring
125. Bhattacharya S, Jang J, Yang L, Akin D, Bashir R (2007) BioMEMS and nanotechnology-based approaches for rapid detection of biological entities. J Rapid Meth Automat Microbiol 15:1–32
126. Seema Yardi A, Kant R, Boolchandani D, Bhattacharya S (2015) High efficiency coupling of optical fibres with SU8 micro-droplet using laser welding process. Lasers in manufacturing and materials processing, pp 1–17
127. Binnig G, Quate C, Gerber C (1986) Atomic force microscope. Phys Rev Lett 56:930
128. Lavrik N, Sepaniak M, Datskos P (2004) Cantilever transducers as a platform for chemical and biological sensors. Rev Sci Instrum 75:2229–2253
129. Tamayo J, Kosaka P, Kosaka J, San Paulo A, Calleja M (2013) Biosensors based on nanomechanical systems. Chem Soc Rev 42:1287–1311
130. Braun T, Ghatkesar M, Backmann N, Grange W, Boulanger P, Letellier L (2009) Quantitative time-resolved measurement of membrane protein–ligand interactions using microcantilever array sensors. Nat Nanotechnol 4:179–185
131. Naik A, Hanay M, Hiebert W, Feng X, Roukes M (2009) Towards single-molecule nanomechanical mass spectrometry. Nat Nanotechnol 4:445–450
132. Kumar V, Boley W, Yang Y, Ekowaluyo H, Miller J, Chiu G (2011) Bifurcation-based mass sensing using piezoelectrically-actuated microcantilevers. Appl Phys Lett 98:153510
133. Gupta A, Akin D, Bashir R (2004) Single virus particle mass detection using microresonators with nanoscale thickness. Appl Phys Lett 84:1976–1978
134. Gupta A, Akin D, Bashir R (2004) Detection of bacterial cells and antibodies using surface micromachined thin silicon cantilever resonators. J Vacuum Sci Technol B 22:2785–2791
135. Davila A, Jang J, Gupta A, Walter T, Aronson A, Bashir R (2007) Microresonator mass sensors for detection of Bacillus anthracis Sterne spores in air and water. Biosens Bioelectron 22:3028–3035

Chapter 3
Manufacturing Methods Overview for Rapid Prototyping

Nikolay Dimov

1 Motivation

Microfabrication is seldom in the focus of a biological studies but it influences the performance and functionality of a miniaturised fluidic platform that affects research outcomes. The scope of this chapter falls on methods, which are accessible, reproducible, and applicable for small-scale, laboratory production. However, fundamental techniques such as photo- and soft lithography are also briefly introduced with references to more detailed sources to facilitate the reader in his choice of fabrication method. It is impossible to cover all the various aspects of microfabrication with extensive details in a single chapter, instead an overview of the rapid and inexpensive methods are conveyed.

2 Selecting a Fabrication Method

Designing a microfluidic platform passes through multiple stages and iterations. Typically, the design process starts with listing all the functionalities, requirements and parameters that overarch the functionality of the device (Fig. 3.1). Then critical design dimensions can be determined using background knowledge about the on-chip operations and the laws of physics that govern them. The critical dimensions serve as a base to select the fabrication process by which the first prototype can be manufactured. With the fabrication method in mind a suitable material for manufacturing prototype device is selected. As a designer you need to determine number of suitable substrates based on those operational criteria, set by target

N. Dimov (✉)
Advanced Centre for Biochemical Engineering, University College London, London, UK
e-mail: n.dimov@ucl.ac.uk

C.K. Dixit, A. Kaushik (eds.), *Microfluidics for Biologists*,
DOI 10.1007/978-3-319-40036-5_3

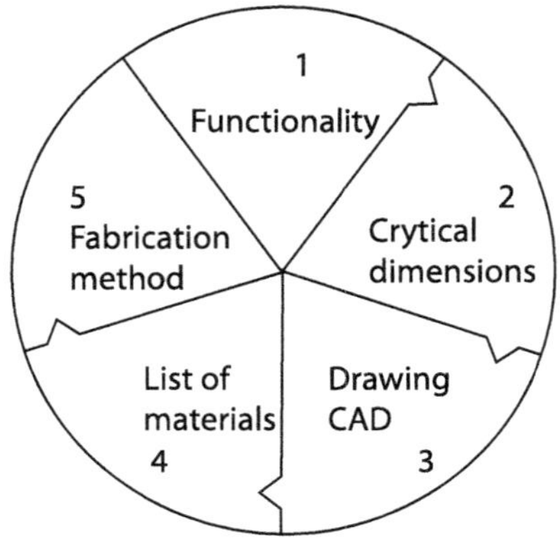

Fig. 3.1 Design iterations. The design cycle begins with list of target functionalities, followed by calculation of the critical dimensions, technical computer-aided drawing, determination of suitable materials and fabrication methods that can obtain those dimensions

usability. Accessibility to equipment and techniques are also indispensable part of the selection process. Scale of production is another factor, which plays role in method selection. Last but not least comes available time for manufacturing and testing complex variations in the device architecture. It is important to think about both experimental and fabrication time as factors that need to be considered when choosing a fabrication strategy. Minimizing the fabrication guarantees more project time on experiments.

Fabrication methods can be grouped in direct: those translating the designed features in one rapid step; and indirect—translation occurs in multiple stages between the creation of the technical drawing and the production of the features (Fig. 3.2). An example for direct method is laser ablation, where a computer-aided design (CAD) file is sent to the laser, and the fabrication can start after a definition of the power, speed and firing rate of the laser head.

The choice of microfabrication methods depends (1) on the functionality that you want to achieve with your device and (2) on the overall operational conditions. A system designer takes in consideration the extreme working range of conditions where the device should remain operational. Typically, for laboratory based biological investigations there are no extremums in temperature as most applications run below 40 °C. But if your microfluidic chip is meant as reusable, autoclavable and resistant to polar aliphatic solvents, such as acetone for instance, then a proper material should be selected (Table 3.1). This means that a material must meet the following criteria: glass transition temperature above 121 °C, so it can be autoclaved; plus, it must have high resistance towards solvents. The suitable candidates, polycarbonate (PC) and cyclic olefin copolymer (COC), are then considered with regards to all available fabrication techniques. Some additional requirement is necessary to narrow down the list of possible fabrication candidates and to select one. Such selection criterion is the maximum possible resolution of the fabrication method. If we return to the hypothetical case study, and for argument sake a minimum feature size of 50 μm is considered for fully-fledged device, then the viable options are milling, hot-embossing casting and moulding. However, economical and time constrains play significant role in the decision making process

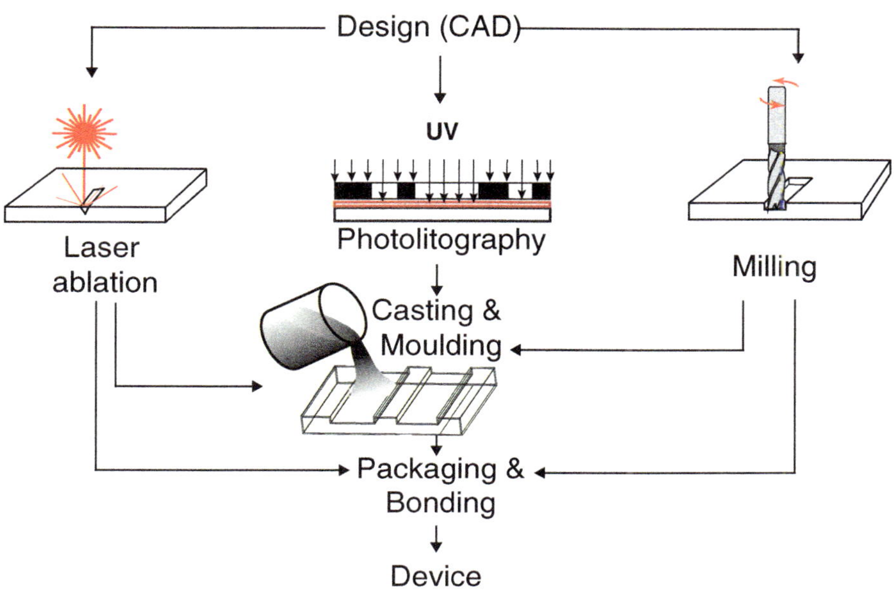

Fig. 3.2 Schematic representation of rapid prototyping techniques, which are commonly implemented in microfluidics. Laser ablation and milling are direct fabrication methods because of fast turnover times between design and ready-to-use device in minimum number of stages. Indirect methods are photolithography, hot embossing, casting and moulding as they all consist of intermediate stages. Tentatively, laser ablation can be used to create moulds for casting, shown here with a dashed line

across all production scales. These constrains define our next criteria—time. From the three enumerated methods milling is the only direct method; it provides a single chip after a short processing time. In case the design stage is more advanced and your device has performed well at the initial tests, then a mould can be made for casting or hot embossing of multiple copies.

2.1 Photolithography and Mask Design

Photolithography is a keystone in the fabrication of microelectronics and microelectromechanical systems (MEMS). A stencil mask is used to transfer the designed patterns onto a wafer (Fig. 3.3). This technique requires dust-free manufacturing conditions plus strict temperature and humidity control. Such conditions are maintained in clean rooms with ISO standardised classes, depending on the content of dust particles per cubic meter of air. Clean room processes are costly because of the high maintenance expenditures; in addition, to gain access one must follow multiple courses that can be also time intensive. Outsourcing the fabrication

Table 3.1 Physical properties of plastics commonly used for microfluidics

Plastic	Properties	Mechanical strength	Chemical resistance	Thermal resistance	Functional groups	Fabrication method	References
Polycarbonate (PC)	Transparent in visible range	Good	Fair	Very good	Carbonate	Milling	[1]
						Hot embossing	
						Injection moulding	
Polymethylmethacrylate (PMMA)	Transparent	Very good	Fair	Good	Acrylate	Milling	[1]
						Hot embossing	
						Laser ablation	
Polyvinyl chloride (PVC)	High flexibility	Good	Very good	Good	Chloride	Milling	
						Hot embossing	
						Injection moulding	
Cyclic Olefin Copolymer (COC)	Transparent in visible and mid-UV	Very good	Good	Excellent	Olefin	Milling	[1–3]
						Hot embossing	
						Injection moulding	
Poly-ether-ether-ketone (PEEK)	Thermal resistance	Very good	Very good	Excellent	Ketone	Milling	[4]
						Injection moulding	
						Laser ablation	

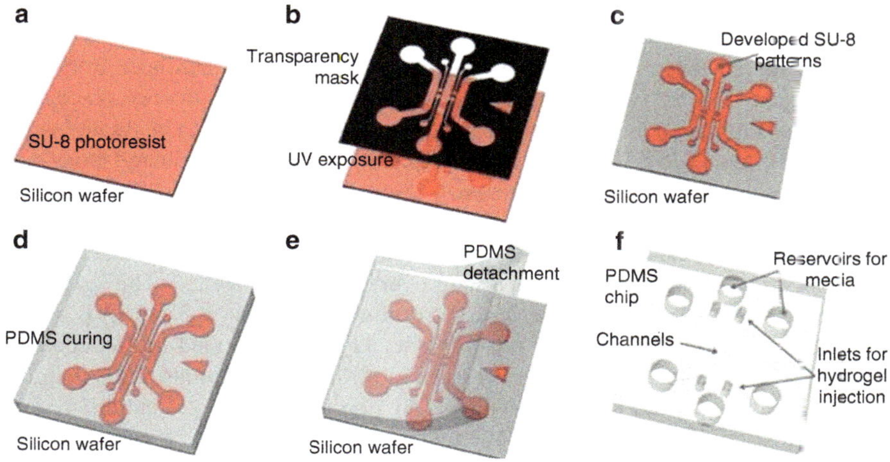

Fig. 3.3 Illustration of the method of photolithography using SU-8 photoresist. As described from (**a**)–(**c**), SU-8 photopolymer is first coated on a isopropanol cleaned silicon wafer, followed by aligning the transparency mask (*black square* in section (**b**)) of the desired microfluidic design. Later, the aligned mask is exposed to UV light, which crosslinks SU-8 in the shape of clear mask region. After mask removal, the SU-8-coated wafer was washed with SU-8 removal solvent. This will remove all the unexposed SU-8; thus, leaving behind desired structures on the mask, as depicted in frame '(**c**)'. On this mould, soft lithography can be used by pouring PDMS and curing it thermally. Upon removal, PDMS will give us the desired dimensioned fluidic network. Reprinted by permission from Macmillan Publishers Ltd: [Nature Protocols, 2012] [5]

is one way of circumventing these apparent obstacles. There are adjacent services and team of professionals who can carry out an operation instead of you. To communicate your plan for photolithography one needs a properly designed mask with definite criteria for fabrication. In the following paragraphs few guidelines for masking will be introduced that will allow you outsourcing of your designs easily.

It is important to know what sizes of masks and substrates are typically used in the cleanroom to which you are outsourcing. Standard 4 in. wafers are processed with 5 in. masks; 3 in. wafers usually require 4 in. masks. Once you know the size of substrate that the process line can handle, maximum lateral features of your design can be determined. Important is to leave free space of about one centimetre over the periphery of the wafer. That space is for handling the wafer manually, using tweezers, without damaging the features close to the periphery. When the design is fully tested and goes to mass cleanroom robotised production the space can be reduced. Photolithography serves as the base for more ubiquitous application, outside of the cleanroom environment. Soft lithography can be achieved in a regular chemical laboratory due to the advances in photopolymerization chemistry and the plethora of commercially available photoresists.

The most common negative photoresist used in microfluidics is SU-8 due to its multiple implementations and well characterised fabrication properties. It is epoxy-based resist that comes in different viscosities, which facilitate the formation of features with various thicknesses, which can be controlled by the spin speed during coating. For exposure i-line (365 nm) is used and exposure time correlates to the thickness of the layer. That information is provided in the datasheet from the manufacturer. The SU-8 has excellent adhesion properties to silicone substrates; demonstrates limited adherence to glass or quartz, the later needs to be treated with HDMS prior to coating. It is important to remember that the illuminated areas are rendered insoluble for negative photoresist. Overexposure is required for complete polymerization, as the process starts from the top, where the light hits first, to the bottom, where the resist adheres to the substrate. Major consequence is that for full polymerization of very thick layers a long overexposure is needed. The scattered light during the extra time diminishes the resolution to about 2–3 μm ([6], p. 7). Multi-step processes are possible when thinner layers are applied in consecutive coating and patterning cycles. However, such multi-layer depositions require alignment and also very thin films can result in non-homogeneous coating in the substrate and pin whole formation. The resulting polymerized SU-8 has excellent optical properties and is suitable for producing integrated optical guides on silicon. As a mould the SU-8 can form undercuts, protruding edges at the top surface that hinder demoulding and can lead to breakage of the silicone master. As shown in Fig. 3.4 the tapered angle can be reversed and such impractical demoulding can be avoided during the fabrication step [3] by using transparent substrate (Pyrex) instead of silicon and by exposing from the bottom.

Typical SU-8 25 series for *vias* and channels follows established protocol [7], which can vary in terms of baking and exposure times related to the thickness (15 μm) of the resist layer. First SU-8 is poured to cover 60 % of the wafer surface.

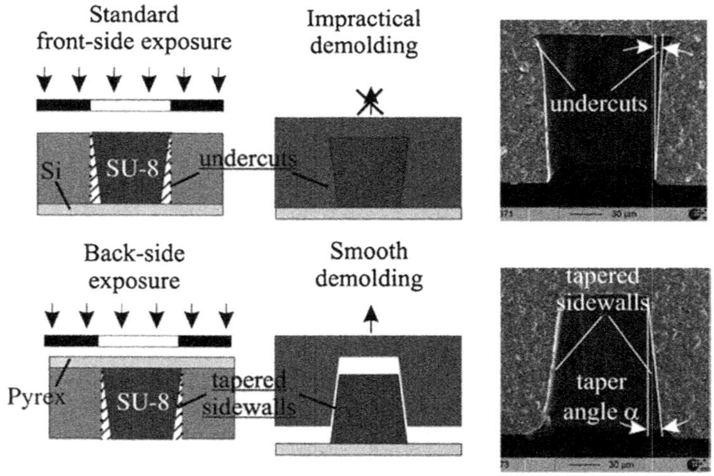

Fig. 3.4 *Top*: standard front side exposure of negative resist tends to form undercut structures at underexposed sidewalls. *Bottom*: backside exposure reversed the effect to form tapered sidewalls (taper angle α = 6°), which allowed facile demolding. Reprinted by permission from Institute of Physics Publishing: [Journal of Micomechanics and Microengineering, 2007] [3]

If it is manually dispensed from a small bottle it is recommended, to minimize the distance between the bottle and the substrate surface. Second, the wafer is spun (e.g., 3000 RPM) to create uniform coating. Third, is soft bake (5 min) at 65 °C, which is followed by (2 min) incubation at 95 °C. Fourth comes the alignment and exposure using hard contact with vacuum for better resolution. A delay before exposure is optionally introduced that lasts from 4 min to 2 hours and serves as stress reducing step in the photoresist. The total exposure time for this particular case (SU-8 25, spun at 3000 RPM) is 60 s on 12 mW/cm^2 EVG mask aligner. Fifth step is post exposure bake, (5 min) at 65 °C followed by 95 °C (2 min). Finally the latent image is developed in REP 600 (5 min) with agitation. The wafer is rinsed with isopropanol and after drying is ready to use. The time values provided in brackets may change as per the datasheet provided by the supplier; depending on the fabrication conditions.

Patterning of positive photoresist (i.e., OiR-, AZ-series or PMMA-based) follows similar processing steps: spin of photoresist; soft bake to remove the solvent from the resulting layer; expose the wafer and resist using UV-light; develop the resulting pattern. To promote adhesion the substrate is treated with HDMS. But the major difference is the polarity of the used mask as after exposure the illuminated areas become soluble. In terms of application, the positive photoresists are not building the channel walls but serve for masking during etching or sputtering followed by lift-off processes.

2.2 *Moulding and Casting*

The principle of moulding is that a master, bearing an inverse image of the targeted structure, is covered with a liquid polymer, or other material that solidifies after exposure to light, temperature or other external factors. As the resin polymerises it mimics the features of the master. At the end of the process the replica is demoulded leaving the master intact, and ready-to-use, which makes the whole process applicable for rapid device generation.

The most encountered moulding material is polydymethylsiloxane PDMS, which is heavily brunched elastomer, that gains its popularity among researchers due to its multiple implementations in cell studies. The properties of PDMS that potentially have adverse effects for microscale cell studies are discussed in a critical review [8] covering the following material characteristics: deformation, evaporation, absorption, leaching of uncrosslinked oligomers, and hydrophobic recovery. It has Young's modulus ranging from 8.7 to 3.6 kPa [9] that is tailored by varying the ratio between the elastomer and crosslinking agent. In this chapter the process flow of PDMS moulding onto SU-8/silicon master is conveyed.

Surface functionalization of the mould is the first step in this process. In the case of silicon substrate a straightforward approach is to use vapour deposition of fluorinated silane to render the surface of the master hydrophobic. For this purpose, the surface of the master is activated using 30 W RF plasma for 1 min. Then the substrate is transferred into a desiccator with 5 μL of Triethoxy(1H,1H,2H,2H-perfluoro-1-octyl)silane. The desiccator is pumped and the master is incubated under vacuum for 4 h. Self-assembly layer of silane is formed on the surface during that process. Next, the master is baked for 1 h to eliminate water and to cross-link the ethoxy-groups thus stabilizing the coating.

The second step is, preparation of mixture of standard 10:1 (w/w) ratio of elastomer and curing agent (Sylgard 184, Dow Corning). PDMS elastomer is very viscous and it adheres to the walls of the beaker, so it is weighted first and in excess. Then the corresponding amount of curing agent is added to that same beaker using a Pasteur pipette. The two components need to be well mixed for a successful polymerisation. Inevitably, bubbles will form while mixing and because of the high viscosity they will remain in the mixture.

To remove the air entrapped in the mixture and to improve the reproducibility between mouldings the mixture is placed inside a desiccator and the chamber is then evacuated. Depending on the amount of mixture the degassing time can vary; the rule of thumb is to keep the degassing until the foaming stops. Typically for 11 g of mixture in 50 mL plastic beaker it would take 4–5 min of degassing. While the mixture is inside the desiccator, the silanized master is placed in a petri dish.

Pouring the degassed mixture onto the master inside the petri dish is the moulding step. Specifically for complex structures air may be trapped onto the master; to avoid it a second degassing step is performed. Once the foaming stops the moulded structure can be placed in an oven for curing at 85 °C for 2.5 h. The final step of the casting process is cutting out the structure from the bulk of the mould and consecutive de-moulding, *i.e.*, detaching the PDMS slap from the master. Holes for connectors are punctured at this stage and the PDMS cast is ready for bonding. In my experience the best way to protect the master from dust or damage is to pour a thin layer of PDMS on top and allow it to cure. This may look like waste of elastomer but actually preserves your masters for future casting and moulding.

This is a protocol for the thermal curing of PDMS (Sylgard 184), other types of PDMS chemistry exists that might be more suitable [10] for the functionality you need. The authors report on hard PDMS with tensile modulus of 8.2 MPa in comparison to 1.8 MPa for the 184-PDMS from Sylgard.

2.3 Hot Embossing

The hot embossing is a rapid fabrication process for thermoplastic material, which is heated above a glass transition temperature, while being pressed against a master. There are three parameters essential for optimisation of hot embossing: the first is holding time, the second temperature on each platen, the third is embossing pressure. We are going to touch upon each of the three in this section. The embossing temperature should exceed by 15–40 °C the glass transition temperature (T_g) of the polymer (Fig. 3.5). Using a higher temperature leads to lower viscosity of the polymer which facilitates filling of the features on the master and high fidelity

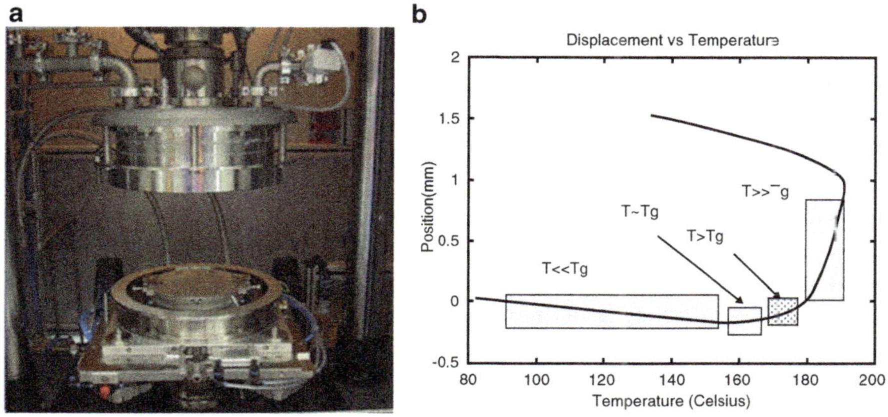

Fig. 3.5 Jenoptik HEX02 hot embossing machine, vacuum chamber and mould insert fixture (*left*). Displacement of the platens as a function of temperature with different states of the polymer material (*right*). Ideal temperatures for bonding is about T > Tg but it can vary depending on the design complexity. Reprinted with the permission of [Springer-Verlag, 2006] [11]

of replication. However, the polymer might degrade if the temperature is too high. Ideally, the temperature of the tool and substrate should be the same within a range of ±2 °C during the holding time. Increasing the holding time gives the polymer more time to flow into the mould resulting in better replication. Also more complex architectures from the master require longer holding times allowing for the molten plastic to fill the features. A higher pressure obviously pushes the polymer faster in to the mould but the disadvantage is that it also increases the risk of damaging the master, especially when using SU-8 on Si wafer. Breaking a master takes you back to photolithography and potentially into the cleanroom, which is time and cost intensive. The parameters are interconnected which makes optimisation nontrivial task.

A conservative approach to optimize the hot embossing consists of isolating a single parameter by keeping the remaining two constant. For instance, hot embossing of PMMA, having glass-transition temperature of 105 °C, will require heating to 130 °C and pressure of 290 N/cm^2 with varying holding times that depend on the complexity of the featured design. Studies exist that develop methodologies for process optimisation [11]. The more advanced optimisation methods include deterministic studies based on numerical simulations and computational fluid dynamic (CFD), which aim to elucidate critical values [12]. In practice the process optimisation targets a specific outcome. For instance to increase the lifespan of the master; first, holding time, then embossing temperature and finally pressure are altered. Such optimisation approach will increase the number of replications per master by reducing the risk of damage, but might also result in prolonged cycle times. In contrast, time efficacy is achieved through changes in embossing pressure, and only then holding time.

An optimised embossing process can use the same master over again without any wearing of its features or compromise of the replica quality. Moreover, the method has short processing times (40 min per device), which makes it suitable for rapid prototyping of devices on small to medium scale [1].

2.4 Micromilling

Micromilling is a subtractive fabrication process based on the mechanical removal of material by a cutting tool from a stock piece, also known as substrate or *workpiece*, resulting in micrometre-sized features. It is a direct method enabling transfer of 3D designs into bulk material, which makes it suitable for machining masters used in hot embossing, injection moulding and resin casting or as an immediate microfluidic device. The micromilling is applied for rapid prototyping in polymer and metal substrates. Resulting architectures can be utilized multiple times as masters for casting resins such as polydimethylsiloxane (*e.g.*, PDMS) and/or can be integrated in the final version of a device. Typically, for microfabrication purposes, dimensions of the features fall in the range from 25 μm; also can decrease down to 3 μm depending on the class and price range [1]. The lower price range machines provide comparable quality and accuracy as their mid-range counterparts (Fig. 3.6a). However, the midrange milling machines come with extra functionalities such as automatic holders and substrate aligners. Variations of milling machine designs exist where additional degrees of freedom are available, *i.e.* rotational axle. The development of milling has shifted the lower limit boundaries even further down (15 μm) lateral obtained with basic functionality CNC, that is available for 15 k USD, thus bringing mechanical methods closer to the level of precision obtained with soft lithography. When automation is required together with high accuracy the price can exceed 200 k but guarantees better quality and high throughput.

2.5 Choice of Instrument and Relation to Material

The choice of cutting tools, substrates and milling parameters is driven by the intended functionality of the final product, also by the design complexity and stage of design development. At the beginning of the design process, when a concept is sought and single devices are investigated direct manufacturing in polymer might be sufficient. Metal substrates (*e.g.*, brass, copper) are used for milling moulds in advanced design stages that intend multiple replications and produce large number of devices. The choice of substrate relates to the final functionality of a microfluidic device and is typically difficult because of the large variety of substrates. Choosing milling-tools, however, is straightforward and depends on the features that need to be fabricated, milled material maximum spindle speed.

A. CNC Mill Comparison

	Manufacturer	Model	Cost (USD)	Work Envelope (X,Y,Z) X in	X mm	Y in	Y mm	Z in	Z mm	Feed ipm	Speed RPM	Power HP	ATC tools	Accuracy µin	Accuracy µm
< $50k	Tormach	PCNC 770	$ 15,500	14.0	356	7.5	191	13.3	337	135	10,000	1	10	600	15
	Minitech Machinery	Mini-Mill/GX	$ 21,750	12.0	305	9.0	229	9.0	229	40	25,000	-	N/A	-	-
	Haas	Minimill	$ 34,000	16.0	406	12.0	305	10.0	254	600	6,000	7.5	10	-	-
$50k - $100k	Mill Tronics	RW15	$ 63,000	25.0	635	25.0	635	20.5	521	800	8,000	12	16	400	10
	Bridgeport	GX 480	$ 68,000	19.0	483	16.0	406	17.0	432	590	10,000	20	20	40	1
	DMG Mori Ellison	Milltap 700	$ 70,000	27.6	700	16.5	420	15.0	380	590	10,000	7.5	15	400	10
	DMG Mori Ellison	DMC 635 V	$ 80,000	25.0	635	20.1	511	18.1	460	1181	8,000	17.4	20	40	1
$100k <	Microlution	363-S	$ 100,000	2.5	63	2.5	63	2.5	63	-	50,000	-	36	40	1
	Atometric	G4-Ultra CNC	$ 130,000	4.0	102	4.0	102	3.0	76	1181	100,000	-	14	20	.6
	Datron	Datron M7	$ 150,000	20.0	508	20.0	508	9.0	229	639	48,000	2.6	15	-	-
	Sodick	TT1-400A	$ 220,000	15.8	400	9.8	250	11.8	300	1440	40,000	-	N/A	120	3

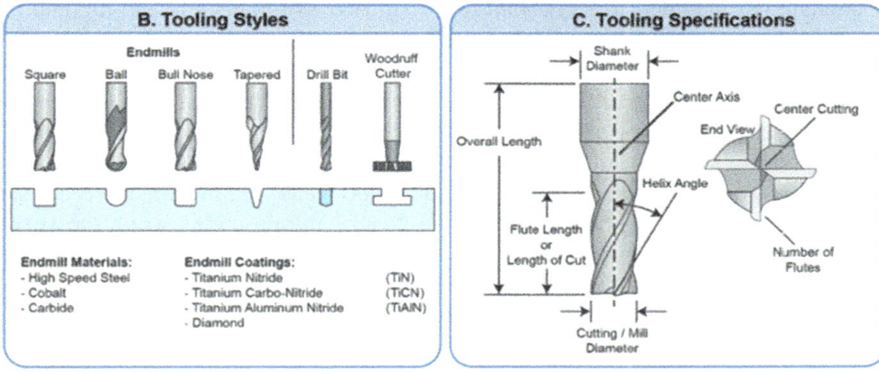

Fig. 3.6 (**a**) CNC mills from several manufacturers are compared and categorized into price ranges. Costs were assessed based on quotes of the lowest level mill from each manufacturer, except for the Tormach mill, which was quoted to be comparable in terms of capabilities to the other CNC mills. Unlisted specifications were not given by the manufacturer. (**b**) Endmills—the most common cutting tool for milling—are available in many profiles, in a variety of materials, and with a variety of coatings. Mills are also compatible with a variety of other cutting tools, some of which are shown. (**c**) Endmills are defined by several characteristics, each of which contributes to the endmill capabilities and feature quality. [© The Royal Society of Chemistry] [1]

Endmills are common tools that remove material from a substrate material. Essentially this is a cutting process powered by high-speed rotation of a motor, a.k.a. spindle, while lateral displacement of the endmill takes place following a machine pathway, called G-code or M-code. As the substrate is immobilized on a flat coordinate table a precise removal of material is achieved. A *pass* is defined by the lateral movement of that table in XY-plane at constant Z-value. Stepper motors control the movements along the axles (X-, Y-, and Z-axle), but the profile of the cut depends on the tooling style (Fig. 3.6b). Square tools result in channels with straight vertical walls with no undercuts that makes them ideal for cutting reservoirs and chambers. It is the most commonly used endmill style, which is the reason it serves as reference for spindle speeds and feed rates. To achieve round bottom channels and a ball style instrument is used. *Bull nose* is different from the square and the ball style tools as it has round edges causing round-offs at the corners of the milled structure. Together with the tapered style tools they are used to obtain slanted, round-off edges that are often required for injection moulding or hot embossing.

Depending on the application various tool types are commercially available, for instance the woodruffer cutter that can create undercuts. High-aspect ratio features, dominant in one dimension, are best achieved with drill bits rather than end mills. The problem is that the *length of cut*, which is defined by the length of the cutting edge, restricts the fabrication depth. Custom tools are available with longer necks for deep cuts; they can be tricky to operate at high-feed speeds though. Tool manufacturers provide recommendations for milling that relate the cutting diameter to spindle speed to feed rate. The purpose of these tables is to provide optimal parameters for milling without tool breaking. It is important to emphasize that most of the data is calculated for cuts shallower than two tool diameters and high chip load. Chip load is the thickness of the removed material by each cutting edge, and it is calculated by dividing the feed rate (meters per minute) by the spindle speed (RPM) times the number of flutes. The flutes are the spiral cutting edges starting at the tip of the tool and protruding towards the shank (Fig. 3.6c). Let us consider that the diameter of a two-fleet tool is large (3 mm) in comparison to the chip load (5 μm) then at extremely low spindle speeds (1000 RPM) the maximum feed rate should not exceed (1000 RPM $\times$ 5e$-$6 m $\times$ 2 $=$ 0.1 m/min). This would be very inefficient, as the milling would take extremely long time. Increasing the feed rate without compensating for the spindle speed could cause dragging of the tool against the substrate. Therefore, spindle speed needs to be corrected accordingly. Higher spindle speeds than those recommended by the tool manufacturer result in friction that wears off the cutting edges. Thus, the optimal milling conditions are defined by knowing the characteristic chip load for a tool with a given diameter and maximal spindle speed.

Optionally, coolant can be added during the fabrication process to improve heat dissipation and prevent melting of the plastic substrate. The cooling agent must be selected such not to interact with the substrate. Increasing the number of passes and reducing the depth per pass diminishes the amount of build-up and hence can prevent melting. Another option is to use air stream as coolant, which is easily achievable by splitting the air supply to the spindle and directing it at the substrate surface.

2.6 Laser Ablation

Laser ablation is another rapid fabrication technique for direct translation of computer-aided laser design (CAD) on to polymer substrates. Laser ablation can also be used for patterning surfaces of photopolymers and /or antiadhesive layers by forming micro- or nanoimprints (Fig. 3.7). These patterns can then be functionalised with proteins, oligonucleotides or whole cells. Power, stability and low operational cost of CO_2 lasers make them suitable for cutting and engraving PMMA. The wavelength of CO_2 laser lies in the far infrared spectrum (10.6 μm) and due to the constraint of light diffraction this limits the focused laser beam diameter [14]. Evidence exist that shorter wavelengths, i.e., ultraviolet laser

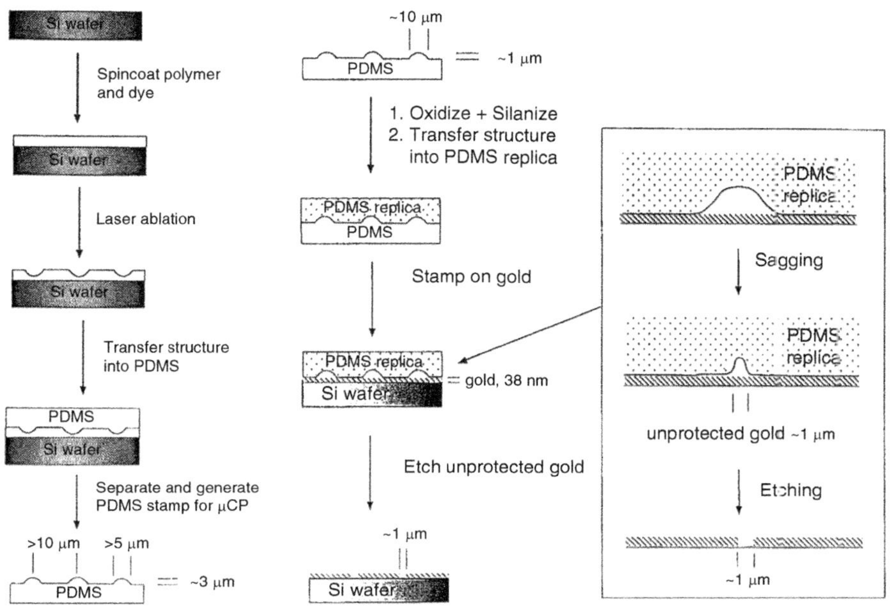

Fig. 3.7 Stamp preparation by laser ablation for microcontact printing. The pattern is transferred from the PDMS stamp to the PDMS replica, which bears a positive image of the originally ablated pattern that has been embossed on its surface. Then the replica is applied for stamping hexadecanethiol onto a gold surface. This replica is used for stamping onto a gold surface and the unprotected gold is removed by etching to give features of sizes of 1 μm. Reprinted by permission from Analytical Chemistry: [Journal of Analytical Chemistry, 1998] [13]

(355 nm), results in better quality PMMA channels [15]. Most organic materials show an intense absorbtion in the ultraviolate region, for instance, excimer laser at wavelengths of about 190–248 nm or third harmonic of Solid State Lasers of 355 nm; remain transparent at the most common semiconductor (800 nm) and Nd-YAG (1064 nm) wavelengths [16]. In this chapter infrared laser fabrication is considered, as it is universal for PMMA fabrication; however the other wavelength can be utilized as well under certain conditions and ablation environments [14].

Among the merits of microfabrication with laser is the capability to engrave and to cut designs on a wide variety of materials. The materials suitable for engraving/cutting cover a wide range including: wood, acrylic, plastic, glass, leather, corian, fabric, coated metals, annodized aluminum, ceramics, mylar, pressboard, and more. Thermal plastics (e.g., PC, COC) are not suitable substrates for laser ablation because they tend to degrade in the process. PVC brings another sort of difficulty it melts and fills the cut groove. Therefore, we look mainly into PMMA. At 1200 dpi, the moving precision of CO_2 laser (e.g., Epilog Mini 18 Laser cutter) is about 0.02 mm (~20 μm). Microdevices are easily made within minutes by engraving channels in PMMA, or by moulding PDMS onto a PMMA master. It has been demonstrated that ablation of mechanically stable material such as PEEK (poly-

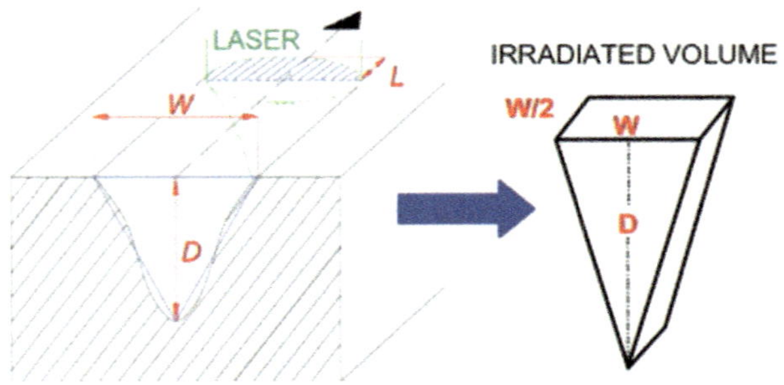

Fig. 3.8 Geometrical definition of the groove profile and irradiated volume. Reprinted by permission from Elsevier [Optics and Lasers in Engineering] [16]

ether-ether-ketone) is possible, and the resulting structure has been implemented for embossing PMMA, PC and PS and COC. However, the technique is restricted by the amount of removed PEEK material, and only negative polarity features up to 10 μm in height are feasible [4].

The CAD file is converted in to a format that is suitable for the machine interface that one uses. Epilog series are compatible with files processed in Illustrator (.AI) or Corel Draw (.CDR). Lasers from Synrad in contrast use their own 2D CAD software to control the movements of the laser head. Despite the different file formats, all lasers have similar setting parameters that define the translation from 2D drawing into 3D feature onto the substrate. The amount of energy that is dissipated by the laser head is determined by the power of the laser. But the absorbed energy is equal to the product of the absorption coefficient α, the incident power P_{in} and the irradiation time Δt:

$$Q_{in} = \alpha P_{in} \Delta t, \quad \text{where} \quad \Delta t = L/v$$

The irradiation time is related to the cutting speed (v) and the effective spot size along the scanning direction L as illustrated in Fig. 3.8. This model is suggested by Romoli et al. [16] who assume the irradiated area to be rectangular and the irradiance distribution of the nearly-Gaussian beam to be triangular, L is represented by half of the groove width W. In the same paper authors assume that the removed material m is proportional to the amount of heat Q available to vaporize the polymer. This heat can be evaluated as the difference between the absorbed energy Q_{in} and the threshold energy Q_{th}:

$$m = kQ = k(Q_{in} - Q_{th})$$

where, k is a constant related to the chemical bond energy of the PMMA polymer and is generally characterized by a nonlinear dependence on temperature. Here Q_{th}, is cumulative value combining two major loss terms: one is the energy needed to heat-up the irradiated area prior vaporization starts, the other energy lost into the surroundings due to conduction.

Based on these formulas and the relation between the material density (ρ), the channel depth (D), channel width (W) and laser spot (L), the groove depth is calculated as:

$$D = (2k\alpha/\rho)(P_{in}/Wv) - 4kQ_{th}/\rho W^2$$

This formula provides a linear correlation between groove depths, incident laser power while reciprocal dependence on the cutting speed is observed. This simple model has shown good agreement with experimental data performed earlier [17]. Essentially, the ratio of the incident power to the scanning speed is the energy density ε [J/m]. As it becomes obvious the second term of the model is the offset in energy density due to dissipation into the material and surroundings.

The predictive power of the model makes it valuable for determination of the laser parameters necessary to attain a feature size. Varying simply the cutting speed and the incident power defines the energy density, hence the physical parameters of the groove. The nominal power of the laser diminishes with time and must be calibrated to make sure that the real values are implemented.

3 Packaging and Bonding

A separate section is dedicated to packaging and bonding as that is the final step of fabrication that facilitates means of connectivity between auxiliary equipment (*e.g.*, pumps, optical cables, signal amplifiers etc.) and the microfluidic device. Several methods are presented that allow bonding in regular chemical lab.

3.1 Thermal Bonding

This technique is very similar to the embossing approach; namely, the layers (2 or more) of the device a are placed in contact with one another and the assembled stack is incubated at temperature above the glass transition temperature of the material. In contrast to the hot embossing, feature alterations need to be avoided. The method can be performed with sophisticated machine as the HEX02 (Fig. 3.5); also using static loads (*e.g.*, weight) and standard lab oven. The process development involves optimisation of the pressures (loads), temperatures and incubation time.

3.2 Chemical Bonding

The uppermost layer of the plastic surface is deplastified using chemical agents, vapours (e.g., acetone, heptane), the substrates are aligned, placed in hard contact incubated at temperature that exceeds the evaporation temperature of the solvent. This allows the extra solvent to evaporate from substrate surface while a bond is formed. However, the incubation temperature should not exceed the Tg of the plastic as this can bring distortion of the existing features and formation of artefacts. The plastic substrates are typically exposed to solvent vapours to minimize feature loss at bulk solvent treatment. Critical parameter for this process is the exposure time that can vary depending on the critical dimensions. A risk exists that miniscule features are damaged when the exposure times are long; too short exposure times can lead to poor bonding quality.

3.3 Bonding Through Pressure Sensitive Adhesive

Bonding through pressure sensitive adhesive (PSA) is a method that requires minimum investment but time, effort and manual operations. The method has been reported for sealing laser-ablated channels on a disc bonded through pressure-sensitive adhesive (86 μm, PSA, Adhesive Research, Inc.) to another PMMA disc, which hosted loading, collection chambers, as well as milled channels and valve beds [18, 19]. The method involves cutting out the silhouette of these structures from the PSA using a precision knife cutter from Graphtec. The extra material is removed manually which can be time consuming and even prohibitive for small dimensions (<20 μm) restricted by the stylus dimensions of the knife cutter. Alignment is achieved using pins or rig to facilitate the adjustment of the PSA relative to the features on the substrate. For designs that include suspended, flimsy, features a transport film needs to be used in order to handle the PSA more easily. After alignment, the protective layers from the foil are removed and the plastic layers are brought together. Pressure is applied to finalize the bonding process. Ideally, a hot/cold press or laminator provide enough pressure; rubber, manual rollers can be used in case resources are limited.

3.4 Plasma Bonding of PDMS

After casting of PDMS structures they must be de-moulded and sealed either with a piece of plastic, glass or another piece from the elastomer. Well-established method is plasma activated bonding of PDMS. In the process RF-generated plasma is used to activate the surfaces of both substrate and PDMS. A radio frequency voltage is applied between two electrodes that cause the free electrons to oscillate and to collide with gas molecules leading to RF plasma generation. Electrons pick-up enough energy during oscillation in the RF field to cause ionization of the gas, which is entrapped in the reactor chamber. Steady flow of gas (e.g., oxygen, nitrogen or air) into the reactor chamber of the plasma generator provides population of molecules that can be ionised. Very importantly the pressure inside the chamber should be stable at 500 mTorr for a successful process. The substrates are placed inside the chamber exposing the binding surfaces to the RF-plasma. Exposure times can vary from one reactor to another, for example 45 s of exposure are sufficient using plasma asher (Harrick Pro) at 30 W nominal power. After exposure the PDMS and substrate are immediately brought in contact, and the left to incubate at 65 °C over 2 h. The incubation step is optional, however, it facilitates water molecule removal from the interface between the substrates and the formation of bonds. Plasma treatment oxidises the siloxane bonds on the surface of the PDMS and renders it hydrophilic, as silanol groups are formed. This process is reversible and takes up to 12 h before PDMS returns to its native hydrophobic state.

References

1. Guckenberger DJ, de Groot TE, Wan AMD, Beebe DJ, Young EWK (2015) Micromilling: a method for ultra-rapid prototyping of plastic microfluidic devices. Lab Chip 15 (11):2364–2378
2. Jeon JS, Chung S, Kamm RD, Charest JL (2011) Hot embossing for fabrication of a microfluidic 3D cell culture platform. Lab Chip 13:325–333
3. Steigert J, Haeberle S, Brenner T, Muller C, Steinert CP, Koltay P, Gottschlich N, Reinecke H, Zengerle R, Ducree J (2007) Rapid prototyping of microfluidic chips in COC. J Micromech Microeng 17:333–341
4. Jensen MF, McCormack JE, Helbo B, Christensen LH, Christensen TR, Geschke O (2004) Rapid prototyping of polymer microsystems *via* excimer laser ablation of polymeric moulds. Lab Chip 4:391–395
5. Shin Y, Han S, Jeon JS, Yamamoto K, Zervantonakis IK, Sudo R, Kamm RD, Chung S (2012) Microfluidic assay for simultaneous culture of multiple cell types on surface or within hydrogels. Nat Protocols 7:1247–1259. doi:10.1038/nprot.2012.051
6. Madou MJ (2002) Fundamentals of microfabrication: the science of miniaturization. CRC Press, Boca Raton
7. Nichols KP, Eijkel JCT, Gardeniers HJGE (2008) Nanochannels in SU-8 with floor and ceiling metal electrodes and integrated microchannels. Lab Chip 8:173–175
8. Berthier E, Young EWK, Beebe D (2012) Engineers are from PDMS-land, biologists are from Polystyrenia. Lab Chip 12:1224–1237
9. Armani D, Chang L, Narayan A (1991) Re-configurable fluid circuits by PDMS elastomer. In: Proceedings of MEMS 1991, Orlando, pp 222–227. doi:10.1109/MEMSYS.1999.746817
10. Kyung M, Rogers JA (2003) A photocurable poly(dimethylsiloxane) chemistry designed for soft lithographic molding and printing in the nanometer regime. J Am Chem Soc 125:4060–4061
11. Datta P, Goettert J (2006) Method for polymer hot embossing process development. Microsyst Technol 13(3):265–270
12. He Y, Fu J-Z, Chen Z-C (2008) Optimization of control parameters in micro hot embossing. Microsyst Technol 14(3):325–329
13. Grzybowski A, Haag R, Bowden N, Whitesides GM (1998) Generation of micrometer-sized patterns for microanalytical applications using a laser direct-write method and microcontact printing. Anal Chem 70:4645–4652
14. Tangwarodomnukun V, Chen HY (2015) Laser ablation of PMMA in air, water, and ethanol environments. Mater Manuf Process 30:685–691
15. Teixidor D, Orozco F, Thepsonthi T, Ciurana J, Rodríguez CA, Özel T (2013) Effect of process parameters in nanosecond pulsed laser micromachining of PMMA-based

microchannels at near-infrared and ultraviolet wavelengths. Int J Adv Man Tech 67 (5–8):1651–1654
16. Romoli L, Tantussi G, Dini G (2011) Experimental approach to the laser machining of PMMA substrates for the fabrication of microfluidic devices. Optics Lasers Eng 3:419–427
17. Romoli L, Tantussi G, Dini G (2007) Layered laser vaporization of PMMA of 3D manufacturing mould cavities. Annals CIRP 56:1
18. Dimov N, Clancy E, Gaughran J, Boyle D, McAuley D, Glynn M et al (2014) Solvent-selective routing for centrifugally automated solid-phase purification of RNA. In: Microfluid and Nanofluid 18:1–13
19. Tsao C-W, De Voe DL (2009) Bonding of thermoplastic polymer microfluidics. Microfluid Nanofluid 6:1–16

Chapter 4
3D Printed Microfluidic Devices

Gregory W. Bishop

1 Introduction

Recently, increasing interest has developed around the use of 3D printing methods for the preparation of microfluidic devices. 3D printing generally involves deposition or curing of materials in a layer-by-layer fashion as determined by a three-dimensional representation of the desired object. While 3D printing was invented and first demonstrated in the 1980s, prohibitive costs and the limited number of commercially available printers and materials restricted applications to rapid prototyping for manufacturing. Present-day enthusiasm surrounding 3D printing can largely be attributed to momentum established about a decade ago by initiatives such as the RepRap and Fab@Home Projects [1, 2]. These endeavors have provided great progress toward democratizing 3D printing by promoting interest in technology, fostering collaboration through communities of enthusiasts, and decreasing barriers associated with high costs and levels of expertise. Improvements in affordability and accessibility of 3D printers have enabled new applications in many fields, including medicine and biotechnology. 3D printing is especially attractive for the fabrication of microfluidic devices due to its rapid prototyping capabilities and simple procedure compared to other previously described methods, which are more time-consuming and typically require a greater amount of expertise and expensive equipment [1, 3].

G.W. Bishop (✉)
Department of Chemistry, East Tennessee State University, Johnson City, TN, USA
e-mail: bishopgw@etsu.edu

C.K. Dixit, A. Kaushik (eds.), *Microfluidics for Biologists*,
DOI 10.1007/978-3-319-40036-5_4

2 3D Printing Techniques

3D printing offers fast design-to-object workflow and typically requires few, relatively simple steps. First, a computer-aided design (CAD) file is generated through the use of CAD software or by computer-assisted scanning of a real object. Free CAD programs and online libraries that allow users to share CAD files have provided means for increased utilization of 3D printing technologies. Printer instructions are generated by a slicer program, which separates the CAD file into sections along the vertical axis to define the composition of each printed layer. After uploading the instructions to the printer, the object is fabricated on a platform via a controlled deposition or curing apparatus interfaced with a precise positioning system. Finally, the printed object is removed from the platform. Depending on the printing method and parameters defined in the slicer program, post-processing may be necessary to remove extraneous material and supports used to bolster the structural integrity of the printed layers during the fabrication process.

Generally, layer fabrication in 3D printing is accomplished by deposition of thermoplastics (fused deposition modeling) or viscoelastic materials (syringe deposition or direct ink writing) through a nozzle or syringe, sintering of powdered materials (selective laser sintering), exposure of photocurable resin contained in a reservoir (stereolithography), or inkjet printing of photocurable inks follow by immediate exposure (PolyJet or MultiJet) [4–6]. Some important aspects for printing methods relevant to fluidic device fabrication are briefly described in this section.

2.1 *Extrusion-Based Methods*

In fused deposition modeling (FDM), a thermoplastic filament (typically 1.75 or 3.00 mm in diameter) is extruded through a heated nozzle (typically ~0.2–0.5 mm in diameter) onto a moving platform (Fig. 4.1). The extruder assembly (Fig. 4.1b) is often mounted on a gantry system that controls XY movement, and the platform or stage moves in the Z direction. Fabrication of single objects composed of multiple materials can easily be accomplished with this method simply by including more than one extruder nozzle in the printer design. Typical filament materials include poly(lactic acid), acrylonitrile butadiene styrene, and poly(carbonate). However, there is a great deal of interest in developing composite materials with improved physical and chemical characteristics for various applications. Composite filaments that incorporate carbon nanotubes, graphene, ceramics, and magnetic materials are commercially available.

Direct ink writing, pioneered by Lewis et al., is another extrusion-based technique for 3D printing [7]. Direct ink writing uses a pneumatically controlled syringe to print viscoelastic materials, such as colloidal suspensions of nanoparticles, polymers, and ceramics, onto a moving platform [7–13]. Rheological properties

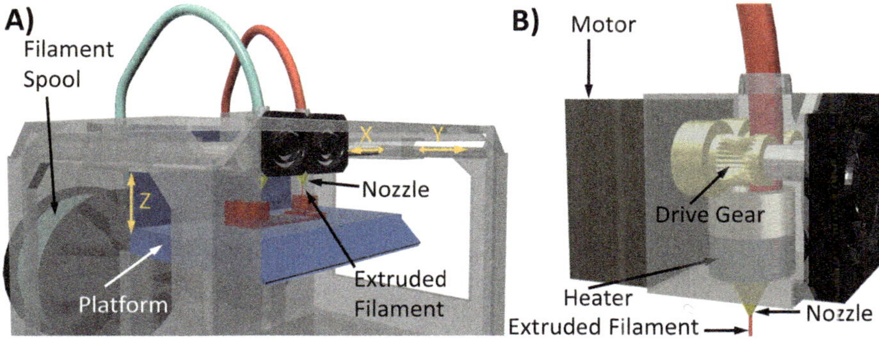

Fig. 4.1 Illustrated representation of an FDM-based 3D printer. (**a**) Object is printed from extruded filament deposited on a moving platform. Extruder assembly is mounted on a gantry system that controls deposition in the X and Y directions. (**b**) Close-up view of the extruder assembly. Adapted from Reference [6] with permission from IOP Publishing

of viscoelastic inks are tailored to facilitate extrusion through syringes with diameters as small as 1 μm [9].

2.2 *Methods Based on Photocuring*

In stereolithography (SLA), photocurable resin held in a reservoir is exposed to a light source. In the traditional form of SLA, a laser is scanned in a pattern defined by the slicer program to cure each layer onto a moving platform (Fig. 4.2). Some SLA printers now incorporate projection systems such as a digital mirror device to facilitate curing of layers with single exposure steps. Digital light processing (DLP) projectors that feature ultra-high performance lamps or light emitting diodes permit fabrication of objects with greater detail since XY resolution is limited by pixel size [14] (~30 μm for 920 × 1140 pixels) rather than laser spot size (often ~100 μm) [15]. However, printing with improved resolution limits the overall object size since DLPs possess a fixed number of pixels. SLA materials are limited to acrylate- and epoxy-based resins [4], and fabrication of single objects composed of more than one material is not easily accomplished.

Recently, Tumbleston et al. described a 3D printer based on a process called continuous liquid interface production (CLIP) [16]. CLIP is similar to SLA in its operation; however, CLIP uses an oxygen permeable membrane between the optical window and cured part. The presence of the membrane leads to a controllable oxygen-containing layer where photopolymerization of the resin is not permitted. This enables the exposure, replacement of resin, and movement of cured part processes to be performed in a continuous manner instead of as discrete steps as they are in traditional SLA. CLIP has been shown to reduce printing times from hours to minutes over traditional SLA.

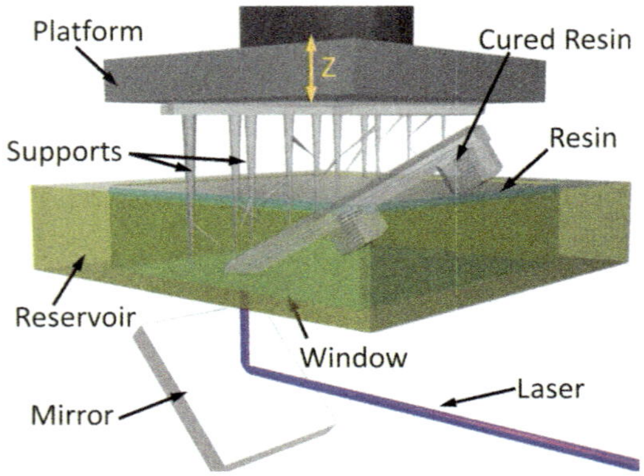

Fig. 4.2 Illustrated representation of a laser SLA-based 3D printer. Adapted from Reference [6] with permission of IOP Publishing

Like SLA, MultiJet and PolyJet technologies also employ photocurable materials. However, these methods rely on inkjet printing and immediate exposure to harden each layer. MultiJet and PolyJet printers are outfitted with multiple printheads so objects composed of more than one material can be produced.

2.3 Cost and Materials

Printers based on each of these techniques are commercially available. Depending on the printing method and capabilities, 3D printers range in price from <\$1000–\$100,000 USD or more. Printers based on FDM are typically among the lowest in cost with some SLA printers also approaching a consumer-grade price-point. Printers based on PolyJet and MultiJet technologies are among the most expensive. There are also several commercial services that enable customers to submit CAD files to receive corresponding printed objects prepared by a high-resolution 3D printer [3, 17–19]. Likewise, commercially available materials also vary in cost with prices for thermoplastic filaments ranging from ~\$30/kg or more and photopolymer materials for SLA and inkjet-based technologies starting at ~\$120/L. Many of the commercially available materials for SLA-based 3D printing are not biocompatible [5]. However, there are a growing number of biocompatible options as interest in materials development continues to be a focus for extending the capabilities of various 3D printing technologies.

3 3D-Printed Microfluidics

Various methods and applications of milli- and microfluidic devices have been described. Due to the inadequate resolution of many 3D printers, the limiting dimensions of the 3D-printed fluidic channels are typically on the order of tens to a few hundred micrometers. Therefore, many 3D-printed "microfluidic" devices described in recent literature are given this distinction to convey the microliter to sub-microliter volumes they contain. Very few reports of 3D-printed "microfluidic" devices correspond to channels with micrometer dimensions characteristic of true microfluidics. While 3D printing can be used to make small features, it is not yet routine for most current technologies to fabricate objects that include design elements of different vastly scale, such as microfluidic devices with submicrometer channels in housings of several centimeters. However, progress continues to push the boundaries of 3D printing towards capabilities necessary for true microfluidic devices.

3.1 *Molds and Scaffolds for Fluidic Channels*

The first applications of 3D printing to fluidics involved the production of molds for elastomer-based microfluidics. Unlike photolithography, which requires the production of a mask to create the master mold, 3D printing enables fabrication of the master directly from the design file. These techniques also facilitate the preparation of complex molds for channels that propagate in all three dimensions. Such molds would be difficult or impossible to produce by two-dimensional photolithography, since they require multiple masks and layer bonding.

Whitesides et al. demonstrated that a printer based on FDM could be used to prepare molds for PDMS channels with limiting dimensions of 250 μm or more [20]. Channel dimensions are limited by nozzle size, and the surface roughness of FDM-printed objects is quite large (~8 μm) since each printed layer is essentially composed of adjacent cylindrical threads of thermoplastic filament. Molds produced by SLA exhibit surface roughness of <1 μm and can be used to produce PDMS channels with dimensions of ~50 μm [21, 22]. Mixing channels and channels with integrated valves prepared from 3D-printed molds have been demonstrated [20–22].

Microvasculature scaffolds for epoxy-based fluidic devices have been prepared by direct ink writing [8]. A pneumatically controlled syringe with diameter as small as 10 μm is employed to produce the scaffold from fugitive organic inks (Prussian blue paste). After curing, the scaffold is removed by heating to 60 °C under light vacuum. Fluidic devices with remarkably smooth cylindrical channels (surface roughness 13.3 ± 6.5 nm) result. Similarly, FDM has been used to prepare scaffolds from acrylonitrile butadiene styrene (ABS) for PDMS-based fluidic devices [23]. ABS dissolves in acetone, permitting its removal from cured PDMS.

3.2 Fluidic Devices Prepared by Direct Printing

Direct fabrication of 3D-printed fluidic devices has also been demonstrated. Channels with limiting dimensions of ~500–800 μm have been produced from poly (propylene), poly(lactic acid), and poly(ethylene terephthalate) by FDM [24]. As previously stated, FDM-printed objects are composed of layers of adjacent cylindrical threads, which results in channels with ridged or scalloped internal surfaces [5]. Additionally, the overlap of adjacent layers makes it difficult to visualize fluids contained within the channel even when clear filaments are used [25]. Fluids in channels are largely obscured when located beneath 14 or more printed layers (0.200 mm) of clear PET filament. Better visualization of fluids within channels can be obtained by printing the bottom layer of the device on a heated platform, which reduces surface roughness.

Printers based on photocurable inks and resins can produce channels with typical limiting dimensions of ~250 μm [26, 27]. Smaller channels are also possible; however, it is often difficult to remove necessary support material from smaller channels. For example, during SLA, the channel is filled with uncured resin, which can be difficult to force from small channels. Also, since some light can reach uncured resin held within the channel during the printing process, complete blockage of the channel can result in some areas. The surface roughness of 500 μm-high channels printed by SLA was reported to be ~2.54 μm [3], and a study of four different 3D printers based on inkjet technologies found surface roughness to range from ~0.09–2.24 μm [28].

Nordin et al. investigated the effect of resin composition on limiting channel dimensions produced by DLP-SLA [14]. Resins for SLA usually consist of monomer material(s), a photoinitiator, and an absorber, which is included to control the penetration depth of the light source. Nordin et al. found that controlling the penetration depth by using higher concentrations of the absorber was crucial for production of small channels. 60 μm-high channels were printed with 100 % yield using a custom-formulated resin that contained poly(ethylene glycol) diacrylate (PEGDA), 1 % photoinitiator phenylbis(2,4,6-trimethylbenzoyl)phosphine oxide (Irgacure 819), and 0.6 % absorber Sudan I. For 100 % success rate, width for these channels must be at least 108 μm (4 pixels) as determined by the limitations of the 912 × 1140 pixel DLP projector.

Due to the availability of clear resins and inks, 3D printed fluidic devices prepared by SLA and inkjet technologies can also permit visualization of fluids within the channels. As-printed, devices are typically opaque and must be processed to improve transparency. For example, the outer layers of SLA-printed devices can be sanded with up to 2000-grit sand paper, polished with a plastic cleaning compound, and coated with clear acrylic spray to enhance clarity [29, 30].

Photopolymer-based 3D-printing methods possess sufficient resolution to produce complex fluidics and modular fluidic devices. SLA has been used to prepare channels with integrated membrane-based valves that can be pneumatically controlled [27, 31]. SLA and MultiJet printing have been employed to yield microfluidic components that can be reversibly connected to construct customizable fluidics [18, 19]. Channels can also be filled with fast-drying, conductive suspensions of colloidal silver to make resistors and inductors for electrical circuits and wireless sensors [32].

4 3D-Printed Fluidics and Bioanalysis

3D-printed fluidic devices have been described for use in cell studies and measurements of biomolecules and biologically relevant species. With relatively high resolution printers based on SLA and inkjet methods, threaded ports can be printed to interface the channel with commercially available fittings and tubing for easy access [33]. Spence et al. have employed a PolyJet printer to produce various fluidic devices that feature access ports for reversibly integrating commercially available membrane inserts and other components into channels for various cell studies [29, 33–35] (Fig. 4.3). Membrane inserts selectively transport small molecules from the channel and prohibit transport of larger species like large biomolecules and cells. In one cell viability study, a layer of bovine pulmonary artery endothelial cells was deposited on a cell culture membrane inserted into the 3D-printed channel [33]. Saponin (a detergent that disrupts the cell membrane) or Hank's balanced salt solution (HBSS) was delivered into the fluidic channel. Sytox Green staining of cells indicated that a larger population of dead cells resulted upon exposure of cells in the membrane to saponin compared to HBSS.

Fluidic devices have also been designed such that a commercial plate reader can be used to perform measurements [34]. Such platforms have enabled investigation of the effects of storage conditions related to transfusion medicine on ATP production by erythrocytes. Erythrocytes stored in FDA-approved additive solution 1 (AS-1) that contains ~110 mM glucose or a modified AS-1 with lower glucose concentration (5.5 mM) were delivered into the 3D-printed channel. Membrane inserts provided reservoirs for collecting and measuring adenosine triphosphate

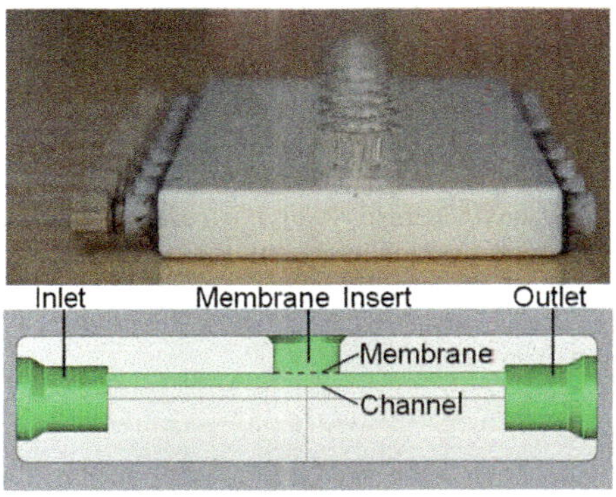

Fig. 4.3 3D-printed fluidic device used in cell studies. Device features 3 mm wide × 1.5 mm deep channels for incorporating membrane inserts. Reprinted with permission from Reference [33], Copyright 2013 American Chemical Society

(ATP) transported from the channel. Erythrocytes stored in the modified AS-1 exhibited greater ATP production than those stored in commonly used AS-1.

In addition to membrane inserts, sensing elements like electrodes and optical fibers have also been incorporated into 3D-printed channels to facilitate bioanalytical measurements [17, 25, 28, 30, 35–37] (Fig. 4.4). Electrodes can be fastened in channels through access holes, deposited on substrates that are bound to open-sided channels to complete the fluidic device, or housed in threaded fittings that are compatible with threaded ports included in the device design. 0.5 mm carbon and platinum electrodes incorporated in 3D-printed channels have been used to detect viruses that were labeled with cadmium sulfide quantum dots [36]. Dopamine, nitric oxide, ATP, and hydrogen peroxide have also been measured using 3D-printed fluidic devices with integrated electrodes [25, 35]. Continuous monitoring of glucose and lactate in human subjects during and after physical activity was accomplished by connecting an FDA-approved microdialysis probe to a 3D-printed channel (375 μm × 508 μm, internal volume 1.9 μL) equipped with needle electrode biosensors [37].

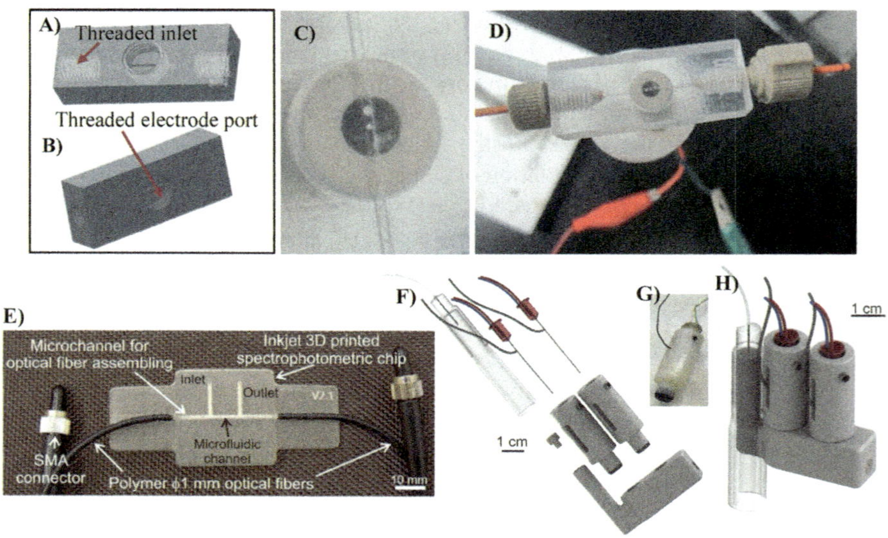

Fig. 4.4 3D-printed channels with integrated devices. (**a–d**) Schematic (**a**, **b**) and photographs (**c**, **d**) of a device that features a threaded port for incorporating disk-shaped electrodes into a 500 μm diameter channel. Reproduced from Reference [35] with permission from the Royal Society of Chemistry. (**e**) Photograph of fluidic channel interfaced with optical fibers for in-channel spectrophotometry. Reproduced from reference [28] with permission from IOP Publishing. (**f–h**) Fluidic device with integrated needle biosensors for continuous monitoring of glucose and lactate from dialysate. Reproduced by the permission from Reference [37]; copyright American Chemical Society

5 Outlook and Prospects

Growing interest in 3D printing techniques and improvements in technological capabilities and materials have resulted in many new applications. Due to the simple and fast design-to-object workflow, 3D printing offers advantages over traditional fabrication techniques for the production of microfluidic devices. Fluidic devices can be directly printed from CAD files that are processed using slicer software, and several free and open-source design and slicer programs are available. The printing process allows more freedom in design than other fabrication techniques to a certain extent due to the ability to produce channels that propagate in various directions. Also, several device designs can be produced and tested relatively quickly, since there is no need to prepare various masks and molds that are required with other techniques. Currently, there are few 3D printing techniques that can produce channels with dimensions <100 μm; however, as the capabilities of 3D printing continue to improve, so too will these boundaries. Applications of 3D-printed fluidic devices have shown their utility and robustness in bioanalytical applications, including cell studies, biomolecule sensing, and immunoassays.

References

1. Waldbaur A, Rapp H, Länge K, Rapp BE (2011) Let there be chip–towards rapid prototyping of microfluidic devices: one step manufacturing processes. Anal Meth 3:2681–2716
2. Ventola CL (2014) Medical applications for 3D printing: current and projected uses. Pharmacol Ther 39:704–711
3. Au AK, Lee W, Folch A (2014) Mail-order microfluidics: evaluation of stereolithography for the production of microfluidic devices. Lab Chip 14:1294–1301
4. Gross BC, Erkal JL, Lockwood SY, Chen C, Spence DM (2014) Evaluation of 3D printing and its potential impact on biotechnology and the chemical sciences. Anal Chem 86:3240–3253
5. O'Neill PF, Azouz AB, Vázquez M, Liu J, Marczak S, Slouka Z, Chang HC, Diamond D, Brabazon D (2014) Advances in three-dimensional rapid prototyping of microfluidic devices for biological applications. Biomicrofluidics 8:052112
6. Bishop GW, Satterwhite-Warden JE, Kadimisetty K, Rusling JF (2016) 3D printed bioanalytical devices. Nanotechnology 27:284002
7. Lewis JA (2006) Direct ink writing of 3D functional materials. Adv Funct Mater 16:2193–2204
8. Therriault D, White S, Lewis JA (2003) Chaotic mixing in three-dimensional microvascular networks fabricated by direct-write assembly. Nat Mater 2:265–271
9. Ahn BY, Duoss EB, Motala MJ, Guo X, Park S-I, Xiong Y, Yoon J, Nuzzo RG, Rogers JA, Lewis JA (2009) Omnidirectional printing of flexible, stretchable, and spanning silver microelectrodes. Science 323:1590–1593
10. Wu W, DeConinck A, Lewis JA (2011) Omnidirectional printing of 3D microvascular networks. Adv Mater 23:H178–H183
11. Sun K, Wei T-S, Ahn BY, Seo JY, Dillon SJ, Lewis JA (2013) 3D printing of interdigitated Li-ion microbattery architectures. Adv Mater 25:4539–4543
12. Compton BG, Lewis JA (2014) 3D-printing of lightweight cellular composites. Adv Mater 26:5930–5935

13. Muth JT, Vogt DM, Truby RL, Mengüç Y, Kolesky DB, Wood RJ, Lewis JA (2014) Embedded 3D printing of strain sensors within highly stretchable elastomers. Adv Mater 26:6307–6312
14. Gong H, Beauchamp M, Perry S, Woolley AT, Nordin GP (2015) Optical approach for resin formulation for 3D printed microfluidics. RSC Adv 5:106621–106632
15. Ho CMB, Ng SH, Li KHH (2015) 3D printed microfluidics for biological applications. Lab Chip 15:3627–3637
16. Tumbleston JR, Shirvanyants D, Ermoshkin N, Janusziewicz R, Johnson AR, Kelly D, Chen K, Pinschmidt R, Rolland JP, Ermoshkin A, Samulski ET, DeSimone JM (2015) Science 347:1349–1352
17. Snowden ME, King PH, Covington JA, Macpherson J, Unwin PR (2010) Fabrication of versatile channel flow cells for quantitative electroanalysis using prototyping. Anal Chem 82:3124–3131
18. Lee KG, Park KJ, Seok S, Shin S, Kim DH, Park JY, Heo YS, Lee SJ, Lee TJ (2014) 3D printed modules for integrated microfluidic devices. RSC Adv 4:32876–32880
19. Bhargava KC, Thompson B, Malmstadt N (2014) Discrete elements for 3D microfluidics. Proc Natl Acad Sci 111:15013–15018
20. McDonald JC, Chabinyc ML, Metallo SJ, Anderson JR, Stroock AD, Whitesides GM (2002) Prototyping of microfluidic devices in poly(dimethylsiloxane) using solid-object printing. Anal Chem 74:1537–1545
21. Comina G, Suska A, Filippini D (2014) PDMS lab-on-a chip fabrication using 3D printed templates. Lab Chip 14:424–430
22. Chan HN, Chen Y, Shu Y, Chen Y, Tian Q, Wu H (2015) Direct, one-step molding of 3D-printed structures for convenient fabrication of truly 3D PDMS microfluidic chips. Microfluid Nanofluid 19:9–18
23. Saggiomo V, Velders AH (2015) Simple 3D printed scaffold-removal method for the fabrication of intricate microfluidic devices. Adv Sci 2:1500125
24. Kitson PJ, Rosnes MH, Sans V, Dragone V, Cronin L (2012) Configurable 3D-printed millifluidic and microfluidic 'lab on a chip' reaction ware devices. Lab Chip 12:3267–3271
25. Bishop GW, Satterwhite JE, Bhakta S, Kadimisetty K, Gillette KM, Rusling JF (2015) 3D-printed fluidic devices for nanoparticle preparation and flow-injection amperometry using integrated Prussian blue nanoparticle-modified electrodes. Anal Chem 87:5437–5443
26. Shallan AI, Smejkal P, Corban M, Guijt RM, Breadmore MC (2014) Cost-effective three-dimensional printing of visibly transparent microchips within minutes. Anal Chem 86:3124–3130
27. Rogers CI, Qaderi K, Woolley AT, Nordin GP (2015) 3D printed microfluidic devices with integrated valves. Biomicrofluidics 9:016501
28. Walczak R, Adamski K (2015) Inkjet 3D printing of microfluidic structures—on the selection of the printer towards printing your own microfluidic chips. J Micromech Microeng 25:085013
29. Gross BC, Anderson KB, Meisel JE, McNitt MI, Spence DM (2015) Polymer coatings in 3D-printed fluidic device channels for improved cellular adherence prior to electrical lysis. Anal Chem 87:6335–6341
30. Bishop GW, Satterwhite-Warden JE, Bist I, Chen E, Rusling JF (2016) Electrochemiluminescence at bare and DNA-coated graphite electrodes in 3D-printed fluidic devices. ACS Sens 1(2):197–202
31. Au AK, Bhattacharjee N, Horowitz LF, Chang TC, Folch A (2015) 3D-printed microfluidic automation. Lab Chip 15:1934–1941
32. Wu S-Y, Yang C, Hsu W, Lin L (2015) 3D-printed microelectronics for integrated circuitry and passive wireless sensors. Microsys Nanoeng 1:15013
33. Anderson KB, Lockwood SY, Martin RS, Spence DM (2013) A 3D printed fluidic device that enables integrated features. Anal Chem 85:5622–5626

34. Chen C, Wang Y, Lockwood SY, Spence DM (2014) 3D-printed fluidic devices enable quantitative evaluation of blood components in modified storage solutions for use in transfusion medicine. Analyst 139:3219–3226
35. Erkal JL, Selimovic A, Gross BC, Lockwood SY, Walton EL, McNamara S, Martin RS, Spence DM (2014) 3D printed microfluidic devices with integrated versatile and reusable electrodes. Lab Chip 14:2023–2032
36. Krejcova L, Nejdl L, Rodrigo MAM, Zurek M, Matousek M, Hynek D, Zitka O, Kopel P, Adam V, Kizek R (2014) 3D printed chip for electrochemical detection of influenza virus labelled with CdS quantum dots. Biosens Bioelectron 54:421–427
37. Gowers SAN, Curto VF, Seneci CA, Vadgama P, Yang G-Z, Boutelle MG (2015) 3D printed microfluidic device with integrated biosensors for analysis of subcutaneous human microdialysate. Anal Chem 87:7763–7770

Chapter 5
The Centrifugal Microfluidic: Lab-on-a-Disc Platform

Brian Henderson, David J. Kinahan, and Jens Ducrée

1 Introduction

Over the past decades microfluidics has proved its game-changing potential where biomedical diagnostics are performed [1, 2]. The field of research focuses on so-called 'Lab-on-a-Chip' or 'Micro Total Analysis Systems' (μTAS) that manipulate and analyse fluid on the microscale. These devices integrate the labour intensive operations of a specific laboratory diagnostic exam onto a chip of the typical size of a credit card. The method offers significant reduction of reagent volumes, reaction times and human intervention through on-chip automation [3]. Since the adaptation of microfluidics towards the biological sciences there has been a surge of interest in the evolution of these two complimentary fields; leading to the adaptation of a huge array of bioanalytical assays onto microfluidic platforms. One of the most promising areas of research in the field of microfluidics is the development of Point of Care (POC) diagnostic devices that are simple to use, cost efficient, fast and effective. By now different modes of operation such as continuous flow, batch-wise discrete volumes and droplet-based liquid handling have been implemented. A range of integrated microfluidic platforms have been developed including passively driven systems propelled by capillary action and paper imbibition, alongside pneumatically, peristaltically, and electrokinetically actuated platforms. This chapter introduces the centrifugal microfluidic "Lab on a Disc" platform [4].

B. Henderson • D.J. Kinahan (✉) • J. Ducrée
Biomedical Diagnostics Institute, Dublin City University, Dublin, Ireland
e-mail: david.kinahan@dcu.ie

C.K. Dixit, A. Kaushik (eds.), *Microfluidics for Biologists*,
DOI 10.1007/978-3-319-40036-5_5

2 Centrifugal Microfluidics

One of the main challenges for improving the efficiency and reliability of the resulting diagnostic device is flow control on the microscale. On the centrifugal microfluidics, the inertial pseudo forces that occur in a rotating reference frame are combined with a purpose-built network of multi-scale channels, chambers and obstacles to pump and manipulate fluids towards through a sequence of laboratory unit operations (LUOs). These LUOs include metering, mixing and particle separation [4]. The centrifugal field depends on the mass density of the fluid and the distance of the liquid volume from the centre of rotation. The centrifugal field squares with the frequency of rotation [4] and so provides a wide range of available force strengths. The disc cartridge functions without connection to an external pump and can be loaded at atmospheric pressure using a simple pipette. Thus there are few or no 'world to chip' interface limitations; similarly the disc cartridges are often relatively cheap and disposable and thus particularly suitable for diagnostics. The inherent capability to centrifuge samples makes the platform particularly suitable for blood processing and also for assays based around particle sedimentation and cell handling [29].

2.1 Centrifugal Hydrodynamics

Within the centrifugal microfluidic platform there are three forces acting on a fluid during rotation an angular spin frequency ω, there are three (pseudo) forces (densities).

Centrifugal Force

$$f_{\omega} = \rho r \omega^2 \tag{5.1}$$

$$\text{Euler Force} \quad f_E = \rho r \frac{d\omega}{dt} \tag{5.2}$$

$$\text{Coriolis Force} \quad f_C = 2\rho\omega v \tag{5.3}$$

acting on the fluid ρ is the mass density of the liquid, r is the radial distance from the centre of rotation, ω is the frequency of rotation and v is the fluid velocity in the plane of the disc (Fig. 5.1).

The effect of changes in these variables is visualised below in Fig. 5.2.

The three forces can be changed by the frequency of rotation to achieve a specific mean flow velocity [4]. Given a fluid plug within a straight channel (segment) with round cross section of diameter d, liquid plug length Δr, mean radial position $\bar{r}$, at a set angular frequency of rotation ω, we obtain the mean velocity

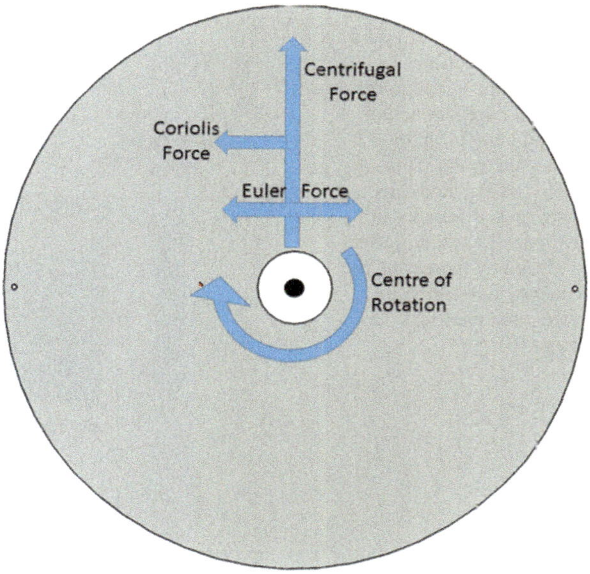

Fig. 5.1 A graphical representation of the direction each force acts in for a give direction and constant spin rate (Centrifugal and Coriolis), Euler Force depends on acceleration which can be positive or negative and can therefore act in both directions

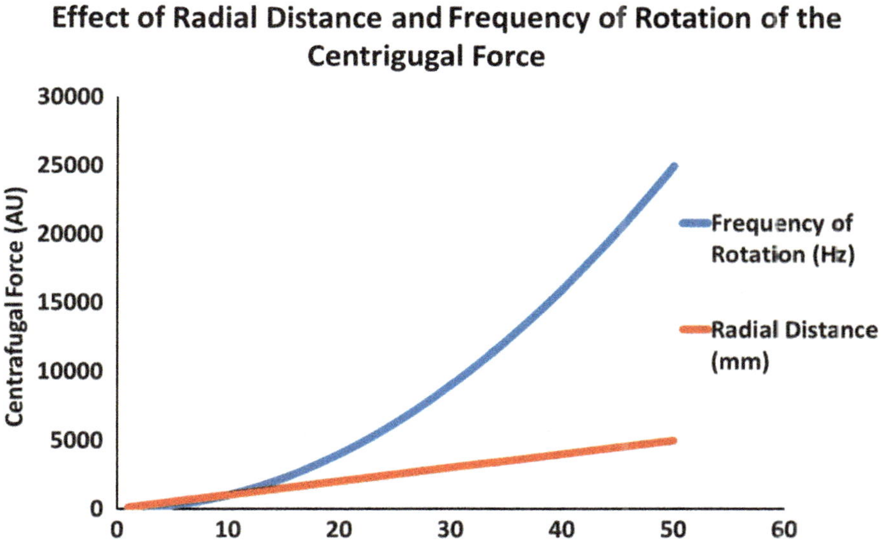

Fig. 5.2 A graphical representation of the effect of radial distance from the centre of rotatior and the frequency of rotation on the centrifugal forcefelt by a particle on the disk

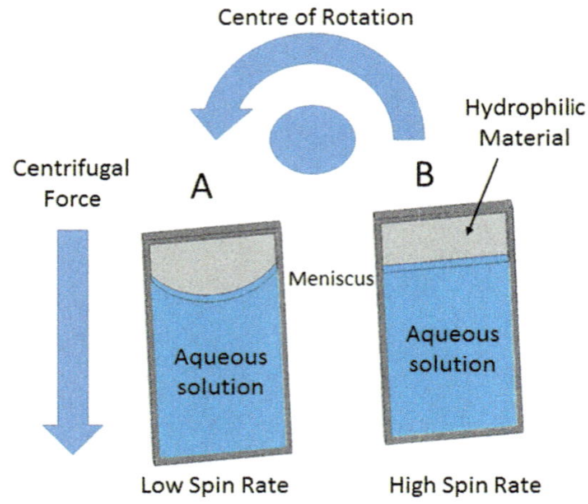

Fig. 5.3 Image a represents a metered volume of liquid within a hydrophilic microchambers, due to the effects of capillary action the liquid creeps up the sides of the walls. This bending of the meniscus causes inconsistencies in metered volumes, this affect is reduced by increasing the frequency of rotation to create a flat meniscus (Image (**b**))

$$\overline{v} = \frac{\rho}{32\eta}\overline{r}d^2\omega^2 \tag{5.4}$$

and the volumetric flow rate

$$Q = \vec{v}A = \frac{\pi\rho}{128\eta}\overline{r}d^4\omega^2 \tag{5.5}$$

with the cross sectional area of the channel $A = 1/4\ \pi\ d^2$. As for pressure-driven systems, centrifugal flow is characterised by a parabolic velocity profile, which peaks in the centre and vanishes at the wall due to no-slip boundary conditions.

2.2 *Metering*

Volume metering on the centrifugal microfluidic platform is often based on a simple principle of centrifugally driven overflow from a chamber with a specific volume below the bottom edge of an upper outlet channel [4].

Some microfluidic devices require hydrophilic materials, which aid the flow against the centrifugal force. This can cause inconsistencies when attempting to handle very small quantities of liquid; the effect can be suppressed at elevated rotation frequencies where by the centrifugal force far exceeds interfacial forces and also flattens the meniscus becomes flat, e.g., to improve accuracy of metering (Fig. 5.3).

The metering chamber is structured with two exits, one will be sealed with a valve that can be opened upon demand and the other is an unobstructed channel that leads to a waste chamber. When the sample or reagent is placed into the metered chamber, the excess will overflow into the waste chamber thereby leaving only the

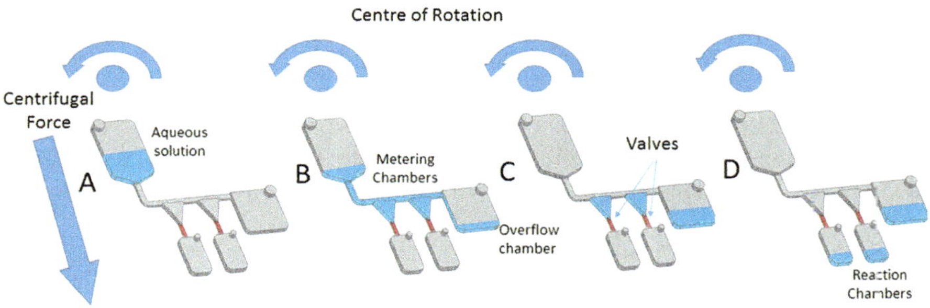

Fig. 5.4 Image (**a**) shows a random quantity of liquid for use in a biochemical reaction, this liquid is required in 10 μL quantities for reactions downstream. The spin rate is increased to force all liquid further from the centre of rotation where it fills all available space on the disk, Image (**b**) show the liquid fill the metered chambers and proceed to the overflow chamber. The dimensions of the metering chambers have been specifically designed to hold only 10 μL of liquid (Image (**c**)), once all excess liquid has passed to the overflow chamber it is possible to open the valving system holding back the metered volume by increasing the frequency of rotation to the "Burst Frequency" of this specific valve

predetermined volume of liquid in the initial chamber. The liquid in this chamber then can be released by opening the valve, valving will be discussed in more detail in further sections (Fig. 5.4).

2.3 *Mixing*

The flow regime of a fluid is described by the Reynolds number

$$\mathrm{Re} = \frac{\rho v D_h}{\eta} \tag{5.6}$$

where ρ is the fluid density, v is the velocity of the fluid, μ is the fluid viscosity and D_h denotes the hydraulic diameter [3]. Turbulent flow begins to form as the Reynolds number surpasses (roughly) 2300. In a typical microfluidic system the values of each term could be $\frac{\varrho}{\mu} \sim 10^{-5}\,\mathrm{cm^2\,s^{-1}}$, $D_h \sim 200\,\mu\mathrm{m}$ and $v \sim 1\,\mathrm{cm\,s^{-1}}$, using the above equation this would result in a Reynolds number ≈ 2 which shows that laminar flow dominates within typical microchannels [3, 5].

Due to the laminar flow regime where advective mixing is absent, mixing primarily occurs through diffusion, which is the process whereby a concentrated group of particles spread out over time by means of Brownian motion to distribute themselves uniformly across a given volume at a time scale

$$t_D = D_h^2/D$$

Determined by the Molecular Diffusion Coefficient D which can range considerably between bio-molecules; for small molecules, the diffusion coefficient would be $10^{-3}\,cm^2\,s^{-1}$ whereas for large molecules (e.g., proteins), the coefficient could be in the range of $10^{-7}\,cm^2\,s^{-1}$. Even though the dimensions are quite small, diffusive mixing is rather slow as the time scales needed to diffuse half the width of a microchannel can range from 10 s for small molecules to 16 min for large proteins.

2.4 Diagnostic Applications

A full-fledged Point of Care (POC) Diagnostic device requires a scope of attributes; first and foremost, POC systems allow patient diagnosis in a more decentralised manor away from the infrastructure of primary health care facilities. With this in mind, a POC device must include a method of sample preparation, biorecognition of target molecule, signal transduction and signal conversion to a readout or display, which can be interpreted qualitatively or quantitatively.

Using this general schematic, a number of biological analyses have been implemented on-disc including DNA hybridisation Microarrays [6], Real Time PCR (qPCR) Amplification [7], Protein Microarrays [8], Immunoassays [9, 10] biochemical analysis [11, 12], cell based screening bioassays [13] (Burger et al. 2012) and biochemical screening of (individualized) cells [14].

2.4.1 Nucleic Acid Amplification

The most established method of nucleic acid amplification is Polymerase Chain Reaction (PCR) which uses enzymes to duplicate strands of DNA as the reaction mix alternates between high ($\approx 95\,^\circ C$) and low temperatures ($\approx 65\,^\circ C$) in the prescience of excess nucleotide bases [15]. This thermocycling requires a method of heating and cooling designated locations of a rotating disc; previously described methods for achieving this have ranged from including battery packs on the disc, slip rings to transfer power to resistors and infrared radiation [15]. Each of these methods has allowed amplification of a targeted DNA sequence which has great relevance in the field of POC molecular diagnostics. PCR can be used to identify the presence of pathogenic microorganisms by including primers specific to the microorganism's DNA, if the DNA is present the amplification will occur whereas no amplification occurs in the absence of the pathogens DNA.

2.4.2 Immunoassays

Automation of immunoassays on disc is an area of increasing interest in microfluidics as they are one of the most widely used biological tests in patient diagnostics, food safety and environmental monitoring. The most frequently used immunoassay on-disk is the Enzyme linked Immunosorbent Assay or ELISA, this technique requires a solid support with bound capture antibody (Sandwich Configuration) or antigen (Direct Configuration) which can then isolate target molecules from solution before impurities are washed away [16, 17]. Using a multi-valve architecture to sequentially deliver reagents [18] at pre-defined times according to an assay protocol, it is possible to implement a benchtop ELISA on the Lab-on-a-Disc platform. More complex versions of protein analyte detection have been integrated on-disk like multiplex immunoassays which utilize individual fluorescent tags for specific targets, and protein microarrays, which require surface modification and precision to efficiently bind and array the proteins on the disc [15]. The most common detection methods utilize either optical or electrochemical set-ups. Such devices could replace conventional blood tests in the future due to their ease of use, speed and overall cost.

3 Valving Technologies

As all disc-based liquids experience the centrifugal field, a set of valves is required to orchestrate the spatio-temporal release sequence of on-board liquids along the test protocol. A number of valving schemes have been devised for the Lab-on-a-Disc platform which typically use the interplay of centrifugal forces with other on-disc forces. These other forces typically depend on statically defined parameters such as microchannel geometries, surface tensions and liquid viscosity.

Valving schemes are usually divided into two distinct types, passive and active valving [19]. With passive valves being controlled by the interplay of both the centrifugal and the capillary counter pressure and active valves require external instrumentation, therefore needing an extra interfaces adding to the overall complexity of the system [19]. Each valving scheme.

3.1 *Passive Valving*

3.1.1 Capillary Valves

On the microscale the surface-to-volume ratio is reduced such that the forces governing the liquid are predominantly surface orientated, this allows the movement of liquid through a narrow channel by capillary action [20]. Capillary valves can take two forms. Geometry based valves use either an expansion or constriction depending on if the surface is natively hydrophobic or hydrophilic. The other form, a hydrophobic patch, locally modifies the channel surface tension to function as a barrier to liquid movement [4, 20]. Surface tension is typically characterised by the surface contact angle, $\varnothing$, which denotes the wettability of the surface to a given liquid. A contact angle $\varnothing < 90°$ indicates a hydrophilic surface whereas a contact angle $\varnothing > 90°$ indicates a hydrophobic surface (Fig. 5.5).

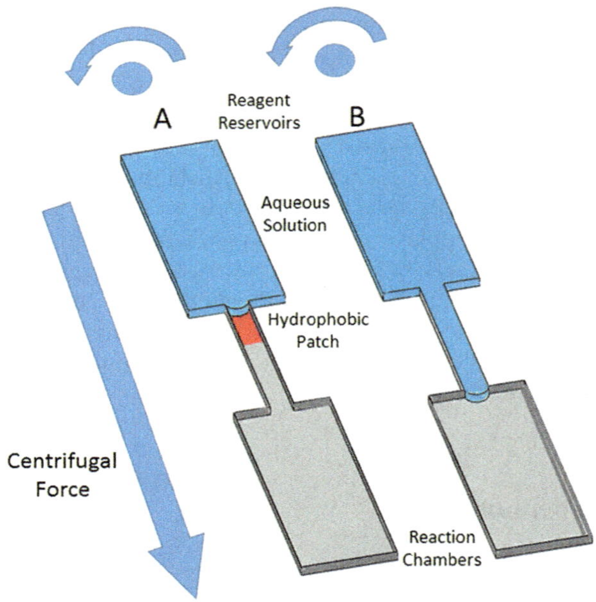

Fig. 5.5 Figure 5.3 showed a metered volume of liquid held back by a "hydrophobic constriction Valve" (Image (**a**)), this type of valve stops the flow of liquid due to the repulsive interactions between the liquid and the surface of the microchannels which have been altered to be hydrophobic. Image (**b**) depicts the concept of "Hydrophilic expansion" whereby the walls of the microchannel diverge rapidly, this halts the contact between the liquid and the walls of the microchannel which is drawing the liquid along the channel. Without the new surface to draw the liquid down the channel the advancing liquid stops at the opening. Both valve types can be opened by increasing the frequency of rotation to the burst frequency

3.1.2 Siphon Valves

The structure of a siphon valve typically utilizes hydrophilic microchannels to ensure capillary action dominates at a low spin frequencies. These structures consist of a reservoir chamber attached to siphon channel that is directed towards the centre of rotation [20]. At a crest point, located radially inwards of the reservoir, the siphon channel turns back away from the centre of rotation to a point radially outward below the original reservoir.

At rest flow can be restricted from entering the siphon valve by a capillary valve (described above) when the rotation of the disc is increase to the "Burst Frequency" the valve will open and fluid will flow up the siphon until it reaches hydrostatic equilibrium, that is when the force of capillary action is counteracted by the centrifugal force [4, 20]. As this frequency of rotation is lowered the capillary force will become more dominant until the fluid passes around the siphon, it is at this point that the siphon is said to be "primed". Next, when the frequency of

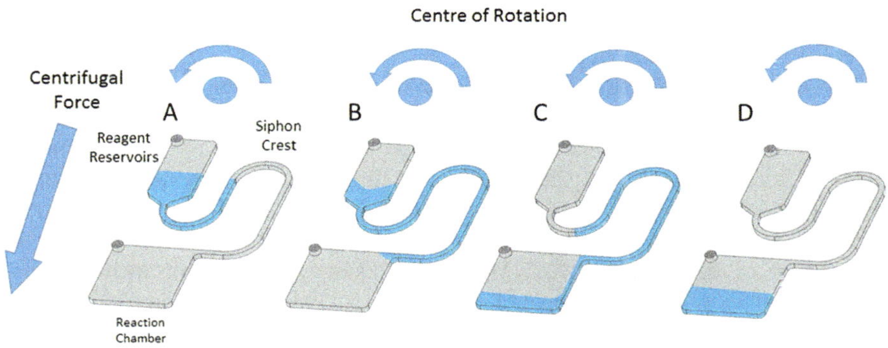

Fig. 5.6 Siphon valves build upon the principle of capillary action as previously described, the walls of the siphon channels draw the liquid from the reagent reservoir back towards the centre of rotation however at a high frequency of rotation (Image (**a**)) the liquid is held back by the counteracting centrifugal force, this causes the liquid to reach "Hydrostatic Equilibrium" at a point beneath the height of the siphon crest. As the frequency of rotation is reduced the adhesive interactions between the liquid and the channel walls begin to dominate over the centrifugal force causing the liquid to advance around the siphon crest and down towards the reaction chamber (Image (**b**)). It is at this point that the siphon is "Primed", the liquid will now flow towards the furthest average radial point from the centre of rotation (Image (**c**) and (**d**))

rotation is increased the fluid will move around the siphon to fill the next chamber. A number of these siphon valves can be placed in series to delay the introduction of different reagents to a reaction chamber [11, 12, 21] (Fig. 5.6).

Both the time taken to prime a siphon valve and the burst frequency for each capillary valve can be calculated in order to create simple changes in the speed of rotation to allow the addition of reagents in an automated manner, the time of each speed of rotation can be elongated in order to allow time for specific reactions to take place, as well as mixing of reagents [4, 20]. The time needed to prime each valve

$$t = \frac{4\eta l^2}{d\sigma \cos\ \theta} \tag{5.9}$$

Depends on the surface tension, the contact angle θ and the viscosity η, the length of the liquid plug l and the channel diameter d. The equation

$$v = \frac{1}{\pi}\sqrt{\frac{\sigma|\cos\ \theta|}{\rho\bar{r}\Delta rd}} \tag{5.10}$$

provides the capillary burst frequency.

3.1.3 Centrifugo-Pneumatic Valving

This type of Lab-on-a-Disc platforms create structures between layers of different materials bonded together to form a continuous network of microchambers and microchannels; due to this feature it is possible to insert dissolvable film between layers of adhesive material to form valves [22]. The valve is be dissolved upon contact with an aqueous solution and also blocks the passage of air meaning that on the addition of liquid upstream a sealed channel of air is created between the valve and the front of the liquid plug [22]. When the frequency of rotation is increased the fluid plug advances down the channel, this is due to the centrifugal force exerted on the liquid as described above; however, due to the now decreasing volume of air there is increasing centrifugally induced pressure which contracts that centrifugal force attempting to advance the liquid [22]. This induced pressure scales according to the following equation [22],

$$\Delta P_{\omega} = \rho \Delta r \bar{r} \omega^2 \tag{5.11}$$

where ΔP_{ω} describes the pressure at a specific frequency of rotation and all other terms are consistent with above descriptions. The advancing liquid plug is halted as $\Delta \rho_{\omega}$ becomes balanced by the air pressure within the microchambers which can be calculated according to the following equation,

$$P = P_0 \frac{1}{1 - \frac{\Delta V}{V_0}} \tag{5.12}$$

where P is the pressure within the microchannel, P_0 is the initial pressure within the microchannel, V_0 is the initial volume of the microchannel and ΔV is the change in Volume of the microchannel [22].

Using the above information, it is possible to design pneumatic chambers with calculated dimensions that will result in a specific burst frequency.

3.1.4 Event Triggered Valving

A key disadvantage of rotationally-actuated capillary burst valves is that their minimum achievable feature size requires a relatively large amount of disk real estate as well as large differences in rotational frequency [23]. This limits the number of discrete LUOs which can be automated on-disk. These specific limitations can be circumvented through the use of so called "Event Triggered Valves". This mechanism of valving builds upon that of dissolvable film valves described above, an extra degree of control is introduced by utilizing a second dissolvable film which controlled the pressure in the pneumatic chamber, in this way the liquid is restricted by the counter acting pressure even at high rotational frequency's [18]. The created valve requires a pneumatic chamber which is sealed by two dissolvable film (DF) tabs, one DF is connected to the reservoir and leads to further unit operations downstream, this is known as the Load Film (LF). The second DF is located within a second chamber which is connected to the pneumatic chamber by a pressurised microchannel, when this chamber fills with an ancillary liquid the DF is dissolved and the pressure holding back the liquid is released, this DF is known as the Control Film (CF) [18]. The main liquid in the reservoir can then advance down the microchannel connecting the reservoir to the LF where it dissolves and releases the liquid [18].

3.1.5 Paper Imbibition Valves

A further improvement with respect to event triggered valving is realised through the use of low cost paper imbibition valving. This iteration utilizes the well-defined mechanism of liquid "Wicking" through absorbent paper, an ancillary liquid wicks from a liquid field chamber through the absorbent paper which is in contact with a number of sequential Control Films (CFs) [23]. As the liquid wicks across the paper at a well-defined rate the control films are dissolved which in turn releases the liquid held in the corresponding reservoir towards further LUOs in the same manner as described above [23].

3.2 Active Valving

3.2.1 Phase-Change Micro-valves

Park et al. introduced a nanocomposite material of 10 nm-sized iron oxide nanoparticles dispersed in paraffin wax which could be used to create a unique phase change microvalve controlled by an external laser diode [24]. The integration and operation of the Laser Irradiated Ferrowax Microvalve (LIFM) is achieved by including a small chamber called an Assistant Valve Chamber (AVC) which is located next to the ferrowax plug [24]. The microvalve can be installed on disk in a Normally Open (NO) or Normally Closed (NC) configuration which can then be reversed by irradiating the iron nanoparticles, on irradiation the iron partials quickly heat causing the paraffin wax to melt thereby opening or closing the channel [24].

The ferrowax plug is located within the microchannel in the case of the NC configuration, as the laser diode transfers heat to the iron nanoparticle the paraffin wax melts and flows into the AVC thereby unblocking the microchannel and opening the valve [24]. In the case of the NO configuration the reverse occurs, the ferrowax is contained within the AVC in a solid state until the laser diode irradiates the iron nanoparticles, this then causes the ferrowax to burst from the AVC and block the microchannel [24].

3.2.2 Electromechanical Pumping

Clime et al. created a microfluidic disk with integrated microelectronics responsible for on-disk pneumatic pumping of liquid through specific chambers [25]. This programmable valving system allows the accurate displacement of fluid by pressurising individual as well as multiple ports of the microfluidic disk. This system has the advantage of being able to pump liquid towards the centre of rotation, thereby giving it a clear advantage over conventional microfluidic designs however its greatest disadvantage is its extensive complexity. The design requires integrated electronics as well as slip-rings for power transfer to the device, this novel technology is currently in its infancy however there is need for this degree of fluid control in the future of centrifugal microfluidics.

4 Valving Design

When developing a new Lab-on-a-disk liquid handling protocol it is necessary to achieve the simplest and most accurate disk design possible, this requires an understanding of the proposed assay or biochemical test to be implemented. Timing of reagent delivery, reagent volumes, incubation times, washes and reagent routing must be taken into account whilst designing the disk architecture, can these unit operations be accomplished using passive valving schemes or do they require more complex integrations. Using passive valving schemes, can the operation be accomplished using frequency dependant valving (Siphon, Capillary, Centrifugo-pneumatic etc.) or should each LUO be triggered by the previous thereby being independent of the rotational frequency (Paper imbibition, Event Triggered). These questions will help in making the simplist architecture possible whilst developing a highly accurate and cost affective Lab-on-a-disk platform.

5 Spin Stand

5.1 Operation

Due to the rapid rate of rotation of microfluidic discs during operation, a special experimental test rig is used to visualise flow. In particular, this development tool provides critical feedback to determine if failures occur (due to manufacturing flaws) and for optimising the fluidic design. This rig is typically referred to as a 'spin stand'[26].

Generally, spin stands use a stroboscopic system to visualise a disc during rotation. Here, a camera and strobe are synchronised such that, as the motor spindle passes through a particular angular location, a pulse is generated which triggers the

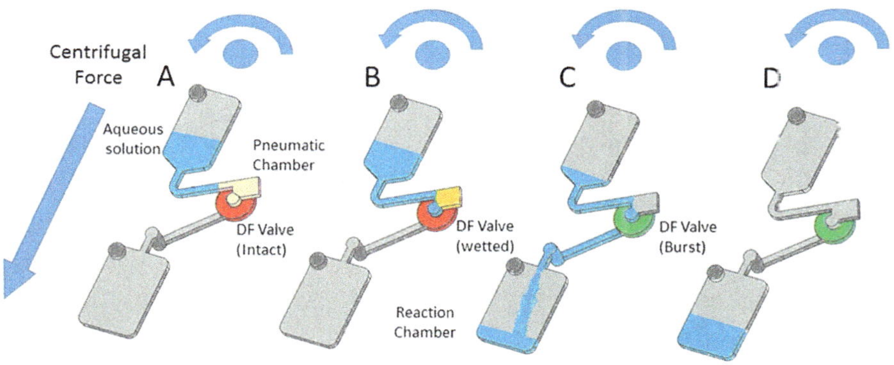

Fig. 5.7 By inserting a piece of dissolvable film between two channels it is possible to create a pressurised channel between the front of the advancing liquid and the dissolvable film, as the frequency of rotation is increased the advancing liquid reduces the volume of air within the channel thereby creating a counteracting pressure gradient that restricts the flow of liquid (Image (**a**)). Only when the frequency of rotations increased to the burst frequency can the liquid advance enough to come in contact with the dissolvable film (Image (**b**)). At this point the film dissolves the pressure is released, thereby releasing the total volume of liquid to pass towards the reaction chamber (Image (**c**) and (**d**))

camera and strobe. Thus, each image of the disc is acquired at the same orientation; a video generated from these still frames shows liquid moving about these apparently still discs. A typical configuration of a spin stand can be seen in Fig. 5.7. Images acquired using a spin stand can be seen in Fig. 5.24.

The most critical aspect for good operation of a spin stand is the correct synchronisation of the motor and strobe. This usually incorporates two steps; generating a signal from the motor as the spindle passes through a particular angular location and filtering this signal before it reaches the camera/strobe.

5.1.1 Signal from the Motor

A stepper or servo-motor should be chosen which has sufficiently high a maximum speed to overcome capillary effects and pump liquids about the disc. Typically centrifugal microfluidics operates using a range of 5–30 Hz but, for some applications, such as blood processing, spin rates of above 60 Hz may be beneficial. The resulting centrifugal force is also influenced by the radial location of the liquid (volume) of the disc. Similarly, a motor should be chosen which can permit high acceleration / deceleration of the disc as this is critical for some applications such as centrifugo-pneumatic siphon valves [27].

The motor should also be chosen which can generate a digital signal once per rotation. If a motor is otherwise suitable, an encoder can be purchased separately and fitted to the motor. Alternatively, an optical sensor can be manufactured using a low-cost LED and photodiode detector [28] (Fig. 5.8).

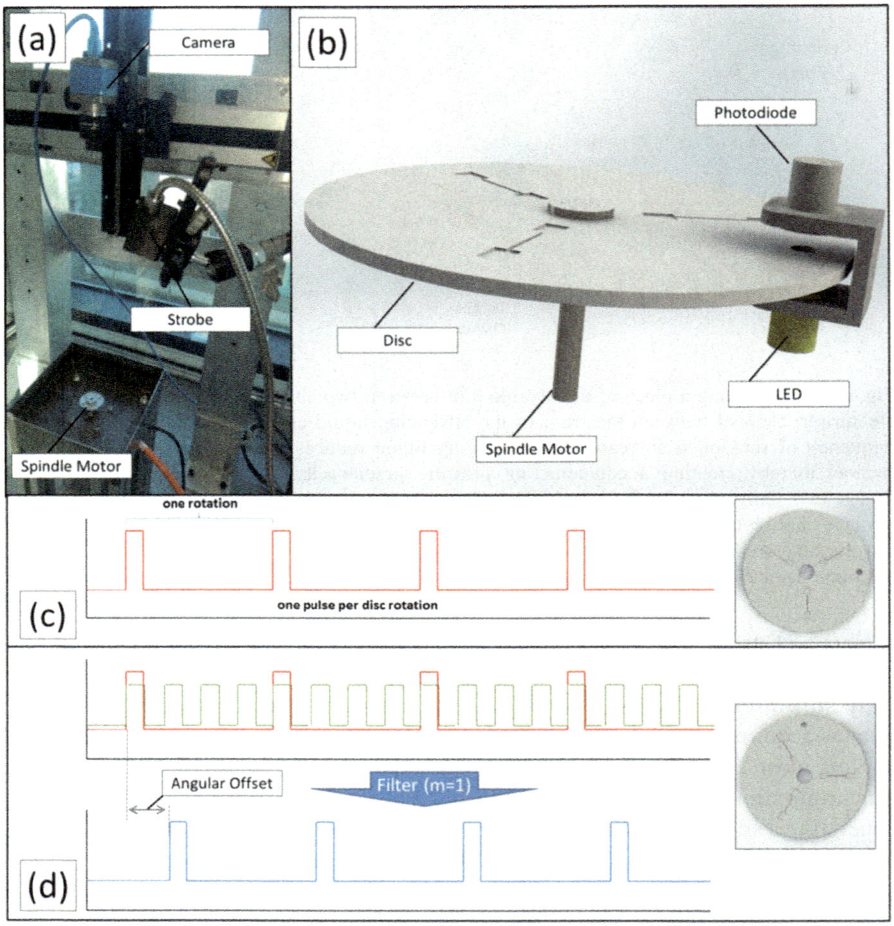

Fig. 5.8 Spin stand Operation (**a**) shows an image of a typical spins stand. The motor is mounted vertically and the camera is mounted above it. A fibre-guided strobe light is mounted on a movable arm. In this case the camera is mounted on an electrical positioning stage (for lateral movement) and a manual positioning stage (to move closer to or further from the disc) (**b**) is a schematic of how a low-cost photodiode and LED can be used to generate one digital pulse per disc rotation. This option is suitable if the motor is not fitted with an encoder that can generate a pulse per revolution (**c**) the typical signal that is generated at one pulse per revolution. Note that in this approach the entire disc must be imaged and images are only acquired at a single angular position (**d**) the typical signal generated by a two channel encoder. The 'home' channel (*red*) shows one pulse per revolution while the second channel (*green*) shows four pulses per revolution. A filter box, with a dial to input value 'm', permits the angular position where the disc is imaged to be changed. Where m = 0 the offset is 0° while m = 1 results in 90° offset. In a typical encoder the channel resolution will be 1000 pulses/revolution and thus the angular resolution will be 0.36°

5.1.2 Positional Triggering

If the motor generates just one digital pulse, at the home position, per rotation then the disc can only be images in at one particular angular rotation. This is also the case if a LED/photodiode based triggering system is used. In this configuration, the entire disc should be imaged. The images acquired can be post-processed (rotated and cropped) using commonly available software.

However, in some cases, particularly if imaging particles or cells, imaging the entire disc is not feasible. In this case, the angular location of the digital pulse must be modified so that the location of interest is imaged during each rotation.

Some motors have software based controls which allow the angular location of the trigger pulse to be varied. An alternative approach is to purchase a motor with an integrated dual-channel encoder. These can also be purchased and fitted separately. These encoders generate one pulse per revolution in Channel 1 and a set number of pulses in Channel 2 can generate a single pulse. A hardware based filter can be applied which counts a specific number of pulses (in Channel 2) after the single pulse in Channel 1 before passing a single pulse to the camera/strobe. Thus, if four pulses are generated per rotation in Channel 2 the system can be rotated in steps of 90° while if 360 pulses are generated the system has a resolution of 1°.

A third approach, when only a single channel is available, is to use a low-cost microcontroller to measure the time between pulses and interpolate between them. However, this approach is less reliable than hardware based timing and is particularly susceptible to jittery imaging during disc acceleration and deceleration.

5.1.3 Filtering the Camera Signal

With the motor signal now filtered to one digital signal per revolution this must now be used to trigger a camera. However, the number of digital signals generated per second is equal to the disc spin rate (up to 100 Hz) while typically low-cost cameras will only acquire ~5 images per second. Therefore the additional signals must be filtered out.

The easiest method is to ensure purchase of a camera with a 'camera ready' output. This digital output from the camera provide information regarding if the camera is ready (or not) to acquire a new frame. This input can be used to filter the incoming digital signals so that a trigger signal only reaches the camera if the camera is ready to receive it. Alternatively, a time delay can be generated so that only one trigger signal reaches the camera in a defined time period; for example if the camera can acquire images at 5 Hz then one trigger signal can be passed to the camera every 200 ms.

5.2 *Application Example: Automation of a Direct Bilirubin Test*

5.2.1 Direct Bilirubin

Direct bilirubin is a measurement of bilirubin which is water soluble and that has been metabolised by the liver for excretion. A decrease in the production of direct bilirubin in conjunction with raised total bilirubin indicates an inherited inability to metabolise bilirubin. Raised direct bilirubin can indicate blockage of the organs of the body which make, store and secrete bile. Here we demonstrate how a standard hospital direct bilirubin test (Randox Laboratories BR411) might be automated on a Lab on a Disc system.

5.3 *Liquid Handling Protocol: Disc Architecture*

The (typical) liquid handling protocol in Fig. 5.9 requires a blood sample to be separated into its different fractions by centrifugation. Once plasma has been

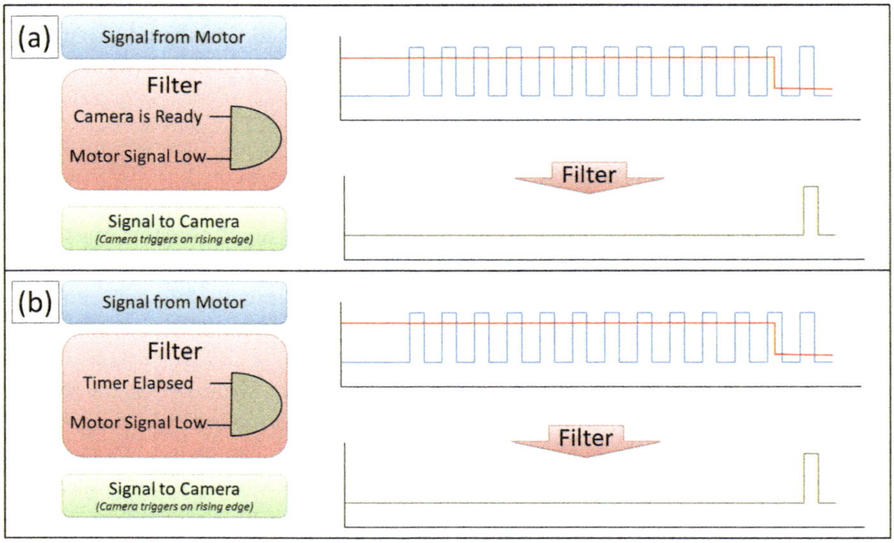

Fig. 5.9 Filtering the Camera Signal (**a**) shows the filtering of the signal using a camera with a 'camera ready out' (*red line*). When the camera is ready this signal line drops. When both this signal and the motor signal are low the filter allows the pulses from the motor to pass through to the camera. Note that when the camera receives a signal it will go into a non-ready state and the subsequent pulses from the motor will be blocked. (**b**) shows the filtering of the signal using a camera and a timer (*red line*). When the timer elapses the signal goes low. When both this signal and the motor signal are low the filter allows the pulses from the motor to pass through to the camera. Note that the timer should be configured such that, when a pulse is passed to the camera, the timer resets and blocks subsequent pulses for the defined time

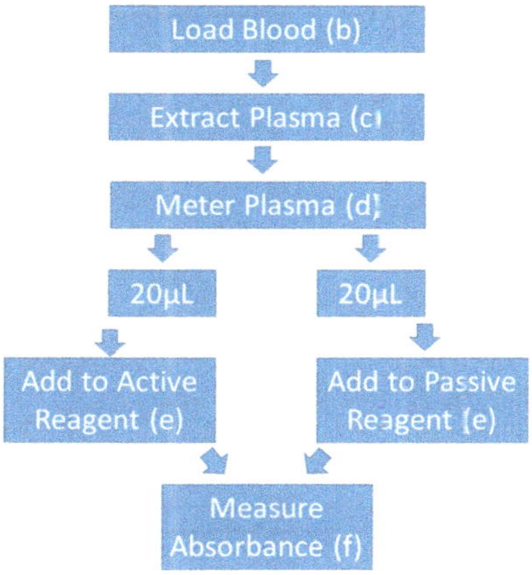

Fig. 5.10 Schematic Diagram of Liquid Handling Protocol

purified, 20 µL is removed and added to each of the two test reagents. The results can then be obtained by measuring the absorbance of the solution.

As shown in Fig. 5.10a, the disc architecture of the Direct bilirubin test requires a blood separation chamber with a DF valve (DF1). This valve leads to two metering structures which are also sealed by DF valves (DF2 and DF3); a waste chamber is included to collect unneeded plasma that overflows from the metering chamber. Each Metering chamber is connected to an individual reaction chamber through DF valves. Prior to running the centrifugal platform, the blood sample must be loaded into the separation chamber through the openings described as Air Vents in Fig. 5.10a, the same procedure is used to load a specific volume of reagent into each of the reagent chambers. Once each chamber is filled (Fig. 5.10b) it is possible to run the test by increasing the spin rate of the disk separate the blood sample.

Accelerating the spin rate of the disk to 30 Hz causes the heavier red blood cells to move towards the edge of the disc without creating enough pressure to burst the connected DF valve (Fig. 5.10c). After the red blood cells have settled further out then the radial distance of the connecting DF valve (DF1), the disk can be accelerated to the burst frequency of DF1. Upon dissolution of the DF film liquid is released and fills each metering chamber with the excess flowing into the overflow/waste chamber (Fig. 5.10d). This step eventually leaves the required volume of 20 µL for each test ready to be added to the reaction chamber (Fig. 5.11).

Once all liquid has settled in the appropriate chambers (Fig. 5.10d) the spin rate of the disk can be increases to the burst frequency of DF valves 2 and 3, this releases the liquid in the above metering chambers and allows it to pass into the reagent chambers. At this point the plasma can react with the preloaded reagents to form a measurable signal which can be measure on disk or if necessary off disk (Fig. 5.10f).

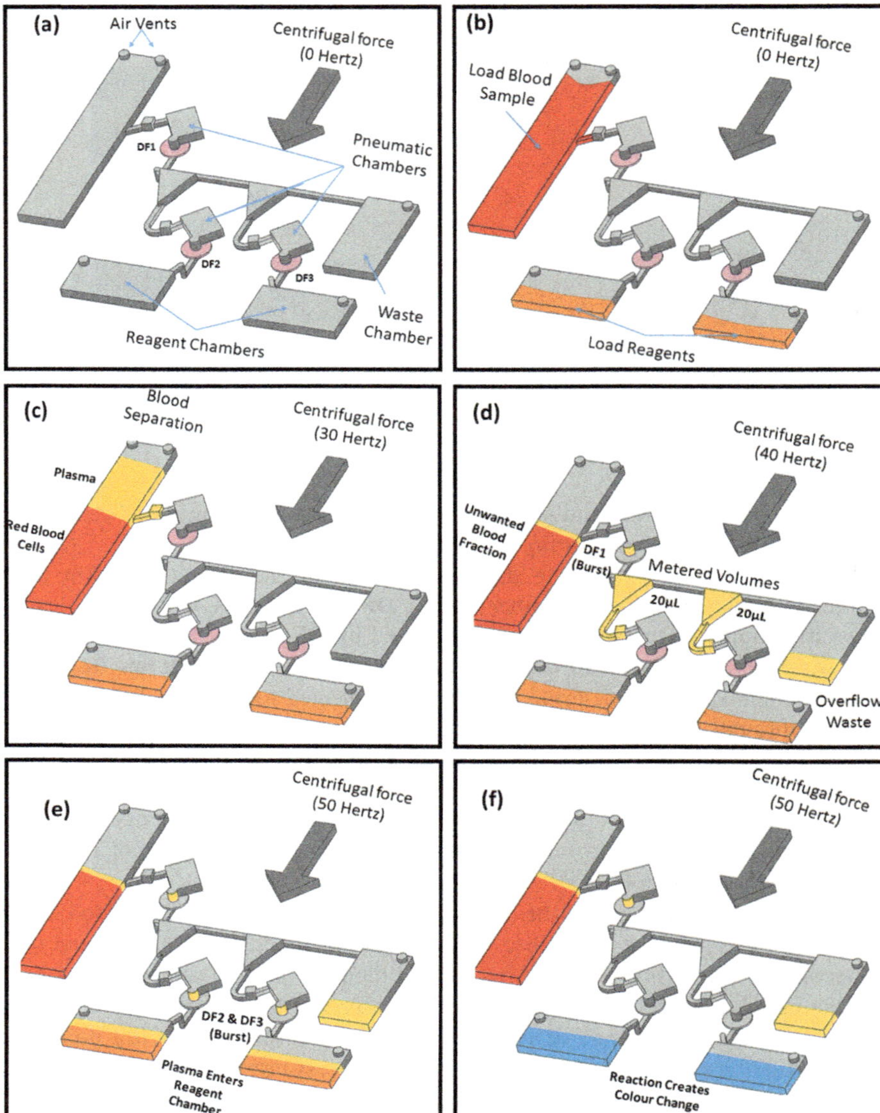

Fig. 5.11 Schematic Diagram representing the disk architecture and liquid handling protocol

6 Manufacturing Protocol

6.1 Designing the Lab-on-a-Disk

The designs used in this protocol were created in the commercial drawing program "SolidWorks"; however, other programs which allow the drawings to be saved as a . DXF file can be used. This Lab-on-a-Disc is fabricated using dissolvable film and multi-layer manufacturing to create a three dimensional internal structure that can control the movement of fluid with a high degree of accuracy (Fig. 5.12).

The SolidWorks programme is well suited for creating the architecture of a Lab-on-a-disk; yet, the code can be difficult to operate at first glance. For this reason, there is a large quantity open source material and instructional video content available online to help beginners learn how to operate the programme. With a rudimentary understanding of the software, it will be possible to create a prototype design in a few hours. This is a key quality of our method of design and fabrication as it allows rapid prototyping of new Lab-on-a-Disc designs (Fig. 5.13)

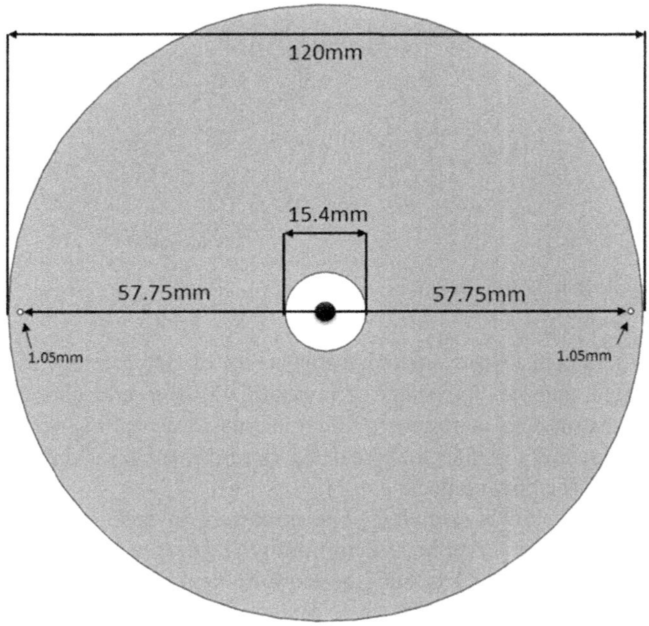

Fig. 5.12 The dimensions of the Lab-on-a-disk platform are similar to that of a compact disk (CD), the outer diameter is 120 mm with the inner diameter set to 15.5 mm. Each disk contains alignment holes used later in the fabrication process to ensure correct alignment of the layers, these holes are 57.75 mm from the centre of the disk (Centre of Rotation) and are 1.05 mm in diameter. The disk itself is 6 mm in width however these dimensions are purely a guideline and may be modified to suit a particular application

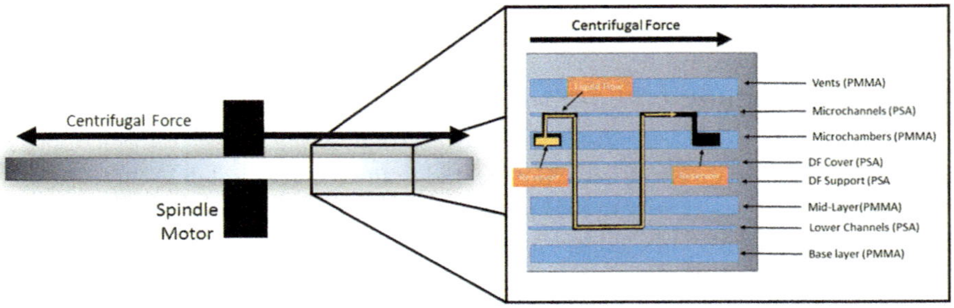

Fig. 5.13 A cross sectional profile of the Lab-on-a-disk shows the location of each as well as the intended material the designs will be cut from. The thin *blue lines* in the image above represent layers of PSA whereas the large blue segments represent layers of PSA. The route between two chambers is depicted by the black voids with the fluid (*orange*) moving outward due to the centrifugal force

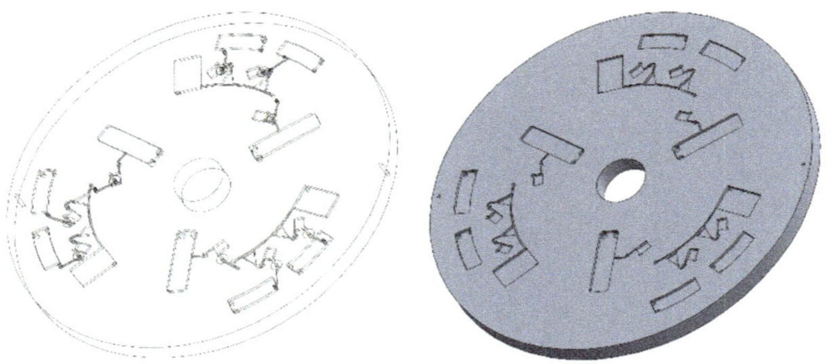

Fig. 5.14 Example SolidWorks design

The microfluidic discs are created using layers of poly(methyl methacrylate) (PMMA) bonded together by layers of pressure sensitive adhesive (PSA). Each layer contains a structural component which is created by cutting voids out of the material, when layered together this creates a three-dimensional network of chambers and channels for flow control.

Once the disk design is complete it is important to ensure that all complex structures requiring the interconnection of multiple layers are perfectly aligned, this can be done by zooming and rotating around the central axis of the disc. If the quality of the design is assured the disk design can be separated into 2D layers which are compatible with the knife cutter and laser cutter (Fig. 5.14).

6.2 *Cutting PSA*

Sheets of pressure sensitive adhesive are cut using a Graphtec automated knife cutter. This device is controlled by the programme "Robo Master-Pro ™ Vision 4.6.0" which accepts the previously attained DXF files (alternative hardware / software can be used) (Fig. 5.15).

6.3 *Dissolvable Film Tabs*

Dissolvable film tabs are used to restrict the flow of fluid from one microfluidic structure to the next by sealing the channel and creating a counteracting pressure which pushes back the liquid. These DF tabs can be wetted and dissolved by increasing the spin rate of the disc until the liquid comes in contact with the DF, this is known as the burst frequency (Fig. 5.16).

Firstly, the tabs are designed in CAD software by patterning an array of inner circles that are equally spaced from each other as shown in Fig. 5.17. Next, a second sketch created to define the outer edge of the tabs; this is drawn with the same pitch as the inner edges.

Fig. 5.15 Graphtec automated knife cutter

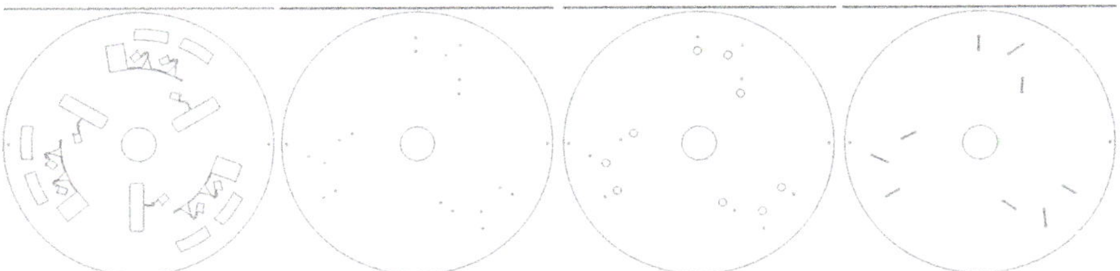

Fig. 5.16 Examples of the four DXF files corresponding to the four layers of PSA

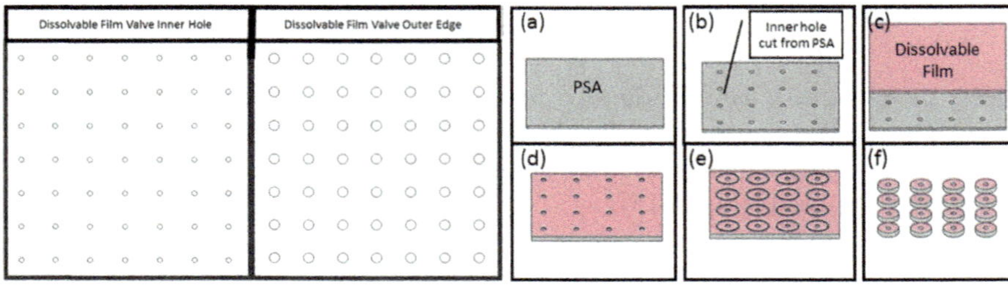

Fig. 5.17 The DXF files showing the inner and outer cuts made in a sheet of PSA (*Left*) and a graphical representation of the procedure required to create dissolvable film tabs

Fig. 5.18 30 W Carbon Laser

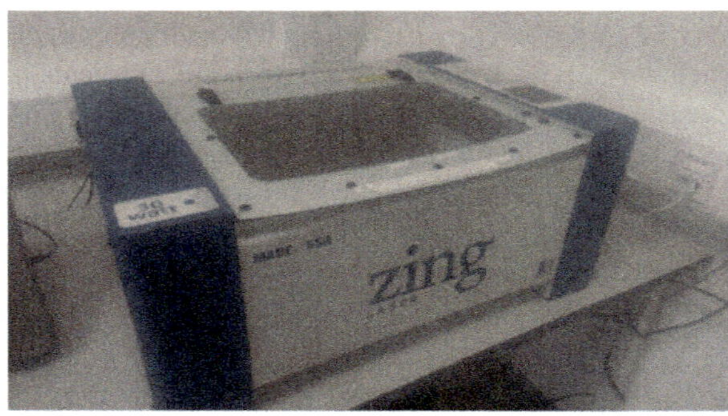

Fabrication of dissolvable film tabs requires an adhesive solid support like PSA (Fig. 5.17). First, the knife cutter is used to cut out the inner holes and the chads are manually removed. Next, the upper layer of PSA is removed exposing the adhesive surface of the PSA. A sheed of dissolvable film is rolled onto this layer and pressure is applied to the film to ensure a good seal over the hole. The outer edge of the DFs is then defined by using the second CAD file above. The force used by the knife cutter should be lower than used for the inner DF cut; this results in the DF tabs remaining mounted on the backing film such that they can be peeled off for placement into the discs (Fig. 5.18).

6.4 *Cutting PMMA*

The cutting of sheets of PMMA is relatively simple in comparison to that of PSA, sheets of PMMA are firstly cut by hand to the size required by the laser cutter being used, in this case the laser cutter is a 30 W Carbon Laser. Once the laser is set up the DXF files, shown in Fig. 5.19, can be opened in the controlling programme Corel DRAW X5 or an alternative programme if using a different piece of equipment. Once this is complete the laser can be activated.

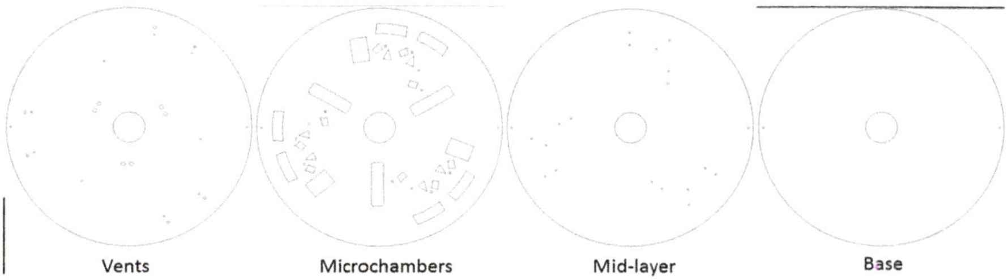

Fig. 5.19 Examples of the four DXF files corresponding to the four layers of PMMA

Fig. 5.20 Apparatus required for disk assembly

7 Assembly

It is necessary to carefully remove all segments of PSA and PMMA that have been cut from the main disk layer, this is done by hand prior to entering the clean room. If left in place these small pieces of PSA and PMMA may interfere with the flow of fluid through the microchannels of the disk during testing (Fig. 5.20).

A custom built alignment housing is used to ensure that that the disks are aligned accurately when assembled, due to the microfluidic nature of the Lab-on-a-disk platform even the smallest deviation during assembly can destroy the functionality of the device. Figure 5.21 shows how this custom assembly jig is used to guide the assembly of the disk. It is for this same reason that assembly is completed within a clean room environment as small particles of dust may inhibit the flow within the device.

As previously stated the PSA serves a dual purpose in the construction and functionality of the disk, it provides channels of narrow dimensions for fluid to pass from one chamber to the next however it also acts as the adhesive that bonds the layers of PMMA together. This requires a great deal of pressure which is provided by a Hot Roll Laminator, or similar device. After each layer is added to the growing Lab-on-a-Disc platform and aligned, it is pressed gently to hold it in place and is

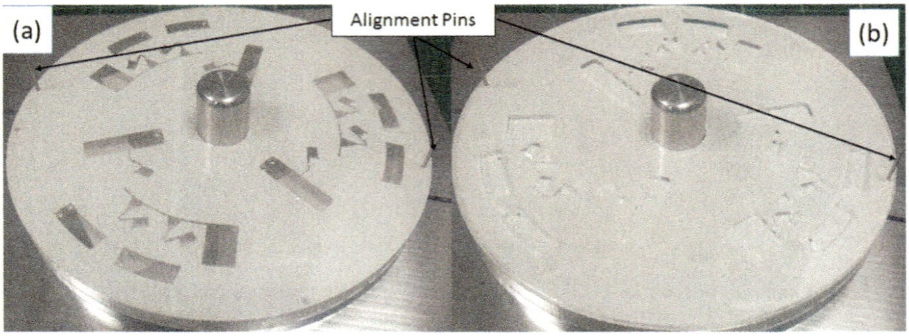

Fig. 5.21 A PSA layer (*Left*) is bonded to a PMMA layer using a pressurised roller

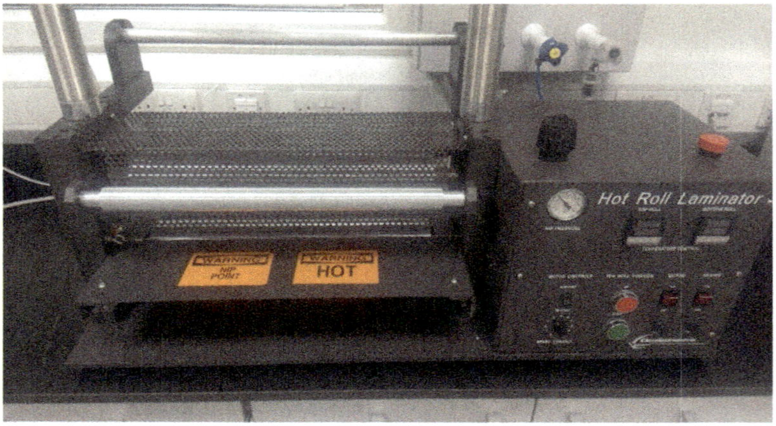

Fig. 5.22 Hot Roll Laminator used to activate PSA

then rolled at room temperature in the hot roll laminator, ensuring that the disk is void of trapped air bubbles (Fig. 5.22).

This procedure is held throughout the assembly process. This Lab-on-a-disk platform is layered as follows:

7.1 Vents (PMMA)

Holes are cut from the a PMMA layer in specific locations to allow air to enter/leave the chamber as fluid enters/leaves to ensure the chamber remains at atmospheric pressure, meaning there is no resistance to the movement of fluid. This layer can also act as a pneumatic seal to create pressurised chambers which restrict the flow of fluid like in the case of dissolvable film valves (Fig. 5.23).

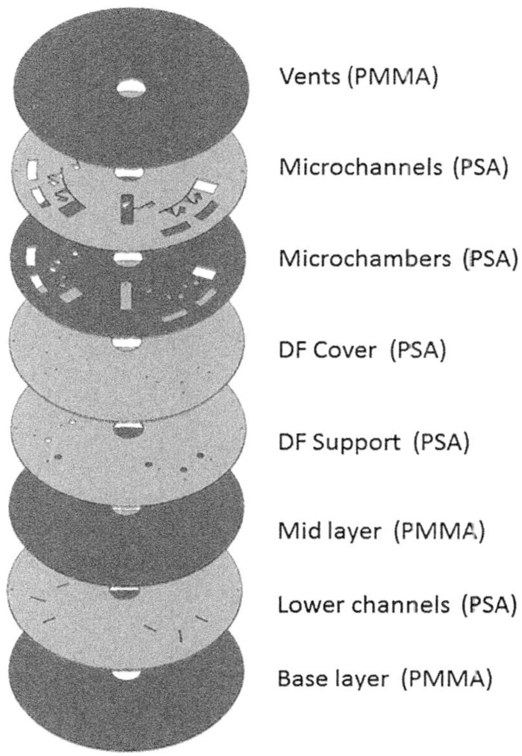

Fig. 5.23 Graphical representation of the order in which the Lab-on-a-disk platform is layered

7.2 Microchannels (PSA)

A layer of PSA directs fluid from each chamber towards further unit operations as the disc spin rate is changed, these microchannels are usually 50 μm in width and maintain a constant height of 86 μm due to the thickness of the PSA. The surfaces of these channels are hydrophilic by virtue of the innate properties of PSA and PMMA.

7.3 Microchambers (PMMA)

Relatively large chambers are cut from the PMMA to form structures responsible for incubation, mixing and liquid storage. This layer can also contain part of the pneumatic chamber responsible for back-pump controlled mixing.

7.4 DF Cover (PSA)

In order to seal each chamber above a layer of PSA was inserted below the microchambers. This layer contains specific cuts to allow the passage of fluid from the above layers to those below as well as creating access points for the fluid to come in contact with the dissolvable tabs contained within the DF support layer.

7.5 DF Support (PSA)

This layer contains the dissolvable film tabs which restrict the movement of fluid by creating a build-up of pressure as the fluid approaches the tab, this resistance to fluid movement can be overcome by increasing the spin rate of the disc thereby allowing release of connected fluid reservoirs. Through cuts were also created to allow the passage of fluid back towards the upper microchannels once valves were burst.

7.6 Mid-Layer (PMMA)

Through channels were cut in a layer of PMMA to allow the movement of fluid between layers in both directions, this layer also forms a support structure and seal for the layers above and below.

7.7 Lower Channels (PSA)

Similar to the microchannel layer the final layer of PSA allows fluid movement connecting the above valves to further microchannels.

7.8 Base Layer (PMMA)

A final PMMA seal is used to complete the structure.

7.9 Testing

Once fabrication and assembly are complete the Lab-on-a-disk design requires testing to ensure that all unit operations (Valving, Metering, routing, etc.) work correctly, this is done using the custom spin stand described above in Fig. 5.7. To test the Lab-on-a-disk design, the reagents intended for use are replaced with water coloured with food dye so that the movement of fluid around the disk can be more easily visualised against the white background of the disk. The spin rate of the disk can be controlled manually to assess the specific burst frequency's required, if the

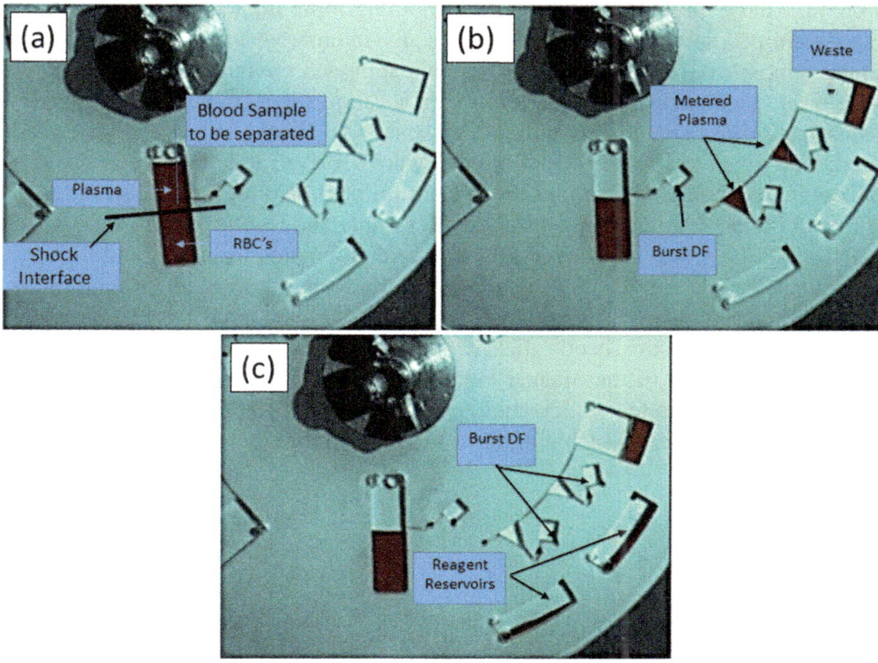

Fig. 5.24 Disc functionality photographed by the custom spin stand described in Fig. 5.7

spin rates are known it is possible to create an automated programme to control the change in spin rate at specific times throughout the process. Figure 5.24 shows a sequence of three real images acquired from a disc designed to isolate blood plasma, meter these samples and distribute them to the periphery of the disc following the liquid handling protocol defined in Figs. 5.10 and 5.11. Coloured water is used as the functional liquid for better visualisation. First, the sample centrifuged at 30 Hz. Once the plasma has been separated the spin rate is increased to 40 Hz to burst the first DF valve, this releases the plasma to flow into the metering chambers and the excess to the overflow waste chamber (Fig. 5.24b). Finally, the spin rate is increased again to 50 Hz which bursts the final two DF valves allowing the metered plasma to reach the prefilled reagent chambers (Fig. 5.24c).

8 Conclusion

The Centrifugal Microfluidic platform offers great versatility in the field of bio-medical diagnostics with applications in environmental monitoring and beyond; this versatility is due to its ease of use and minimal hardware requirements. In the best case scenario, a Lab-on-a-Disc can run a diagnostic test with a minimal impute of blood, can be entirely automated using nothing more than a spindle motor and preprogrammed spin rate, leading to a visually interpreted result or can at times require simple external/on-disk sensors. This approach obviates the need for specially trained technicians as the operation is as simple as turning on a CD player, contamination risks are greatly reduced as the sample to answer process is a closed system and given the portability of the hardware involved it is possible to use this concept in decentralised locations away from primary health care facilities, this in turn would remove any cost associated with sample and results transport.

Described here is a basic overview of the requirements to understand, design, fabricate, assemble and test the Lab-on-a-disc platform, the equipment required is common in academic and some industrial settings with the main fabrication material being cheap plastic sheets, this means the main cost of implementing this procedure is the cost of the manual labour required. This protocol is intended to allow those with limited understanding of centrifugal microfluidics to gain a foothold within the field which will allow them to explore this promising platform.

References

1. Manz A, Graber N, Widmer HM (1990) Miniaturized total chemical analysis systems: a novel concept for chemical sensing. Sensors Actuators B Chem 1(1–6):244–248. doi:10.1016/0925-4005(90)80209-I
2. Whitesides GM (2006) The origins and the future of microfluidics. Nature 442(7101): 368–373. doi:10.1038/nature05058
3. Beebe DJ, Mensing GA, Walker GM (2002) Physics and applications of microfluidics in biology. Annu Rev Biomed Eng 4:261–286
4. Ducrée J et al (2007) The centrifugal microfluidic Bio-Disk platform. J Micromech Microeng 17(7):S103–S115
5. Ottino JM, Wiggins S (2004) Introduction: mixing in microfluidics. Philos Transact Series A Math Phys Eng Sci 362(1818):923–935
6. Barathur R et al (2002) New disc-based technologies for diagnostic and research applications. Psychiatr Genet 12(4):193–206. http://www.ncbi.nlm.nih.gov/pubmed/12454524. Accessed 17 Nov 2015
7. Focke M et al (2010) Centrifugal microfluidic system for primary amplification and secondary real-time PCR. Lab Chip 10(23):3210–3212, http://pubs.rsc.org/en/content/articlehtml/2010/lc/c0lc00161a. Accessed 17 Nov 2015
8. Kido H, Maquieira A, Hammock BD (2000) Disc-based immunoassay microarrays. Anal Chim Acta 411(1–2):1–11
9. Czilwik G et al (2015) Magnetic chemiluminescent immunoassay for human C-reactive protein on the centrifugal microfluidics platform. RSC Adv 5(76):61906–61912, http://pubs.rsc.org/en/content/articlehtml/2015/ra/c5ra12527h. Accessed 13 July 2015
10. Mishra R et al (2015) Lipophilic-membrane based routing for centrifugal automation of heterogeneous immunoassays. http://doras.dcu.ie/20442/1/Mishra_et_al_LIPOPHILIC-MEMBRANE_BASED_ROUTING_FOR_CENTRIFUGAL_AUTOMATION_OF_HETEROGENEOUS_IMMUNOASSAYS_MEMS2015.pdf. Accessed 21 Aug 2015
11. Nwankire CE et al (2013) At-line bioprocess monitoring by immunoassay with rotationally controlled serial siphoning and integrated supercritical angle fluorescence optics. Anal Chim Acta 781:54–62, http://www.sciencedirect.com/science/article/pii/S0003267013005138. Accessed 25 Nov 2014

12. Nwankire CE et al (2013b) Full integration of a liver assay panel on a centrifugal microfluidic platform. In: 2013 Transducers & eurosensors XXVII: the 17th international conference on solid-state sensors, actuators and microsystems (TRANSDUCERS & EUROSENSORS XXVII). IEEE, pp 377–379. https://www.researchgate.net/publication/236971194_Full_integration_of_a_liver_assay_panel_on_a_centrifugal_microfluidic_platform. Accessed 11 Jan 2016
13. Chen QL et al (2012) An integrated lab-on-a-disc for automated cell-based allergen screening bioassays. Talanta 97:48–54. http://www.ncbi.nlm.nih.gov/pubmed/22841046. Accessed 23 Nov 2015
14. Espulgar W et al (2015) Centrifugal microfluidic platform for single-cell level cardiomyocyte-based drug profiling and screening. Lab Chip 15(17):3572–3580, http://pubs.rsc.org/en/content/articlehtml/2015/lc/c5lc00652j. Accessed 11 Jan 2016
15. Gorkin R et al (2010) Centrifugal microfluidics for biomedical applications Lab Chip 10:1758–1773
16. Cox KL (2012) Immunoassay methods. Assay Guidance Manual, pp 26–28
17. Ng AHC, Uddayasankar U, Wheeler AR (2010) Immunoassays in microfluidic systems. Anal Bioanal Chem 397(3):991–1007
18. Kinahan D, Kearney S, Dimov N (2014) Event-triggered logical flow control for comprehensive process integration of multi-step assays on centrifugal microfluidic platforms. Lab Chip 14(13):2249–2258. http://pubs.rsc.org/en/content/articlehtml/2014/lc/c4lc00380b
19. Strohmeier O et al (2015) Centrifugal microfluidic platforms: advanced unit operations and applications. Chem Soc Rev 44(17):6187–6229, http://pubs.rsc.org/en/Content/ArticleHTML/2015/CS/C4CS00371C. Accessed 25 Jan 2016
20. Siegrist J et al (2010) Serial siphon valving for centrifugal microfluidic platforms. Microfluid Nanofluid 9:55–63
21. Kitsara M et al (2013) Spin coating of hydrophilic polymeric films for enhanced centrifugal flow control by serial siphoning. Microfluid Nanofluid 16(4):691–699, http://link.springer.com/10.1007/s10404-013-1266-x. Accessed 11 Jan 2016
22. Gorkin III R et al. (2012) Centrifugo-pneumatic valving utilizing dissolvable films. Lab Chip 12(16):2894–2902. http://ovidsp.ovid.com/ovidweb.cgi?T=JS&PAGE=reference&D=emed10&NEWS=N&AN=2012436014
23. Kinahan DJ et al (2015) Paper imbibition for timing of multi-step liquid handling protocols on event-triggered centrifugal microfluidic lab-on-a-disc platforms. RSC Adv 5(3):1818–1826, http://www.researchgate.net/publication/268989122_Paper_Imbibition_for_Timing_of_Multi-Step_Liquid_Handling_Protocols_on_Event-Triggered_Centrifugal_Microfluidic_Lab-on-a-Disc_Platforms. Accessed 28 May 2015
24. Park J-M et al (2007) Multifunctional microvalves control by optical illumination on nanoheaters and its application in centrifugal microfluidic devices. Lab Chip 7(5):557–564
25. Clime L et al (2015) Active pneumatic control of centrifugal microfluidic flows for lab-on-a-chip applications. Lab Chip 15(11):2400–2411. http://pubs.rsc.org/en/Content/ArticleHTML/2015/LC/C4LC01490A. Accessed 25 Jan 2016
26. Grumann M et al (2005) Visualization of flow patterning in high-speed centrifugal microfluidics. Rev Sci Instrum 76(2):025101. http://scitation.aip.org/content/aip/journal/rsi/76/2/10.1063/1.1834703. Accessed 6 Jan 2016
27. Godino N et al (2013) Comprehensive integration of homogeneous bioassays via centrifugo-pneumatic cascading. Lab Chip 13(4):685–694. http://www.ncbi.nlm.nih.gov/pubmed/23250328. Accessed 6 Jan 2016
28. Madou M et al (2006) Lab on a CD. Annu Rev Biomedical Eng 8:601–628
29. Burger R, Ducrée J (2012) Handling and analysis of cells and bioparticles on centrifugal microfluidic platforms. Expert Rev Mol Diagn 12(4):407–421. doi:10.1586/erm.12.28
30. Ducrée J, Haeberle S, Lutz S, Pausch S, von Stetten F, Zengerle R (2007) The centrifugal microfluidic Bio-Disk platform. J Micromech Microeng 17(7):S103–S115. doi:10.1088/0960-1317/17/7/S07

31. Gorkin R III, Nwankire CE, Gaughran J, Zhang X, Donohoe GG, Rook M, O'Kennedy R, Ducrée J (2012) Centrifugo-pneumatic valving utilizing dissolvable films. Lab Chip 12(16):2894–2902. doi:10.1039/c2lc20973j
32. Nwankire CE, Czugala M, Burger R, Fraser KJ, O'Connell TM, Glennon T, Onwuliri BE, Nduaguibe IE, Diamond D, Ducrée J (2014) A portable centrifugal analyser for liver function screening. Biosens Bioelectron 56:352–358. doi:10.1016/j.bios.2014.01.031
33. Siegrist J, Gorkin R, Clime L, Roy E, Peytavi R, Kido H, Bergeron M, Veres T, Madou M (2010) Serial siphon valving for centrifugal microfluidic platforms. Microfluid Nanofluid 9(1):55–63. doi:10.1007/s10404-009-0523-5
34. Strohmeier O, Keller M, Schwemmer F, Zehnle S, Mark D, von Stetten F, Zengerle R, Paust N (2015) Centrifugal microfluidic platforms: advanced unit operations and applications. Chem Soc Rev 44(17):6187–6229. doi:10.1039/C4CS00371C

Chapter 6
Materials and Surfaces in Microfluidic Biosensors

Pandiaraj Manickam, Jairo Nelson, and Shekhar Bhansali

1 Introduction

Microfluidics is the science of designing, manufacturing, and formulating processes to generate devices that are capable of analyzing small sample volumes, usually in the range of microliters (10^{-6}) to picoliters (10^{-12}). Microfluidic techniques have emerged as a promising alternative to conventional laboratory assays since they allow complete laboratory protocols to be performed on a single chip, merely a few square centimeters in size. Applied microfluidics have a number of significant advantages in biomedical research and in creating clinically useful technologies. For example, microfluidics enable the fabrication of new cost-effective biosensing technologies for clinical diagnostics. This decrease observed in cost stems from the small scale of the device's architecture, which reduces sample volumes, processing times, and reagent consumption when compared to conventional methods. At such a small scale, material selection is a crucial part of microfluidic system development as it impacts its processing, functionality, application, and the disposability of the sensor strips and the fluidic manifold. This chapter reviews the most common types of materials that are currently used to fabricate microfluidic devices. Methods used for their fabrication, physical and chemical properties of the materials, and advantages they provide to the biosensor configuration are also summarized. Special consideration was also given to the selection of ideal prototyping materials for specific applications based on their cost, mechanical, and biocompatible properties.

P. Manickam (✉) • J. Nelson • S. Bhansali
Department of Electrical and Computer Engineering, Bio-MEMS and Microsystems Laboratory, Florida International University, Miami, FL, USA
e-mail: pmanicka@fiu.edu

C.K. Dixit, A. Kaushik (eds.), *Microfluidics for Biologists*,
DOI 10.1007/978-3-319-40036-5_6

1.1 Biosensors-Involvement of Microfluidics

According to a recent National Institute of Health (NIH) report, point-of-care (POC) testing has the potential to introduce a paradigm shift into personalized medicine by creating a link between disease diagnoses and the ability to tailor therapeutics to the individual [1]. POC testing promotes a shift away from traditional diagnostic tests in the clinical laboratory setting to near-patient settings, providing physicians with timely diagnostic details to make appropriate decisions regarding diagnosis and treatment. The global POC market for in-vitro diagnostics is poised to grow at a CAGR of 9.3 % from 2013 to 2018 and is said to reach $27.5 billion by 2018 [2]. Examples of POC testing have become a familiar topic within the research community in recent years. Successful examples of these widely adopted POC systems are the glucometer, used for managing diabetes mellitus and the disposable lateral flow immuno-strip, used for pregnancy testing. To qualify as a successful candidate for POC diagnostics, the sensitivity, specificity, portability, and cost of the biosensor system must be better than that of centralized laboratory assays [3, 4]. In order to achieve these requirements, most of the POC sensors are built around the idea that they should operate as lab-on-chip (LOC) devices; implying that they are miniaturized automated laboratories. The only way to achieve this device goal and out-compete the need for cumbersome lab techniques that require culture bottles, petri dishes, and microtiter plates is to employ microfluidics. Analysis rates for POC devices integrated with microfluidic channels are usually shorter and several assays can be integrated in a single system without extending the size and complexity of the device. In addition to this, several steps of the analytical procedure can be integrated and automated within the system. An idealized concept of a POC device [5] is shown in Fig. 6.1.

Based on George Whitesides' definition [6], microfluidics is "the science and technology of systems that process or manipulate a small volume of fluids, typically (10^{-9}–10^{-18} L), using channels with dimensions of tens to hundreds of micrometers". The high surface-area-to-volume ratio of microfluidic devices leads to enhanced heat and mass transfer. In addition to the latter, interfacial phenomena that are not usually observable at the macroscale, such as the domination of surface forces instead of inertial and body forces can be elucidated. While microfluidics hold great promise in POC medical diagnostics, limitations to its applicability still exist. Fabrication costs and material compatibility are two major concerns in material development; novel materials and processes that overcome these limitations are addressed in this chapter.

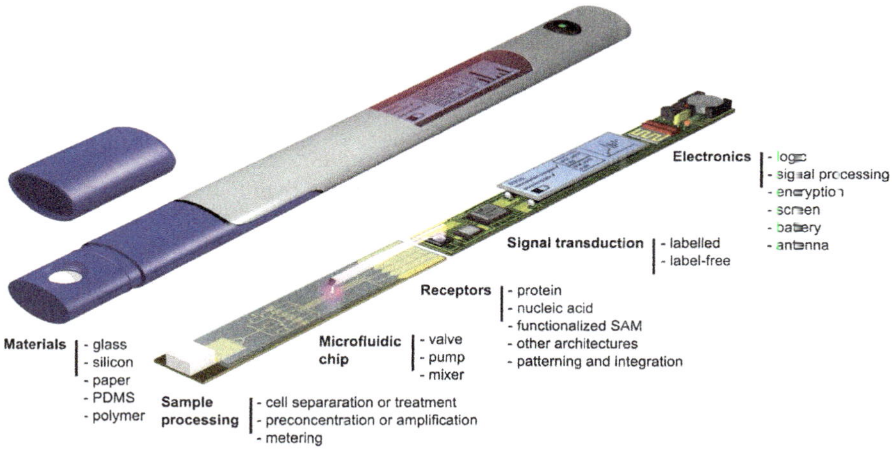

Fig. 6.1 Idealized POC device [5] reproduced with permission from John Wiley & Sons, Inc. The POC device consists of a disposable part, with a loading port for sample introduction, a sample preparation and metering unit (for pre-concentration, amplification, cell lysis), a microfluidic processing unit (for splitting, moving, and mixing samples and reagents), a sensor unit (to target recognition receptors) for labeled or label-free detection, and a signal transduction unit with electronic readout circuit

2 Material Development for Microfluidic Systems

The choice of materials for microfluidic sensor fabrication is a significant factor in performing chemical and physical operations inside the microchip [7, 8]. The surface chemistry and mechanical stability of substrate materials are two major considerations for microfluidic device fabrication [9]. The surface modification and handling of fluids within the chip are primarily controlled by the surface properties of the material while the mode of detection is governed by the transducers employed. Although a variety of transducer methods have been feasible toward the development of biosensor technology, the most common methods are electrochemical and optical followed by piezoelectric. A broad range of materials from silicon, glass, polymers, and paper have been demonstrated as viable compatible materials for creating advanced and low-cost microfluidic devices, with embedded micrometre-sized elements, to provide unique and often combined functionalities for microfluidic processing (Fig. 6.2).

2.1 Inorganic Materials

2.1.1 Silicon

The first generation microfluidic devices were fabricated using glass and silicon-based materials due to their widespread availability in the semiconductor industry

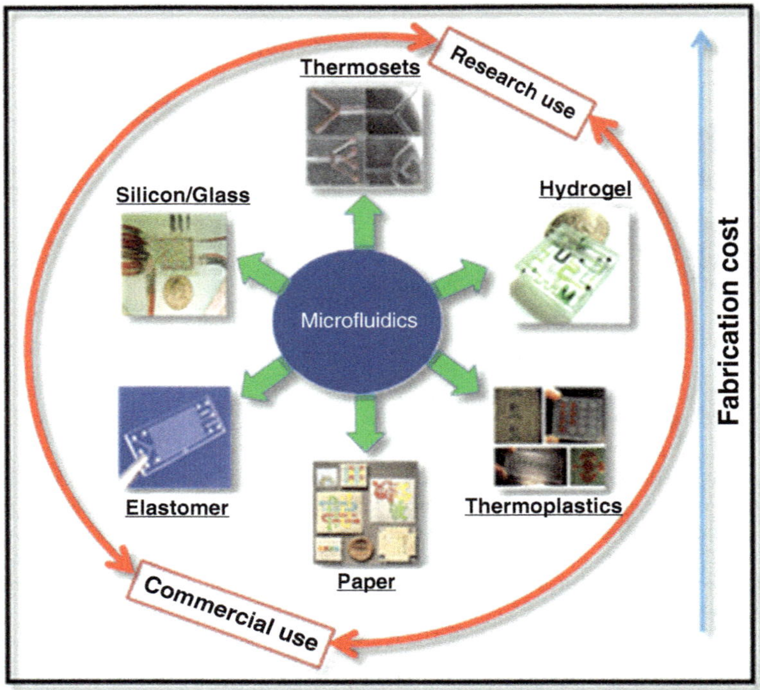

Fig. 6.2 Schematic representation of different types of materials used for microfluidics design

and well-established micromachining processes [10, 11]. Furthermore, both the materials can be chemically modified and functionalized to provide biocompatible platforms for sensing and cell culture applications. However, since silicon is a semiconductor, usually a dielectric SiO_2 layer is thermally grown on the surface of silicon as an insulating layer to separate the electrode surface from the substrate of most silicon-based sensors that require electrical isolation [12]. This is particularly important for sensors that employ electrically-based detection methods since electrical effects in the bulk silicon substrate could potentially interfere with detection or provide stray noise in the sensor. The silicon's surface chemistry, based on the silanol group (—Si—OH) is well developed and modification can be easily accomplished via silanes [9, 13]. For example, nonspecific adsorption can be reduced or cellular growth improved through chemical modification of the surface.

Due to its intrinsic properties, silicon is transparent to electromagnetic wavelengths in the infrared spectrum but not to wavelengths in the visible light spectrum, making typical fluorescence detection or fluid imaging challenging for embedded structures. This issue can be overcome by having a transparent material (polymer or glass) bound to silicon in a hybrid system. Such hybrid devices have led to a renaissance in Si-based detectors for microfluidic systems [15, 16]. For example,

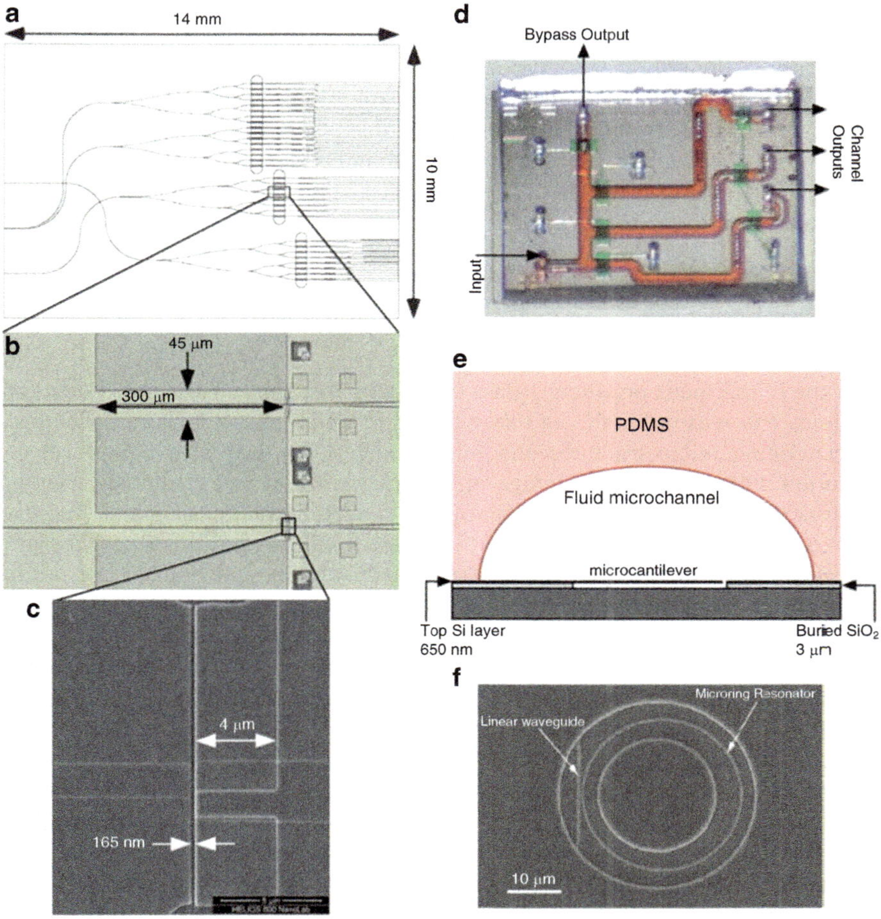

Fig. 6.3 Microcantilevers (**a–e**) and a microring resonator (**f**) made from silicon. (**a**) Schematic diagram of waveguides and microcantilever array layout on die. (**b**) Optical image of two microcantilevers in a fabricated array. (**c**) Close up scanning electron micrograph (SEM) image of the unclamped end of a microcantilever (*left* of 165 nm gap) and the differential splitter capture waveguide (*right* of gap). (**d**) Photograph of complete integrated device showing the fluid microchannels (*red*) and control valves (*green*). (**e**) Cross-section of fluid microchannel at a microcantilever array. (**f**) *Top-view* SEM image of a microring resonator and linear waveguide, visible through an annular opening in the fluoropolymer cladding layer [14]. Reproduced with permission from American Chemical Society

Si microcantilevers and optical resonators have recently been interfaced with microfluidics for optical sensing (Fig. 6.3). Silicon possesses good thermal conductivity and is resistant to high temperatures; therefore, it is suitable for applications requiring a relatively high operating temperature, such as for a polymerase chain reaction (PCR) and for bioreactions. However, silicon substrates are relatively expensive when compared with other materials, such as glass and polymers. Furthermore, the fabrication process for silicon-based microfluidic devices involving substrate cleaning, resist coating, photolithography, development, and wet/dry etching are relatively time consuming and costly. These limitations hinder its practical applications in commercial immunoassays. Another drawback is silicon's optical opacity, which limits its direct applications in real-time optical detection.

2.1.2 Glass

After the initial focus on silicon, glass has emerged as the substrate of choice for the time. The reason for this is that wet and dry etch methods were developed for generating microstructures and microchannels in glass at a high level of resolution. Glass has the advantage of being transparent and chemically stable, which is particularly attractive when optical detection methods such as fluorescence or surface plasmon resonance (SPR) are used. Several types of glass are used in microfluidic devices such as soda lime, quartz, and borosilicate. Soda lime glass is one of the cheapest and most commonly used forms of glass, but it can contain a large amount of contaminants such as aluminum. Quartz is the most suitable glass material for optical sensing devices since it is transparent between the ultraviolet and infrared wavelength spectrum [17]. Unfortunately, the etching rate and cost of the quartz substrates are 2–3 time higher than that of soda-lime glass. Borosilicate glass or pyrex is the most commonly used material in microfluidic and nanofluidics devices because of its optical characteristics (transparent from approximately 350 through 700 nm) and its physical properties (annealing temperature of 640 °C, resistant to most chemicals). However, borosilicate glass is also more expensive than soda lime glass. In the end, selecting the material will affect either the device cost or the device function. The key to selecting an adequate material for a microfluidic device comes down to the device's application. In many cases, many materials need to work together to yield the desired functionality. An example of this can be seen with glass, whose elastic modulus is highly dependent on the glass's composition; as a result constructing active components with more dynamic valves and pumps require multiple materials to form hybrid devices such as the one seen in Fig. 6.3 [18]. Glass has many other favorable traits that become paramount to a microfluidic device's function. One of the most important and well-known traits is its compatibility with biological samples. Since glass has the property of relatively low nonspecific adsorption and is not gas permeable, it is an ideal material for working with biologics where cell kinetics and gas incubation need to be controlled.

One pitfall of utilizing silicon and glass substrates is the costly conditions under which the micromachining of channels and structures need to take place. Most of these microstructures require the use of cleanroom facilities and equipment to maintain a small structural resolution; which includes the use of wet and dry etching, photolithography, electron beam lithography, and a variety of other techniques that can only be done properly in a particle-free environment. The high cost involved in processing glass and other materials like it, will likely limit their usage as disposable devices. In addition to this, glass may not be a very suitable material for more complex multi-layered devices. This challenge stems from the fact that bonding glass-based device layers to create sealed channels is a difficult to do. Often high temperatures and/or large electric fields may be needed to achieve the desired outcome and therefore new fabrication methods for biomaterials have to be adapted to meet these device-bonding requirements.

2.1.3 Ceramics

Ceramics are another inorganic material that offer many novel structural and functional capabilities in microfluidic device fabrication. Ceramics such as, Low-Temperature Cofired Ceramic (LTCC) are a commonly used aluminum oxide based material that comes in laminated sheets that can be patterned, assembled, and then fired at elevated temperatures to construct microfluidic platforms (Fig. 6.4). LTCC has inherent properties that make it another viable material for microfluidic structure construction. It has high-temperature stability, chemical inertness, biocompatibility, low thermal conductivity, excellent dielectric properties, mechanical strength, packaging capabilities, and three-dimensional structuring characteristics that are ideal for developing multi-layered microfluidic devices coupled with electronic components [19, 20]. A representative LTCC device consists of a multilayered stack of sintered ceramic tapes, each of which contains passive electronic elements such as resistors, capacitors and inductors buried within it. The various layers are interconnected via channels filled with conducting materials such as gold (Au) or silver (Ag); topical gold or silver paste can also be utilized to bridge device components. LTCC-based fabrication aids rapid prototyping with a significantly low turn-around time in a semi-cleanroom environment and with minimal use of expensive tools when compared to conventional cleanroom based microfabrication techniques (Fig. 6.5).

2.2 *Elastomers and Plastics*

Polymer-based microfluidic chips were introduced several years after silicon/glass chips. The vast variety of polymers offers great flexibility in choosing a suitable material with specific properties for any microfluidic application. Compared to inorganic materials, polymers are easy to access and inexpensive, which is why

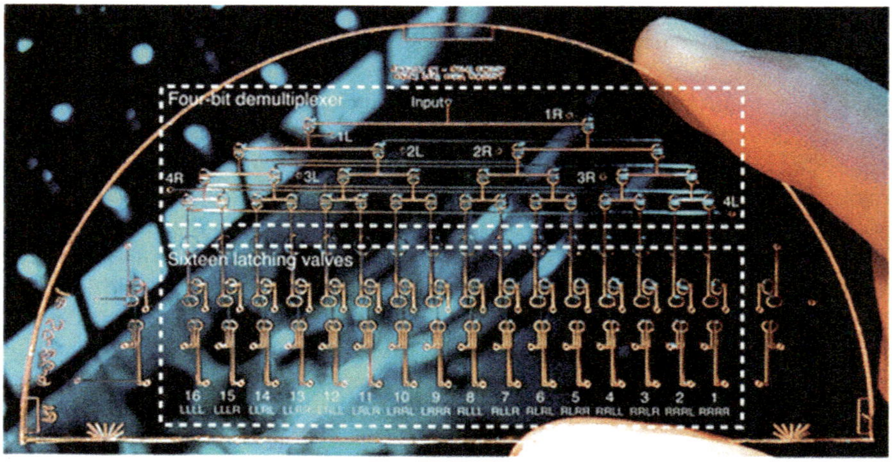

Fig. 6.4 Microfluidic valve network. Photograph of the multiplexed latching valve test device, with a four-bit demultiplexer [18] Reproduced with permission from Royal Society of Chemistry

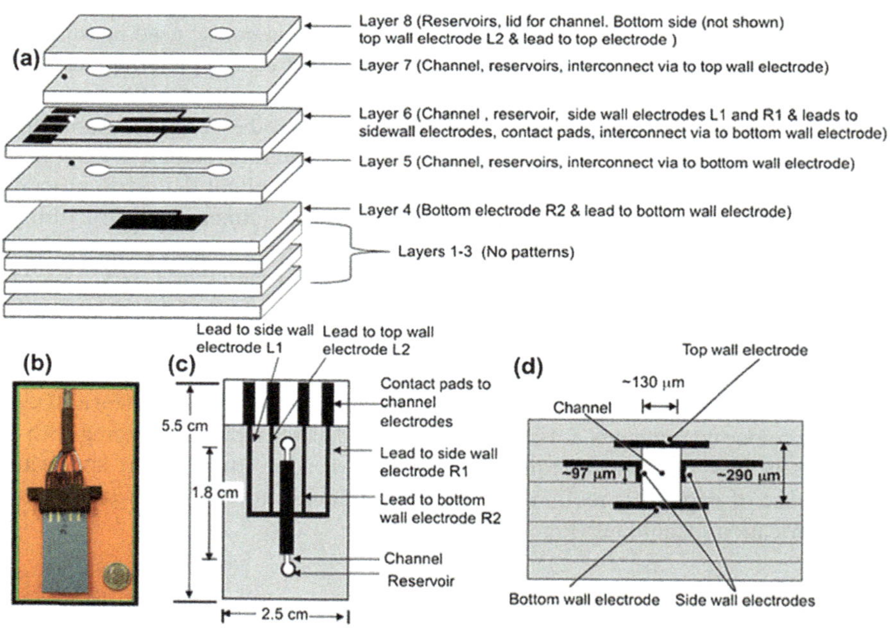

Fig. 6.5 Microfluidic device made from LTCC. (**a**) Multilayer stacking to create a screen-printed gold/LTCC microchannel device. (**b**) Photograph of a microchannel device with an edge connector. (**c**) *Top down view* of a device. (**d**) Cross-section view of the microchannel showing the four gold electrodes. The width of the chip in panel d is not drawn to scale [21]. Reproduced with permission from Springer

they have become the most commonly used microfluidic materials. Based on their physical properties, polymers can be classified into three groups: **elastomers, thermoplastics, and thermosets**.

2.2.1 Elastomers

Elastomers consist of cross-linked polymer chains that can stretch or compress when external force is applied and return to their original shape when the external force is removed. Elastomers, particularly PDMS-based microfluidic systems have been used extensively in the control and manipulation of different liquids because of their remarkable biocompatibility and ease of fabrication. One of the most common PDMS fabrication processes is soft lithography [22, 23], which allows the patterning of sub-micron sized channels. The process is fast, simple, and does not require expensive facilities. PDMS offers a number of unique and attractive features compared to former inorganic materials. These features are as follows: (1) PDMS has a shear modulus of 0.25 MPa and a Young's modulus of roughly 0.5 MPa (characteristic of a moderately stiff elastomer). This elastomeric characteristic allows it to conform to a surface and achieve atomic-level contact, a feature that is useful in forming and in sealing microfluidic systems; (2) PDMS is readily available from commercial sources at decent prices (~\$80/kg); (3) It is optically transparent which facilitates the observation of fluid transfer and content in the micro-channels visually or through a microscope; (4) The surface of the PDMS is hydrophobic (with a water contact angle of ~110°) and can be modified to be hydrophilic (with a water contact angle around 10°) by brief exposure to oxygen plasma; (5) This material can sustain a large temperature range, from 100 to 300 °C, without obvious changes in the property. Figure 6.6 shows the commercially available elastomer microfluidic devices.

Due to its advantageous features, PDMS has been extensively employed in the designing of microfluidic devices including. Examples include the detection of tumor markers [24], anticancer activity evaluation [25], the diagnostic of influenza virus [26], and HIV-1 infection [27]. These flexible materials have also been investigated for use in pressure sensors and wearable healthcare monitoring devices. For example, using a carbon nanotube (CNT)-PDMS composite, Lee *et al.* developed flexible and biocompatible dry electrodes that exhibited good long-term performance in wearable electrocardiographic (ECG) monitoring when connected to traditional ECG devices [28]. Although PDMS has many merits, its hydrophobicity (due to the repeating $OSi(CH_3)_2$) challenges its applications in biochemical sensing because of the nonspecific adsorption of proteins and other molecules it exhibits. Furthermore, the PDMS polymer network sometimes absorbs small molecules, leaches uncured monomers, and swells in solvents. Therefore, applications for PDMS devices are restricted to aqueous solutions. This disadvantage, however, can be easily overcome by bulk/surface modification and well-developed functionalization techniques.

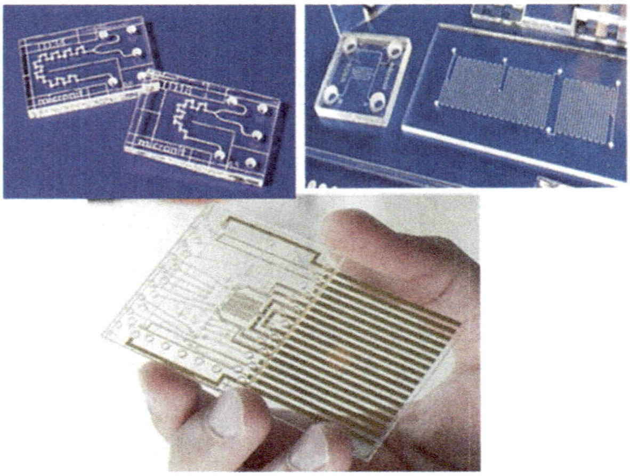

Fig. 6.6 Some commercial elastomer based microfluidic devices

2.2.2 Thermoplastics

Thermoplastics are a class of synthetic polymers that exhibit softening behavior above a characteristic glass transition temperature (Tg) resulting from long-range motion of the polymer backbone, while also allowing them to return to their original chemical state upon cooling. Thermoplastic polymers differ from elastomers by their ability to be softened, fully melted, and reshaped upon heating, while remaining chemically stable over a wide range of operational temperatures and pressures [29]. Polymethylmethacrylate (PMMA), polycarbonate (PC), and cyclic olefin polymers (COP) or copolymers (COC) are most common examples of thermoplastics used in the microfluidic design. A short list of other engineered thermoplastics which have been used for microfluidic chips include polystyrene (PS), polyetheretherketone (PEEK), polyesters, polyethylene terephthalate (PET), polyethylene (PE), polyvinylidene chloride (PVDC), polyvinyl chloride (PVC), polypropylene (PP), polysulfone (PSU), and fluoropolymers such as polytetrafluoroethylene (PTFE) and fluorinated ethylene propylene (FEP). The microchannel fabrication process with thermoplastics involves a different variety of replication methods including hot or cold embossing, injection molding, or thermoforming. Early developments in thermoplastic microfluidic systems were largely focused on using PC and PMMA due to their wide availability in a variety of grades, good solvent and chemical compatibility, and well-characterized molding parameters. A chart summarizing several important material properties for common microfluidic thermoplastics is provided in Table 6.1.

PMMA, a commonly used thermoplastic, is widely known under the commercial names of Plexiglas and Lucite. It has an elastic modulus of 3.3 GPa and good optical clarity from the visible into the UV [30]. Other advantages of this material include biological compatibility, gas impermeability, and ease of micromachining at relatively low temperatures (~100 °C). Yang et al. demonstrated that α-fetoprotein can be quantified in blood serum using immuno-affinity extraction

Table 6.1 Summary of physical properties for common microfluidic thermoplastics [29]

Polymer	Tg (°C)	Tm (°C)	Water absorption (%)	Solvent resistance	Acid/base resistance	Optical transmissivity	
						Visible	UV
COC/COP	70–155	190–320	0.01	Excellent	Good	Excellent	Excellent
PMMA	100–122	250–260	0.3–0.6	Good	Good	Excellent	Good
PC	145–148	260–270	0.12–0.34	Good	Good	Excellent	Excellent
PS	92–100	240–260	0.02–0.15	Poor	Good	Excellent	Excellent
PEEK	147–158	340–350	0.1–0.5	Excellent	Good	Poor	Poor
PET	69–78	248–260	0.1–0.3				
PE	−30	120–130	0.01	Excellent	Excellent	Fair	Fair
PVDC	0	76	0.10	Good	Good	Good	Poor
PVC	80	180–210	0.04–0.4	Good	Excellent	Good	Poor
PSU	170–187	180–190	0.3–0.4	Fair	Good	Fair	Poor

Note: The qualitative metrics shown in this chart are neither comprehensive nor definitive and are provided only as a general guide for material evaluation
In particular, note that solvent resistance can be highly dependent upon the solvent type, e.g. hydrocarbons, alcohols, ketones, etc

coupled with electrophoretic separation in a PMMA-integrated microdevice [31]. Chen et al., reported a self-contained, integrated, disposable, PC microfluidic cassette for nucleic acid—based detection of pathogens at the point of care[32] (Fig. 6.7). The system, in its current state, successfully and reproducibly detected down to 1000 pathogen particles in the sample.

Covalently modified surfaces are generally more stable in thermoplastics than in PDMS. For example, after treatment with oxygen plasma, the surface of PMMA retains hydrophilicity for up to a few months [33]. Also, they can be easily integrated with electrodes for flexible circuits; one related application is digital microfluidics, which can manipulate droplets by electrowetting. In general, they show a slightly better solvent compatibility than the PDMS elastomer. Unfortunately, they are incompatible with most organic solvents, such as ketones and hydrocarbons. In contrast to elastomers, thermoplastics are normally purchased solid and fabricated by thermomolding. Thermomolding can produce thousands of replicas at a high rate and low cost, but it requires templates in metal or silicon for use at high temperatures (to allow ample plastic flow); it is excellent for commercial production but not economical for prototypic use.

From a manufacturing point of view, the main advantage of thermoplastics is their stability and ability to be melted and reshaped against a mold, enabling production of thermoplastic parts with high throughput. From a lab-on-chip perspective, the availability of many commercial, medical grade formulations is a great advantage. Their stiff mechanical properties also provide structural support and protection for the sensor and the microfluidic network. However, many solvents common in chemical analysis and separation dissolve thermoplastics. Nevertheless, most commercial lab-on-chips are made of thermoplastics.

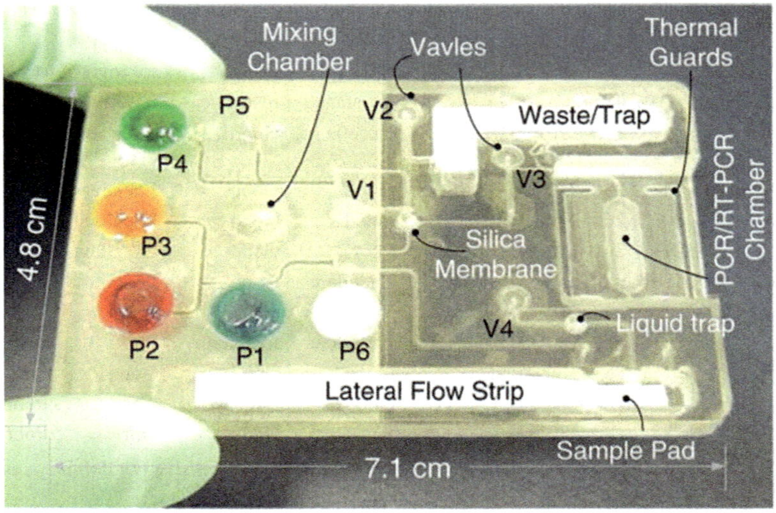

Fig. 6.7 Photograph of the assembled, nucleic acid cassette in its storage state [32] Reproduced with permission from Springer

Thermosets or thermosetting polymers, are covalently cross-linked polymers and thus *do not* melt. From a manufacturing point of view, thermosets are shaped during the polymerization and cross-linking process. Because of the covalent bond formation, thermosets exhibit higher residual stress, shrinkage and crack-formation compared to thermoplastics. From a lab-on-chip perspective, the main advantages of the thermosets are their geometrical stability and solvent resistance. Common thermosets used in microfluidics are poly(dimethylsiloxane) PDMS (an elastomer), the hard resist SU-8 (MicroChem, USA), and the optical glue NOA81 (Norland Products, Inc, USA), that has recently been used for solvent resistant microfluidics [34].

2.3 *Hydrogel*

Hydrogels are a class of crosslinked hydrophilic polymer networks and are able to change their volumes reversibly and reproducibly by more than one order of magnitude; what's even better is that this volume expansion can be accomplished with very small changes of certain environmental parameters [35]. The volume change of smart hydrogels can be induced in response to a variety of inputs such as pH, glucose temperature, electric field, light, as well as by the carbohydrates and antigens present. Hydrogels can be natural or synthesized in a laboratory. Natural hydrogels are proteins that are extracted from mammalian or non-mammalian cells (e.g., collagen, gelatin, and fibrin, and polysaccharides). Most of the natural hydrogels form networks through ionic or physical interactions under physiologically relevant conditions. Synthetic hydrogels (e.g., polyethylene glycol, polyacrylic acid, and polyvinyl alcohol) are hydrophilic polymers and synthesized covalently by radical chain polymerization or step-growth polymerization. Studies by *Mirzabekov* and his collaborators showed that immobilizing nucleic acid probe molecules into hydrogel networks rather than grafting them directly onto solid surfaces (such as in a traditional microarray) led to significant detection advantages; with respect to specificity and sensitivity [36, 37]. *Jinseok et al.* describe a microfluidic biosensor that uses an array of hydrogel-entrapped enzymes to quantitatively determine the concentration of an analyte and simultaneously detect multiple analytes [38]. The hydrogel, poly(N-isopropylacrylamide) (PNIPAAm) is a thermosensitive polymer that exhibits a reversible phase transition from a swollen hydrated state to a dehydrate state. The reversible phase transition can also be described theoretically as a gas-liquid (hydrated to dehydrated state) phase transition [39]. Microvalve actuators are the simplest hydrogel-based components in microfluidic systems. *Beebe et al.* presented a hydrogel-based microvalve concept for autonomous flow control inside microfluidic channels corresponding to different pH values [40]. The hydrogel components were fabricated inside the microchannels using a liquid phase *in situ* photopolymerization process. Owing to low density at the macromolecule scale (and low strength), hydrogels support only lower resolutions (micrometer scale) in microfabrication than other polymers (nanometer scale). While there is no complete study of hydrogel as microfabricated material, numerous methods have been developed to integrate hydrogel into microfabricated devices. The applications of hydrogel devices are mostly cell related.

2.4 Paper

Microfluidic paper-based analytical devices (μPADs) is a growing research field first described by Whitesides and his collaborators in 2007 [41]. At the initial stage, microfluidic channels were designed on chromatography paper to allow the simultaneous colorimetric detection of glucose and proteins in the same sample. Paper has become an alternative starting material to inorganic or polymer materials for fabricating low-cost micro-analytical devices for following three reasons: (1) it is a ubiquitous and inexpensive cellulosic material; (2) it is compatible with many chemical/biochemical/medical applications; and (3) it transports liquids using capillary forces without the assistance of external forces.

Filter paper (Whatman Grade 1) and chromatography paper are the most widely used substrates for μPADs. They are composed of pure cellulose, while many other papers contain structure-reinforcing additives that are potentially detrimental in analytical assays. For example, surface coatings can prevent the analyte from diffusing into the paper matrix. Cellulose is the primary component which has abundant hydroxyl groups (—OH) and a few carboxylic acid groups (—COOH) on the fiber surface [43]. Since cellulose has a slightly anionic surface, it can serve as a scaffold for immobilizing positively charged bio-receptors (enzymes, antibodies, and nucleic acids). Covalent bonding of bio-receptors is often exploited for its robust attachment of biomolecules through EDC/NHS chemistry.

There are many techniques reported in the literature for fabricating paper-based microfluidic devices including: (1) photolithography, (2) plotting with an analogue plotter, (3) ink jet etching, (4) plasma treatment, (5) paper cutting, (6) wax printing, (7) ink jet printing, (8) flexography printing, (9) screen printing, and (10) laser treating. The fundamental applications of all these techniques is to create hydrophobic barriers onto the sheet of hydrophobic cellulose matrix, which will constitute the micron-sized capillary channels. To prevent leakage of the applied solution and keep it in the fluidic channels, the paper materials are often coated with a polypropylene layer (or other plastics). A variety of hydrophobic materials such as photoresist SU-8, wax, or alkyl ketene dimer (AKD) are widely used to create fluidic channels inside the paper [42]. Depending on the hydrophobic agents employed, the paper pores can be either blocked (after using SU-8) or covered by layer of physical deposition (polystyrene or wax); in some cases the cellulose fibers can be chemically modified (after using AKD).

In general, four detection methods have been reported for the detection of analytes in paper-based microfluidics: (1) colorimetric detection, (2) electrochemical detection, (3) chemiluminescence detection, and (4) electrochemiluminescence detection. Colorimetric detection protocols are typically related to enzymatic or chemical color-change reactions. In most cases, the analysis of results can be visually assessed, which is adequate when a yes/no answer or a semi-quantitative detection method is sufficient for diagnosis [41]. Electrochemical detection has higher sensitivity, enabling the detection and quantification of analytes even in the nano-Molar (nM) range. Most studies to date have been focusing on exploiting the colorimetric detection and the electrochemical detection of biomarkers in paper-based microfluidics. Chemiluminescence and electrochemiluminescence are the most common optical detection methods in microfluidics. They are performed in the dark and therefore exposure to ambient light will yield inaccurate readings. However, they have not been widely used in paper-based microfluidics; only a few studies investigated using these two detection methods for detecting analytes in paper-based microfluidic devices or paper-based microarray plates [43, 44].

Although paper-based microfluidics are predominantly used in recent days, they still have issues. They are:

1. Low efficiency of sample delivery to the sensing surface due to the retention of samples within the paper fluidic channels and sample evaporation during transport. In most of the cases, the volume that reaches the detection zones verses the total volume within the device is usually less than 50 %. This makes the application of paper microfluidics difficult to utilize in clinical diagnosis situations where the sample quantity is tiny or the sample is costly.
2. The micro-channels (barriers) patterned with hydrophobic materials are not hydrophobic, or strong, enough to withstand samples of low surface tension.

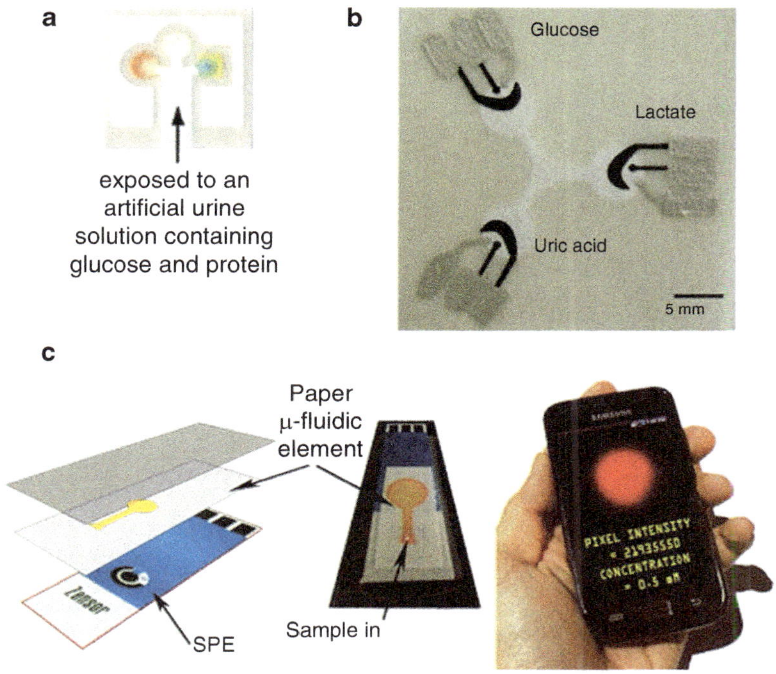

Fig. 6.8 Examples of three detection methods for bioassays on paper-based microfluidic devices. (**a**) Colorimetric detection for simultaneously detecting glucose and protein in an artificial urine sample; (**b**) EC detection on a three-electrode paperfluidic device. (**c**) ECL detection of a sample solution (2-(dibutylamino)-ethanol (DBAE)) [42]. Reproduced with permission from AIP publishing LLC

For example, the hydrophobic channels of a paper-based microfluidic device that is fabricated with wax or AKD will not allow the liquid to leak unless the surface tension of the liquid is lower than a critical value. When the surface tension of a liquid is lower than that value (e.g., biological samples with surfactant), liquid can penetrate not only within the hydrophilic channels but also in the hydrophobic areas of the device, making the device ineffective for proper liquid transport.

3. Another challenge faced by researchers is how difficult it is to employ multiplex analysis using paper microfluidics. Though, a few multiplexed sensing studies have been performed; see Fig. 6.8b for an example. However, as the design and shape becomes more complicated, there is the potential for cross-talk as some of the signal reporters may diffuse to neighboring channels.

A comparison of properties of different type of microfluidic materials is given in Table 6.2.

Table 6.2 Major components of the microfluidic platforms are the materials they are made of and the methods used to control the fluid flow [8]

Property	Si/Glass	Ceramics	Elastomers	Thermoset	Thermoplastics	Hydrogel	Paper
Young's (tensile) modulus (GPa)	130–180/50–90	~120	~0.0005	2.0–2.7	1.4–4.1	Low	0.0003–0.0025
Common technique for microfabrication	Photolithography	Low temperature co-firing technology	Casting	Casting, photo-polymerization	Thermo-molding	Casting, photo-polymerization	Photo-lithography, printing
Smallest channel dimension	<100 nm	~100 μm	<1 μm	<100 nm	~100 nm	~10 μm	~200 μm
Channel profile	Limited 3D	3D	3D	Arbitrary 3D	3D	3D	2D
Multilayer channels	Hard	Easy	Easy	Easy	Easy	Medium	Easy
Thermo-stability	Very high	Very high	Medium	High	Medium to high	Low	Medium
Resistance to oxidizer	Excellent	Excellent	Moderate	Good	Moderate to good	Low	Low
Solvent compatibility	Very high	Very high	Low	High	Medium to high	Low	Medium
Hydrophobicity	Hydrophilic	Hydrophilic	Hydrophobic	Hydrophobic	Hydrophobic	Hydrophilic	Amphiphilic
Surface charge	Very stable	Very stable	Not stable	Stable	Stable	N/A	N/A
Permeability to oxygen	<0.01	<0.01	~500	0.03–1	0.05–5	>1	>1
Optical transparency	No/high	No	High	High	Medium to high	Low to medium	Low

3 The Ideal Microfluidic System for POC Sensor Design

The ideal prototyping material is a tunable material that can provide structural stability and external interface, such as manifold integration. It has to be chemically inert, compatible with chemical and biological samples without absorbing them, and allows stable and patternable surface modifications to be employed to control the wetting and biological functionalization properties of the device. Finally, the material must allow for biocompatible bonding to a wide range of substrates. These properties are concretized in the list below:

1. *Tunable mechanical properties*: The mechanical properties of an ideal prototyping material are somewhat contradictory. On one hand, it must mirror the stiffness of commercial thermoplastics to produce geometrically stable microfluidic devices with robust external chip-to-world interfaces. On the other hand, it must be soft enough to allow for the pneumatically actuated valves and pumps to operate, commonly used in PDMS.
2. *Chemical inertness and low interaction with the sample*: To be able to analyze low concentration samples, the ideal material must not absorb small molecules, such as proteins or DNA from the sample, react with the sample, or leach uncured components that may interact with the sample or the sensor.
3. *Solvent resistance*: Critical to many chemical reactions and separation processes is the use of harsh solvents. The ideal material must not dissolve or swell in the presence of these solvents.
4. *Direct, patternable, and stable surface modifications*: The spatial control of surface properties is instrumental for controlling liquids and immobilizing biological receptors on the chip. The ideal material allows for spatial control of surface modifications, preferably without the need to first activate the polymer surface by plasma or solvents.
5. *Fast, scalable, and utilizing inexpensive materials and processes*: The prototyping method has a fast-curing micro-structuring step and uncomplicated back-end processes, which enables rapid device development. To be useful in academic research, the ideal prototyping method relies on inexpensive materials and does not require access to expensive/technically-complicated facilities. To allow a fast transition to commercial production, the method is possible to scale up to medium or large-scale production.
6. *Three-dimensional microfluidics*: Advanced lab-on-chips must be able to handle multiple liquids, something that often requires 3D microfluidic channels with under- and overpasses. The ideal prototyping method therefore allows for efficient fabrication of multiple vertical interconnections between channel layers.
7. *Biocompatible bonding*: Essential for labs-on-chip is a simple and biocompatible bonding method to surfaces that are functionalized with proteins and DNA. The ideal prototyping method must form a strong bond to a wide number of materials under biocompatible conditions.

4 Conclusion

The first generation of microfluidic POC sensors is already available in the market: blood glucose tests, rapid streptococcal tests, and pregnancy tests, just to name a few. The crucial aspects for commercializing microfluidic devices are powerful microscale technologies and low-cost materials for portable analyses. Although, the development of these technologies is currently the focus of many research groups, the selection of material and expensive manufacturing process bottleneck limiting the transfer of technologies from research to commercial products. Silicon/glass and PDMS are the commonly used materials in research laboratories while plastics and paper are more promising for commercializing microfluidic devices. Each material has its pros and cons (Table 6.2). Silicon, glass, and Teflon represent the most inert materials to chemicals and solvents; PDMS is easy to prototype, making it easy to fabricate complicated fluid circuits with various integrated components; normal thermoplastics are excellent for commercial mass production of standard microfluidic devices; hydrogels are more suitable for biological applications; and paper is highly promising for commercial disposable bioassays. Finally, all the materials can be modified or combined to fabricate more powerful devices for specific aims.

References

1. NIH Fact Sheets—Point-of-care diagnostic testing (n.d.) https://report.nih.gov/nihfactsheets/ViewFactSheet.aspx?csid=112. Accessed 22 Mar 2016
2. Point-of-Care Diagnostic Market Worth $27.5 Billion by 2018 (n.d.) http://www.prnewswire.com/news-releases/point-of-care-diagnostic-market-worth-275-billion-by-2018-274885521.html. Accessed 22 Mar 2016
3. Vashist SK, Luppa PB, Yeo LY, Ozcan A, Luong JHT (2015) Emerging technologies for next-generation point-of-care testing. Trends Biotechnol 33(11):692–705. doi:10.1016/j.tibtech.2015.09.001
4. Kaushik A, Vasudev A, Arya SK, Pasha SK, Bhansali S (2014) Recent advances in cortisol sensing technologies for point-of-care application. Biosens Bioelectron 53:499–512. doi:10.1016/j.bios.2013.09.060
5. Gervais L, de Rooij N, Delamarche E (2011) Microfluidic chips for point-of-care immunodiagnostics. Adv Mater 23:H151–H176. doi:10.1002/adma.201100464
6. Whitesides GM (2006) The origins and the future of microfluidics. Nature 442:368–373. doi:10.1038/nature05058
7. Fong Lei K (2014) Microfluidics in detection science. Royal Society of Chemistry, Cambridge. doi:10.1039/9781849737609
8. Ren K, Zhou J, Wu H (2013) Materials for microfluidic chip fabrication. Acc Chem Res 46:2396–2406. doi:10.1021/ar300314s
9. Iliescu C, Taylor H, Avram M, Miao J, Franssila S (2012) A practical guide for the fabrication of microfluidic devices using glass and silicon. Biomicrofluidics 6:16505–1650516. doi:10.1063/1.3689939

10. Harrison DJ, Fluri K, Seiler K, Fan Z, Effenhauser CS, Manz A (1993) Micromachining a miniaturized capillary electrophoresis-based chemical analysis system on a chip. Science 261:895–897. doi:10.1126/science.261.5123.895
11. Manz A, Harrison DJ, Verpoorte EMJ, Fettinger JC, Paulus A, Lüdi H et al (1992) Planar chips technology for miniaturization and integration of separation techniques into monitoring systems. J Chromatogr A 593:253–258. doi:10.1016/0021-9673(92)80293-4
12. Tjerkstra RW, De Boer M, Berenschot E, Gardeniers JGE, van den Berg A, Elwenspoek M (1997) Etching technology for microchannels. In: Proceedings IEEE The Tenth Annual International Workshop on Micro Electro Mechanical Systems. An Investigation of Micro Structures, Sensors, Actuators, Machines and Robots. pp 147–152. doi:10.1109/MEMSYS.1997.581790
13. Wu Z, Chen H, Liu X, Zhang Y, Li D, Huang H (2009) Protein adsorption on poly(N-vinylpyrrolidone)-modified silicon surfaces prepared by surface-initiated atom transfer radical polymerization. Langmuir 25:2900–2906. doi:10.1021/la8037523
14. Nge PN, Rogers CI, Woolley AT (2013) Advances in microfluidic materials, functions, integration, and applications. Chem Rev 113:2550–2583. doi:10.1021/cr300337x
15. Anderson RR, Hu W, Noh JW, Dahlquist WC, Ness SJ, Gustafson TM et al (2011) Transient deflection response in microcantilever array integrated with polydimethylsiloxane (PDMS) microfluidics. Lab Chip 11:2088–2096. doi:10.1039/c1lc20025a
16. Washburn AL, Gunn LC, Bailey RC (2009) Label-free quantitation of a cancer biomarker in complex media using silicon photonic microring resonators. Anal Chem 81:9499–9506. doi:10.1021/ac902006p
17. Harz S, Schimmelpfennig M, Tse Sum Bui B, Marchyk N, Haupt K, Feller K-H (2011) Fluorescence optical spectrally resolved sensor based on molecularly imprinted polymers and microfluidics. Eng Life Sci 11:559–565. doi:10.1002/elsc.201000222
18. Grover WH, Ivester RHC, Jensen EC, Mathies RA (2006) Development and multiplexed control of latching pneumatic valves using microfluidic logical structures. Lab Chip 6:623–631. doi:10.1039/b518362f
19. Ibanez-Garcia N, Mercader MB, Mendes da Rocha Z, Seabra CA, Góngora-Rubio MR, Chamarro JA (2006) Continuous flow analytical microsystems based on low-temperature co-fired ceramic technology. Integrated potentiometric detection based on solvent polymeric ion-selective electrodes. Anal Chem 78:2985–2992. doi:10.1021/ac051994k
20. Zhang W, Eitel RE (2012) Biostability of low-temperature co-fired ceramic materials for microfluidic and biomedical devices. Int J Appl Ceram Technol 9:60–66. doi:10.1111/j.1744-7402.2010.02581.x
21. Fakunle ES, Fritsch I (2010) Low-temperature co-fired ceramic microchannels with individually addressable screen-printed gold electrodes on four walls for self-contained electrochemical immunoassays. Anal Bioanal Chem 398:2605–2615. doi:10.1007/s00216-010-4098-5
22. Wolfe DB, Qin D, Whitesides GM (2010) Rapid prototyping of microstructures by soft lithography for biotechnology. Methods Mol Biol 583:81–107. doi:10.1007/978-1-60327-106-6_3
23. Qin D, Xia Y, Whitesides GM (2010) Soft lithography for micro- and nanoscale patterning. Nat Protoc 5:491–502. doi:10.1038/nprot.2009.234
24. Zhu Q, Trau D (2012) Multiplex detection platform for tumor markers and glucose in serum based on a microfluidic microparticle array. Anal Chim Acta 751:146–154. doi:10.1016/j.aca.2012.09.007
25. Ziółkowska K, Stelmachowska A, Kwapiszewski R, Chudy M, Dybko A, Brzózka Z (2013) Long-term three-dimensional cell culture and anticancer drug activity evaluation in a microfluidic chip. Biosens Bioelectron 40:68–74. doi:10.1016/j.bios.2012.06.017
26. Wang C-H, Lien K-Y, Hung L-Y, Lei H-Y, Lee G-B (2012) Integrated microfluidic system for the identification and multiple subtyping of influenza viruses by using a molecular diagnostic approach. Microfluid Nanofluid 13:113–123. doi:10.1007/s10404-012-0947-1

27. Wang J-H, Cheng L, Wang C-H, Ling W-S, Wang S-W, Lee G-B (2013) An integrated chip capable of performing sample pretreatment and nucleic acid amplification for HIV-1 detection. Biosens Bioelectron 41:484–491. doi:10.1016/j.bios.2012.09.011
28. Jung H-C, Moon J-H, Baek D-H, Lee J-H, Choi Y-Y, Hong J-S et al (2012) CNT/PDMS composite flexible dry electrodes for long-term ECG monitoring. IEEE Trans Biomed Eng 59:1472–1479. doi:10.1109/TBME.2012.2190288
29. Tsao C-W, DeVoe DL (2008) Bonding of thermoplastic polymer microfluidics. Microfluid Nanofluidics 6:1–16. doi:10.1007/s10404-008-0361-x
30. Tsao CW, Hromada L, Liu J, Kumar P, DeVoe DL (2007) Low temperature bonding of PMMA and COC microfluidic substrates using UV/ozone surface treatment. Lab Chip 7:499–505. doi:10.1039/b618901f
31. Yang W, Sun X, Wang H-Y, Woolley AT (2009) Integrated microfluidic device for serum biomarker quantitation using either standard addition or a calibration curve. Anal Chem 81:8230–8235. doi:10.1021/ac901566s
32. Chen D, Mauk M, Qiu X, Liu C, Kim J, Ramprasad S et al (2010) An integrated, self-contained microfluidic cassette for isolation, amplification, and detection of nucleic acids. Biomed Microdevices 12:705–719. doi:10.1007/s10544-010-9423-4
33. Yu H, Chong ZZ, Tor SB, Liu E, Loh NH (2015) Low temperature and deformation-free bonding of PMMA microfluidic devices with stable hydrophilicity via oxygen plasma treatment and PVA coating. RSC Adv 5:8377–8388. doi:10.1039/C4RA12771D
34. Bartolo D, Degré G, Nghe P, Studer V (2008) Microfluidic stickers. Lab Chip 8:274–279. doi:10.1039/b712368j
35. Peppas NA, Khare AR (1993) Preparation, structure and diffusional behavior of hydrogels in controlled release. Adv Drug Deliv Rev 11:1–35. doi:10.1016/0169-409X(93)90025-Y
36. Fotin AV, Drobyshev AL, Proudnikov DY, Perov AN, Mirzabekov AD (1998) Parallel thermodynamic analysis of duplexes on oligodeoxyribonucleotide microchips. Nucleic Acids Res 26:1515–1521. http://www.pubmedcentral.nih.gov/articlerender.fcgi?artid=147416&tool=pmcentrez&rendertype=abstract. Accessed 23 Mar 2016
37. Pevzner PA, Lysov YP, Khrapko KR, Belyavsky AV, Florentiev VL, Mirzabekov AD (1991) Improved chips for sequencing by hybridization. J Biomol Struct Dyn 9:399–410. doi:10.1080/07391102.1991.10507920
38. Heo J, Crooks RM (2005) Microfluidic biosensor based on an array of hydrogel-entrapped enzymes. Anal Chem 77:6843–6851. doi:10.1021/ac0507993
39. Cheng C-J, Chu L-Y, Zhang J, Wang H-D, Wei G (2008) Effect of freeze-drying and rehydrating treatment on the thermo-responsive characteristics of poly (N-isopropylacrylamide) microspheres. Colloid Polym Sci 286:571–577. doi:10.1007/s00396-007-1817-3
40. Beebe D, Moore J, Bauer J, Yu Q, Liu R, Devadoss C et al (2000) Functional hydrogel structures for autonomous flow control inside microfluidic channels. Nature 404:588–590. doi:10.1038/35007047
41. Martinez AW, Phillips ST, Butte MJ, Whitesides GM (2007) Patterned paper as a platform for inexpensive, low-volume, portable bioassays. Angew Chem Int Ed Engl 46:1318–1320. doi:10.1002/anie.200603817
42. Li X, Ballerini DR, Shen W (2012) A perspective on paper-based microfluidics: current status and future trends. Biomicrofluidics 6:11301–1130113. doi:10.1063/1.3687398
43. Delaney JL, Hogan CF, Tian J, Shen W (2011) Electrogenerated chemiluminescence detection in paper-based microfluidic sensors. Anal Chem 83:1300–1306. doi:10.1021/ac102392t
44. Chitnis G, Ding Z, Chang C-L, Savran CA, Ziaie B (2011) Laser-treated hydrophobic paper: an inexpensive microfluidic platform. Lab Chip 11:1161–1165. doi:10.1039/c0lc00512f

Chapter 7
Paper Microfluidics

Elizaveta Vereshchagina

1 Introduction

Low cost, paper-based point-of-care diagnostics, often also referred as, paper microfluidics and paper lab-on-a-chip is a new, emerging class of microfluidics devices. This chapter summarizes recent advances in the field of paper microfluidics, describes the underlying physical concepts, design and fabrication guidelines, and presents various examples of application of this technology. In this chapter, we mainly focused on the works published in the period 2010–2016. We apologize in advance if any original works and reviews were not mentioned or overlooked due to the large amount of material that has been published in this topic.

Multiple reviews have been published: reviews giving overview over all aspects of paper devices [1–5], reviews which are mainly focused on paper-based bio-sensors with EC detection [6], functionality of paper devices [7], detection [7, 8], chemical measurements using microfluidic paper-based analytical device (μPAD) [9], bio-sensing techniques and, specifically, use of nanoparticles [10], working principles and reaction mechanisms [11], history of development [12], thorough review of properties of paper material [13], paper devices for biomarkers and bacterial detection [14], infectious disease [15], paper devices in resource-limited settings for diagnostics and education [16].

Traditionally paper microfluidics has been seen as a low cost diagnostic tool for the developing world. There are still big hopes associated with this technology, mainly that it can help people in low and middle-income countries in areas such as public health, environmental monitoring, agriculture, water and food safety and many others [17]. At the moment, according to the World Bank data, more than 5 billion people are living in the areas that can be characterized as low and middle-

E. Vereshchagina (✉)
Department of Microsystems and Nanotechnology, SINTEF ICT, Oslo, Norway
e-mail: Elizaveta.Vereshchagina@sintef.no

C.K. Dixit, A. Kaushik (eds.), *Microfluidics for Biologists*,
DOI 10.1007/978-3-319-40036-5_7

income. These people might live in low-resource, hard-to-reach areas with a very limited access to primary healthcare and nearby medical institutions. Access to affordable diagnostic tools can be a very positive change in these communities. Another aspect of the same problem: people might lack basic medical education in these areas. If diagnostic tools are easy to use and interpret, many people would be able to apply them to real life problems it e.g. a community health officer at a remote rural village with limited medical education and not only highly educated practitioners could do the initial diagnostic screenings. With time, when reaching technological maturity, this type of technology could be very useful for rapid diagnostics of e.g. non-communicable diseases (e.g. sickle cell, or daily monitoring of blood glucose levels in diabetic patients), but also for communicable (infectious) disease diagnosis (e.g. malaria or ebola). Monitoring quality of drinking water is another application area that still needs attention worldwide, as some regional discrepancies as well as differences between rural and urban regions persist. According to the World Health Organization, 8 out of 10 people living in rural areas do not have access to high quality, clean drinking water sources. Integration of low cost solutions for water quality monitoring would have positive economic effect and improve well-being of people in those regions. Managing the above-mentioned challenges will be more important over the years as the world population will continue to grow. These ideas are few of the core motivating factors behind the development of paper-based microfluidic diagnostic devices.

As the research community on this technology rapidly developed, the field also started receiving a commercial attention as well as interest from non-profit organizations. This is seen following the numerous patents filed in this area in recent times with a few examples referred to here [18–20]. DFA (Diagnostics for All), for example, is a non-profit organization aiming to deliver low-cost, easy-to-use, point-of-care diagnostic devices designed specifically for the developing world [21], and is one of the successful examples of bringing this technology into use. If paper-based devices will be able to overcome such shortcomings of the lateral flow based devices as lack of quantification, specificity and sensitivity, they may potentially revolutionize the field of low cost diagnostics.

The paper microfluidics device works by wicking liquid sample between separate compartments containing assay reagents. The device may consist solely of patterned chromatography paper or may combine several other materials (e.g. polymers, conductive materials, functional nanoparticles, etc.) used in point-of-care. Typically, a paper device has hydrophobic patterns, and hydrophilic areas which are used as chambers and channels, and performing various kinds of fluidic operations. If, for example, a colorimetric assay has been adapted, the result can be read by a phone, or photographed and sent for further analysis to a centralized lab, where a specialized medical practitioner will be making an informed judgement. Time duration of assays on paper is mainly limited by the time needed to wick paper channels of given dimensions and design and by the inherit properties of the assay itself, and typically is within 30 min, which is well sufficient for many analytical applications. Some applications are particularly suitable for paper diagnostics. These are situations where e.g. qualitative results are of primary interests, the analytes are fairly easy to obtain, some device-to-device variations due to fabrication processes are acceptable and assays can be easily transferred for use on paper substrates [22].

Paper-based devices are so attractive due to these factors:

- Low cost,
- Operating on low volumes,
- Abundant supply,
- Easy to construct devices,
- Suitable for multiplexing, thus multiple samples can be analyzed simultaneously,
- Reagent storage in dry form,
- Facile interpretation of the results (produced signal can be read by eye),
- Do not require additional instrumentation for liquid transport (without pump),
- Can be designed not to require additional instrumentation to read test results [23],
- Possible to combine with portable readout systems [24, 25], smart phones [26–28]
- Highly-skilled staff is not required,
- Light-weight material,
- Does not produce bio-hazardous waste,
- Devices can be mass-produced,
- Portable format of devices, and many others.

Factors that are often mentioned in a positive context can be, however, seen as drawbacks when applied in different settings. In fact, this depends on the specifications to device and the application area.

The following drawbacks are often mentioned:

- Limited sensitivity of many assays,
- Lack of quantitation in existing assay formats,
- Limited number of detection methods can be applied, i.e. colorimetric, electrochemical,
- Often not self-sustained, i.e. to be able to obtain quantitative results paper-based devices need to be integrated with external read-out system, which increases the costs and complexity [22].

Significant progress has been made in increasing functionality of paper-based fluidic devices. At the moment, many fluidic operations are possible including fluidic timers, valves, sophisticated detection units and even power sources (batteries)—therefore, potentially it is possible to design complex assays in single paper-based device. If, some of the promising design concepts could be mass-produced in paper at a price comparable to commercially available rapid tests, they may be commercialized in the near future.

2 Application Areas

The idea of low-cost microfluidics responding to global public health needs has been around before paper microfluidics started developing as an individual field, and, undoubtedly, there was and still is an actual need for monitoring devices used in resource-poor settings [29]. The world Health Organization (WHO) provides the guidelines ("ASSURED" concept) to how devices should be better suitable for being used in developing countries. Paper devices meet such requirements as being affordable, sensitive (sufficient sensitivity for some applications), specific, user-friendly, rapid and robust, equipment-free, portable and target-delivered (to those who in need of the technology) [1]. Essentially, paper devices can be applied to analysis of various bodily fluids and their synthetic substitutes (whole blood [30], human serum [31, 32], artificial serum, synthetic urine, urine, saliva). Detection of a variety of biomarkers has been demonstrated i.e. uric acid and glucose [24, 31], cholesterol [33], simultaneous detection of glucose, lactate and uric acid in urine, ketones [34], salivary nitrite [34], proteins [34], lactate [35], triglyceride [36], nitrates [34].

An overview of the history of the development of this field can be found elsewhere [1]. An important step in growing awareness of this technology was introduction of paper-based ELISA and the first demonstration of the colorimetric glucose assay on paper, which is still used by many researchers as a model application system [27, 37].

While a killer-application, that is only possible on paper, is still missing, a wide range of interesting applications has been reported:

- Medical diagnostics

  Paper fluidics can be successfully implemented in medical diagnostics. Especially it is useful for systematic, routine diagnosis, analysis of patients' samples in places distant from hospitals, analysis of asymptomatic diseases, and evaluation of disease progression [1]. Examples include early cancer detection [38] using multiple biomarkers, among those are r-fetoprotein (AFP), carcinoma antigen 125 (CA125), carcinoma antigen 199 (CA199) and carcinoembryonic antigen (CEA) [32], isolation of extracellular vesicles [39], blood typing [30, 40], drug monitoring (example of induced liver failure [41]), diagnostics of non-communicable diseases (including cardiovascular disease and cancer) [42].
- Veterinary diagnostics

  This application area mainly addresses infectious and viral diseases passed from animals to humans. A wide array of communicable diseases can be transmitted from livestock or wildlife to humans in various ways. Some of these disease can potentially become pandemics [1].
- Food safety and control, agricultural field

  Determination of toxic agents with main examples including salmonellosis and campylobacteriosis infections (via eggs, poultry, and unpasteurised milk), enterohaemorrhagic Escherichia coli (O157:H7 serotype) and cholera (via water, rice, vegetables, seafood) [1], water analysis by electronic tongue device [43], detection of pesticides [44, 45], foods analysis e.g. wine and beer [46].

- Environmental monitoring
  Today, very few developing countries have adapted testing routines for water supplies in rural regions. Contamination due to industrial and agricultural activities, as well as environmental pollution monitoring is still challenging and needs to be tackled. Examples of monitoring of these factors include detection of heavy metals [47], low cost monitoring of environmental pollutants in air [48], measuring the metals (Fe, Cu, Ni, and Cr) content in welding fumes [49, 50], biosensor for organic pollutants in water (e.g. L-DOPA and catechol [51]), bacterial detection (Salmonella and Escherichia Coli) [28].
- Energy storage and generation
  In fact, paper as a substrate has entered other fields, besides traditional point-of-care diagnostics. Batteries and other types of energy storage devices can be constructed based on paper [52], fuel cell harvesting electricity from bacterial metabolism [53], devices generating power when liquid sample is applied [54].
- Pharmaceuticals
  Detection of pharmaceuticals has been demonstrated on paper [55].
- Forensics
  Detection of explosives is a new area of application [56].

3 Physical Principles

The ability to control the flow rates of liquids in paper channels is necessary for successful operation of the device. This section gives an overview of basic physical principles behind flow in paper. For more details, one can refer to the informative articles on this topic [57, 58].

3.1 Flow Through Paper

Darcy's law describing flow through porous media is the basis for estimation of liquid flow through a paper channel. For the constant width of channels fluid flow can be expressed as

$$Q = -\frac{kWH}{\eta L}\Delta P$$

Where Q is the volumetric flow rate, k—permeability of the paper, η—viscosity, WH—area of the channel perpendicular to flow, ΔP—pressure difference over the length of channel L.

When the width of channel is varied:

$$Q = -\frac{\Delta P}{\left(\frac{\forall}{k}\sum_{i=1}^{N}\frac{L_i}{W_i H_i}\right)}$$

$W_i H_i$ is the area perpendicular to flow, L_i is the length in the direction of flow, and ΔP is the pressure difference across the length of the channel.

Using parallels between fluidic and electrical resistance, i.e. liquid flow Qi is an equivalent for current, ΔPi pressure drop along the channel is equivalent to voltage drop, and μLi/(kWiHi) is an equivalent to fluidic resistance of each individual channels, one can estimate total flux. If several fluidic elements are connected in series or in parallel, the total flux through this network will follow analogy of Ohm's law, i.e. sum of individual fluxes when connected in series, and reciprocals when connected in parallel.

3.2 *Spreading of Wax and Width of Patterned Channel*

Spreading of the molten wax and width of final channels can be predicted using the Washburn equation:

$$L = \sqrt{\frac{\gamma D t}{4\eta}}$$

Where L—distance covered by the wax front, η—viscosity (function of time and temperature to which device was exposed during bake), γ—effective surface tension, D—average pore diameter, t—time. The same equation can be used to predict transport of the fluid front in the channel.

The final inner width of the channel formed by wax can be defined as

$W_C = W_P - 2L$, where

W_C—inner width of the hydrophobic channel, W_P—inner width of the printed channel, L—the additional distance that the wax spreads perpendicular to the length of the channel. L is a function of time, heat and the structural properties of paper.

3.3 *Transport Time*

Transport time through a multi-segment geometry can be calculated using the modified Darcy's law equation:

$$t = \frac{V R_{eq}}{\Delta P}$$

Where V is the volume of the geometry, ΔP is pressure difference and R_{eq} expressed as

$$R_{eq} = \frac{\mu}{k} \sum_{i=1}^{N} \frac{L_i}{W_i H_i}$$

If we assume that permeability, viscosity, and pressure difference are constant, differences in the transport times in two channels would be only due to geometrical differences

$$t = \frac{\mu}{k\Delta P} L^2$$

This means that all channels of the same length would have the same transport time regardless to other geometrical dimensions.

Of course, in reality, channels often have several sections characterized by different length and width. For a channel that has two sections, both characterized by constant length and width (and same heights), the equation for transport can be modified to

$$t = \frac{\mu}{k\Delta P} (L_1 W_1 + L_2 W_2) \left(\frac{L_1}{W_1} + \frac{L_2}{W_2} \right)$$

3.4 Signal Visibility

The visibility of the signal is a function of the thickness and the opacity of paper:

$$\%\ Visible\ signal = \frac{Visible\ depth}{Thickness}$$

4 Main Formats of Paper Devices

Many design concepts in paper microfluidics are inspired by bio-sensing principles and structure of the lateral rapid tests and dipsticks [59]. Examples of the commercial point-of-care tests are shown in Fig. 7.1. If focus in lateral flow assays and dipsticks was their robustness and ease of interpretation (often yes/no answer), paper microfluidics explored various ways to enhance the functionality, increase design complexity while still trying to keep the costs down. Current status on fluidic operations in paper can be found elsewhere [60]. Design concepts of paper fluidic devices are constantly under development. Currently several groups succeeded in integration of sample preparation with simultaneous detection of a biomarker of interest using the same paper device [61].

Transport of liquid through paper is equivalent to transport of liquid through the porous media. It occurs as soon as paper is brought in contact with liquid reagents and therefore has to be confined and directed. To gain some understanding on

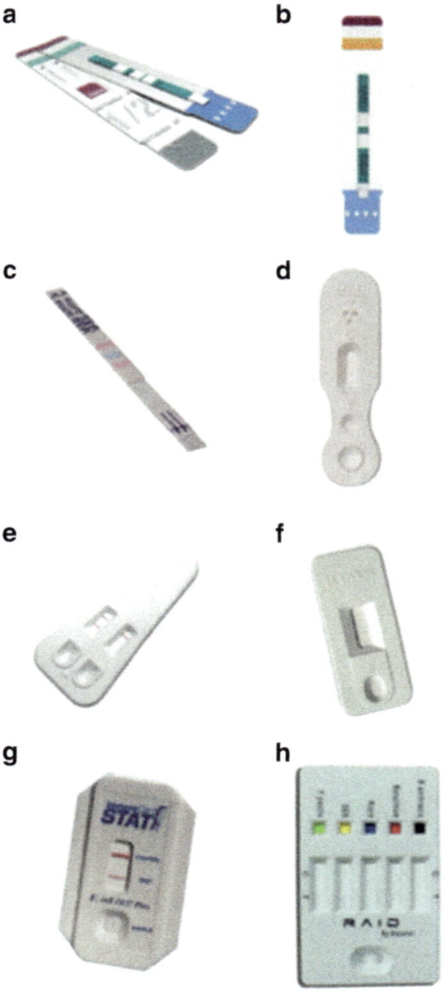

Fig. 7.1 Commercial point-of-care tests. (**a**) DetermineTM HIV 1/2 Ag/Ab Combo, (**b**) DetermineTM TB LAM Ag test, (**c**) QuickVue Influenza A + B test, (**d**) Clearview1 Malaria P.f. Test, (**e**) DirectigenTM EZ Flu A + B, (**f**) ICON HP (**g**) ImmunoCard STAT!1 E. coli O157 Plus, (**h**) A multiplex lateral-flow assay. RAIDTM 5 for the determination of biological threat agents. Adapted from [1] with permission of Royal Society of Chemistry

correlation between the shape of channels and liquid transport, one can also refer to [58]. Most of the channels have rectangular shape, but there is also experimental work with channels with varied width [58]. Typical width of the hydrophilic channel is 0.5–4 mm, hydrophobic barrier can be about 1 mm.

All paper devices can be classified by these three main formats:

- 2D format for paper networks

  Two-dimensional paper networks were the first microfluidic designs. A few typical examples are shown in Fig. 7.2. These are mostly channels fully made out of paper, which can either be cut or patterned with a liquid-repelling material to a required shape. The channels may vary in width, length, may contain several

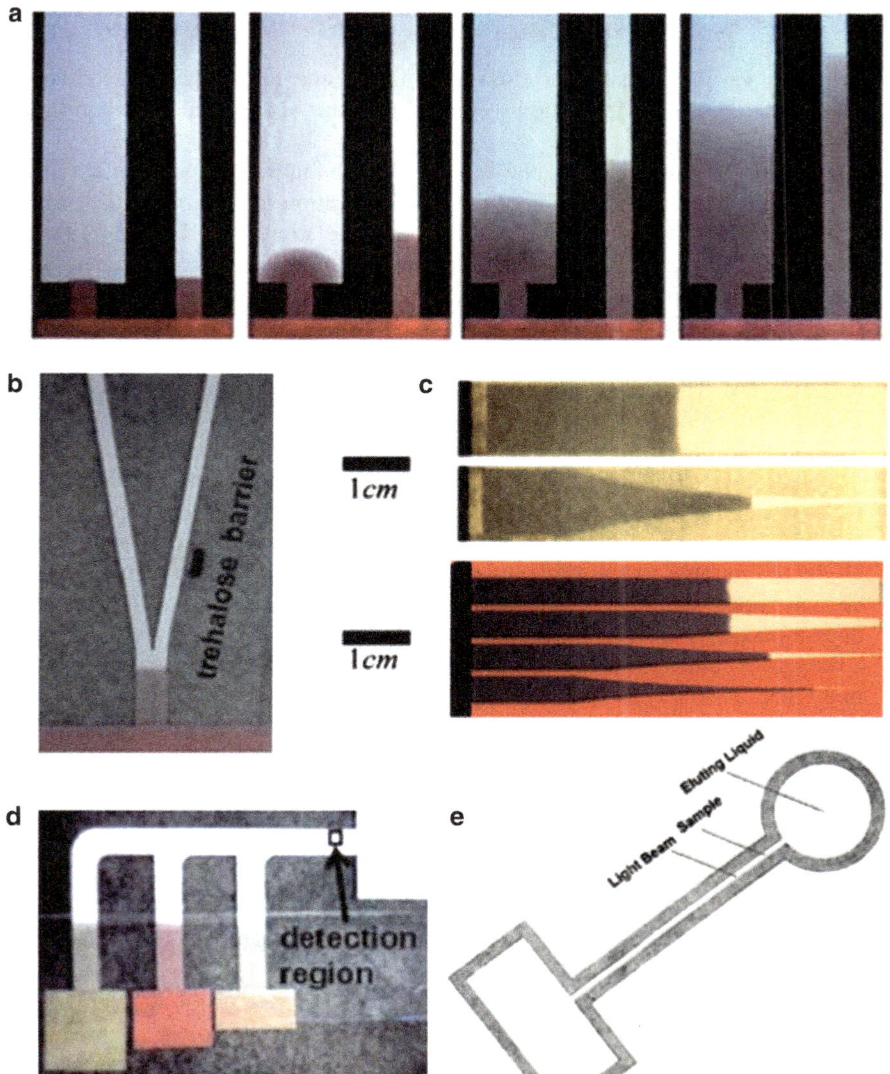

Fig. 7.2 2D paper channels of various designs: (**a**) channels of different width (liquid introduced via the same inlet of the same width) shows faster transport in the strip of smaller width. Images were taken at 2, 10, 50, and 210 s after introduction of fluid to the inlet (Adapted with permission [65] Royal Society of Chemistry); (**b**) Y-shaped channel with dissolvable barrier can be used to create a delay in the transport time of a fluid along the channel (Adapted with permission [65] Royal Society of Chemistry), (**c**) Slit-shaped channels and correponding differences in liquid transport (Adapted with permission [58] Royal Society of Chemistry); (**d**) A simple 2D paper network. The arrival time of multiple reagents at the "detection region" of the paper strip is staggered by placing 3 inlets along the common segment of the device. The geometry of the inlets results in the sequential arrival of fluid from each of the three inlets. (Adapted with permission [65] Royal Society of Chemistry); (**e**) One of the first paper microfluidic designs with inlet, transport channel and reaction zone. (Adapted from [1] with permission of Royal Society of Chemistry)

segments varying in size for storage of liquid and dry reagents. Modifying channel length and width allows controlling time and spatial distribution of reagents and samples.

To prevent contamination between individual fluidic compartments hydrophobic barriers, absorbing pads or physical separation of reacting zones can be applied. Flow in these channels follows the relationships discussed in the previous sections. Some of the early publications discuss various scenarios of liquid transport in 2D networks such as: Y-shaped and T-shaped devices, structures for hydrodynamic focusing, size-based separation, mixing and dilution [57, 62–65], hydrophobic barriers for time-controlled transport of liquid reagents [66]. The majority of the newly published works are using this design principle probably due to its simplicity.

- 3D format for paper networks,

  Three-dimensionality was an important step in developing device complexity. In these devices, liquid transport occurs in both vertical and lateral directions [32, 67–71]. 3D devices, shown in Figs. 7.3 and 7.4, have two main advantages (1) better suitable for multiplexing, a higher number of tests can be simultaneously integrated; (2) more complex assays can be integrated as three-dimensionality (liquid flow can now be controlled in both lateral x-y and vertical z directions) allows for more complex fluidic operations and more suitable for multi-step assays[72, 73]. Since liquid flow can be transported vertically, the distribution times between various reaction zones can be significantly reduced, so the required sample volume.

- Centrifugal, i.e. paper-disc format,

  Flow in 2D and 3D devices are governed by capillary force. Centrifugal paper-based systems are operating on principles of interplay of centrifugal and capillary forces [36, 74, 75]. This adds more possibilities to time and spatial time control as reagents can be recirculated within the same channel multiple times and the flow rate through the paper can be well controlled. Compared to this, in 2D and 3D networks, the flow through channel is typically constant and reverse flow is not possible. Two examples of centrifugal systems are shown in Fig. 7.5.

- Various hybrid formats, that cannot be strictly assigned to either of the above mentioned categories.

  Combination of paper devices with other materials and the advances in other areas of microfluidic research gave rise to some interesting hybrid device concepts. Various efforts for hybrid integration of electrodes into paper has been demonstrated: electrical circuit used for electrical readout can be attached externally [53], functionalized paper can be placed on top of screen-printed electrode [76], or electrodes can be incorporated as a part of the fluidic network directly in paper [43, 77]. Since patterning of electrodes on paper or hybrid integration with external devices are relatively established process, paper microfluidics can be combined with other areas such as e.g. digital microfluidics, which allows development of some complex assays [78]. These areas can potentially profit strongly from each other as low power and flexibility of operational control of digital microfluidics can be well combined with suitability

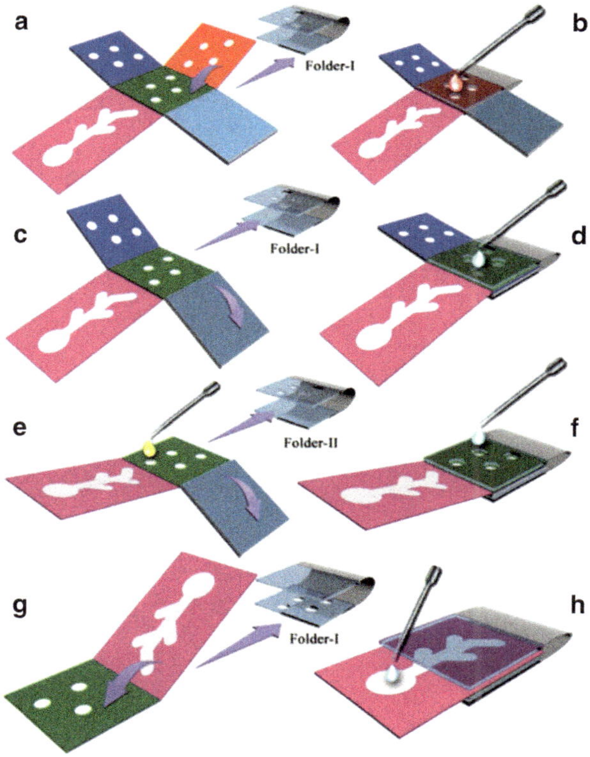

Fig. 7.3 Example of an assay based on 3D origami immunodevice. (**a**) The filter tab was folded above the test pad; (**b**) the blood sample was added into each paper microzone on the folded filter tab with the aid of Folder-I; (**c**) the filter tab was removed and one waste tab was folded below the test pad; (**d**) the washing buffer was added into each immunozone on the test pad to wash the immunozones with the aid of the used Folder-I; (**e**) the used waste tab was removed and a solution of AgNPs-luminol/Ab2 was added into each immunozone on the test pad; (**f**) the remaining waste tab was folded below the reversed test pad and the device was washed again with the aid of Folder-II; (**g**) the used waste tab was removed and the reagent tab was folded above the test pad; (**h**) the hydrogen peroxide solution was added into the reagent cell to trigger the CL reactions with the aid of a new Folder-I. Reproduced from [72] with permission of from Royal Society of Chemistry

for rapid prototyping of paper fluidics. Fabrication of electrodes for digital microfluidic devices in a classic clean room would be significantly more expensive. Another example is a combination of paper with hydrogels, where hydrogels are used for liquid storage and upon application of external stimulus release liquid into paper [79]. Sugars were applied to paper for programmable time-delays, that can be used for sequential reagent delivery [80].

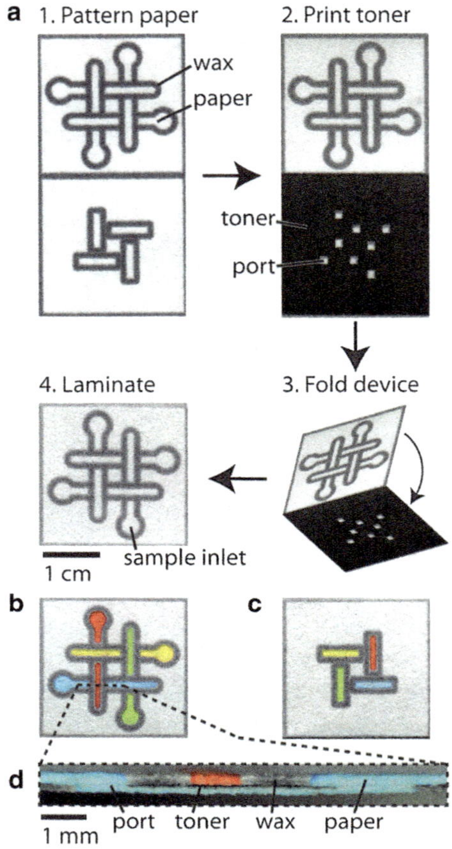

Fig. 7.4 Example of a process flow for fabrication of 3D microPADs using a toner. (**a**) Schematic of the four steps of device fabrication. (**b**) The *top* of the device after addition of aqueous dyes to the sample inlets and wicking across the channels for 2 min. (**c**) The *bottom* of the device after addition of aqueous dyes. (**d**) Cross-section of the device showing the layers of paper and toner. The toner is used for bonding of the layers of paper together, it prevents fluids in channels from mixing. Ports in the toner layer allow wicking between layers of paper. Reproduced from [73] with permission of Royal Society of Chemistry

5 Types of Paper and Its Functionalization

Traditionally, rapid point-of-care tests use nitrocellulose membranes (e.g. membranes available from Millipore). The term "membrane" is exclusively used to describe nitrocellulose in the lateral flow format [1]. Nitrocellulose-based technologies are in close relationship to paper microfluidic devices. Attempts to replace nitrocellulose in the diagnostics industry with other materials (nylon, polyvinylidene fluoride and Fusion 5 from Whatman) were not successful due to high-costs, manufacturing challenges and need for additional optimization of chemistry. In order to replace current manufacturing practices, to which industries are typically reluctant, paper-based devices have to compete in manufacturing price, signal-to-noise ratio, robustness and functionality. The term "paper" refers to cellulosic materials (i.e. filter paper and chromatography paper used in microfluidic devices). 90 % of paper is produced from wood stock, however for diagnostic purposes a paper from cotton is desired (to eliminate interferences from

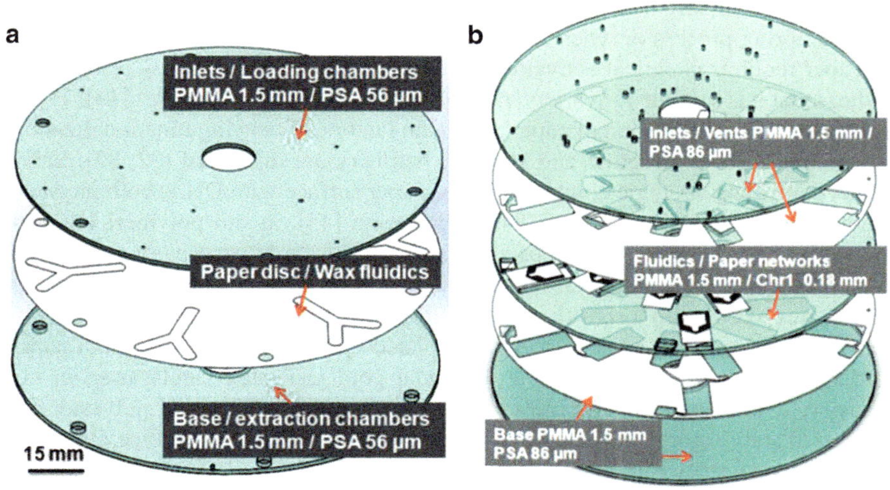

Fig. 7.5 Centrifugal paper-disc devices: (**a**) whole paper disc is used in this device assembly and fluidic structures are patterned in paper by wax, (**b**) paper inserts with wax patterns are cut and integrated into polymeric parts. Adapted from [75] with permission of IEEE

lignin coming from wood). Unlike nitrocellulose membranes exhibiting hydrophobic properties due to its cellulose acetate blends, filter and chromatographic papers are hydrophilic and do not require the deposition of surfactants to improve wetting properties. Additionally, there are several other requirements to paper material: suitability for processing of biological samples in small volumes and within short times (e.g. wicking time between reaction zones), specificity and sensitivity comparable to commercial rapid tests [1].

Another crucial factor is sufficient protein binding to allow for the formation of sharp and intense capture zones while keeping the signal from nonspecific background low.

There is no universal type of paper that will suit all applications. Choice of design is highly dependent on the type and structure of paper. Structural properties, insights into physics and chemistry of paper [81] can be found in dedicated reviews. Utilization of different papers, and different channel width were explored by some researchers earlier [36, 82]. Some experimental work on papers of different grades can be found in [75]. Commercial paper grades differ in flow rate, pore size and porosity, thickness, color, particle retention and other properties. Ideally, in paper microfluidic devices there should be high consistency (within single device as well as batch-to-batch) in surface pore size, thickness, protein binding capacity (irreversible capture of reagents at the detection zones), flow characteristics, magnitude of obtained signal during detection and stability during storage. Binding characteristics of selected paper should be tested during the development stage of device. Protein binding capability will dependent on the paper surface area available for capture. It should also be verified that binding capabilities are not altered during the manufacturing process or when assay is run [13].

Paper requires chemical activation to immobilize antibodies. Many researchers address that it is suitable for variety of (bio)functionalization procedures [13]. Techniques for functionalization of paper, and main factors influencing functionalization procedures such as structure and surface chemistry are discussed [82, 83]. Early works describing strategies for treatment of paper surface with DNA, both physical adsorption and covalent binding [84], salinization [45], in situ polymerization of molecular imprinted polymer on paper [45], modification of paper with poly (vinyl pyrrolidone) and polyaniline [33]. Functionalization with polymers gives numerous active sites to build up a sensitive detection method [85, 86].

There are, however, some concerns associated with possible integration of paper in diagnostic devices. For example, colloidal gold and latex labels, used in the industry, require more open pore materials such as glass fibre and polyester for optimized stabilization and release from it during the assay run. Surface quality is another key parameter in optimizing the performance of paper-based microfluidics. Paper has relatively rough surface characteristics, which might cause various challenges in reproducibility of quantitative measurements. Recently, a novel class of materials, so-called synthetic paper, with controlled porosity characteristics, has been introduced and may have potential for integration in point-of-care devices [87].

Finally, note that it is possible to combine nitrocellulose, filter, and chromatographic paper in one device where positive sides of each type of these materials can be utilized. All these materials can be cut by laser or PC-controlled knife plotter, and assembled in a single process.

6 Existing Fabrication Technologies

This section reviews technologies that are applied for fabrication of paper-based devices. Because paper is a very flexible material, the following techniques are often applied (Fig. 7.6): inkjet, wax, flexography, or screen-printing all use non-toxic reagents. In the majority of the works a paper device needs to go through two main fabrication stages, i.e. patterning of hydrophobic channels for liquid confinement and assembly [88]. On the laboratory level, some deviations from the described low-cost methods are possible, but due to the limited space we only provide general descriptions.

6.1 Technologies for Patterning of Hydrophobic Barriers

Paper is a flexible material; therefore, various printing techniques are well suited to form a pattern. Physical blocking of pores in paper with hydrophobic material is a

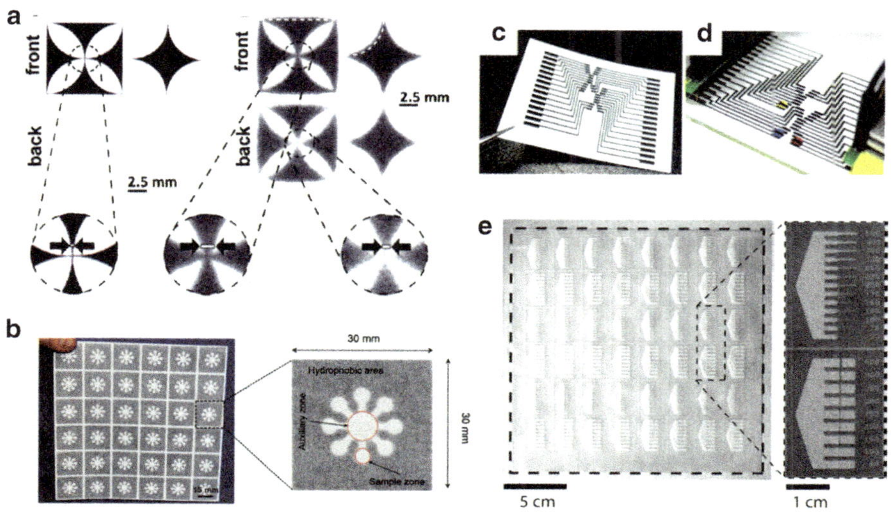

Fig. 7.6 Examples of paper-based structures fabricated by (**a**) wax-printing (Adapted from [92] with permission of Royal Society of Chemistry), (**b**) screen-printing (Adapted from [99] with permission of Royal Society of Chemistry), (**c**) inkjet-printing (Adapted from [104] with permission of John Wiley and Sons) and (**d**) photolithography assisted method (FLASH) (Reproduced from [108] with permission of Royal Society of Chemistry)

widely used method. The rise of this technology goes back to 2007 when a description of the process for creating hydrophobic pattern in paper and using hydrophilic channels for pumpless liquid transport was reported by Whitesides and co-workers [89]. Hydrophobic barriers can be made using wax, SU-8 and other photoresist materials, PDMS, alkenyl ketene dimer [5], polystyrene, ethyl cellulose, silicones, rosin, paraffin, printer varnish, cellulose esters, hydrophobic gels [90] and possibly others. All these materials are impervious to water and allow for implementation of various structures for transport and storage of liquids. Width of channel structures that can be achieved by these methods varies and can go down to 200-300 μm with some optimization.

6.1.1 Wax Printing

Wax printing [91, 92] is the most commonly used technique. It is easy, fast, and low cost, and can be easily applied for small-scale manufacturing. Typically, the channel width is in the range of 1–5 mm. This technology is very straightforward: a design is sent to a printer and printed directly on a sheet of paper, this paper can be sandwiched between a metal foil and placed on a hot-plate at ca. 70 °C for 1 min. For better penetration of wax into paper it can be baked from both sides. Baking temperature, baking time and printing mode are parameters that can be used for optimization in each specific case. Speading of wax in paper after bake can be predicted (see description in the earlier section). Reproducibility of channel dimensions and printing resolution will depend on quality of the printer and homogeneity of heat applied during the bake. Wax printing has been widely applied to 2D, 3D fluidic devices as well as in centrifugal paper-based systems. Details of the method for fabrication of 3D μPADs by wax printing can be found elsewhere [93]. Permeability of the paper is another parameter describing flow through it. Permeability has been alternated using papers that were impregnated with wax [94]. Accurate control over the penetration depth of melted wax, printed on both sides of a paper substrate allows formation of multilayers of patterned channels in the substrate [95].

6.1.2 Screen-Printing

Screen-printing [96, 97] is a versatile technique where liquid material is transferred onto substrate via a screen (a grid with a stencil attached or formed directly on it) manually or using an automatic tool, which regulates pressure applied on substrate and amount of printed material. After applying the material, the paper substrate is allowed to dry, and maybe subjected to heat or other treatments. What is characteristic for this method is that much thicker layers of printed material can be applied compared to wax printing by a commercial printer. There are also less strict requirements to size, planarity, shape and thickness of the selected paper substrate. Advantages of screen-printing technique [98] are fast fabrication times, low costs, flexibility (different materials can be printed) and capability of mass production. Several instrument-free, single-step screen-printing methods on chromatographic paper were demonstrated, e.g. patterning of PDMS solution with minimum channel width of ca. 600 μm [99], patterning of polystyrene through a patterned screen [100] with minimum channel width of ca. 670 μm (minimum width of hydrophobic barrier was ca. 380 μm). Another variation of this technique utilizing spraying of material through a pre-defined micro stencil instead of squeezing through a screen as in conventional screen-printing, i.e. a mask containing pattern to be transferred [101]. This technique is also widely applied to form electrodes in paper [77].

6.1.3 Inkjet-Printing

Inkjet-printing is also one of the early methods applied for fabrication of paper-based devices [102–104]. Review of inkjet-printed technologies applied on paper can be found elsewhere [105, 106]. In short, this technology is based on transferring material directly into paper via a nozzle (nozzle is activated e.g. piezo-electrically or thermally). Material is jetted in a close proximity from paper surface and follows a required pattern. Printed lines are formed from hydrophobic inks, e.g. PDMS, which was also adapted to roll-to-roll technology [103, 107]. Besides the setup of the inkjet-printer (jetting pressure applied via the nozzle), the chemistry and wetting properties of the printed material is probably the most crucial optimization parameter to achieve good quality of the printed image.

6.1.4 Flexographic Printing

Flexographic printing is method utilizing a flexible plate containing a pattern which is transferred by applying a liquid ink. It is another variation of the screen-printing technique, which is widely applied in industrial settings. It is suitable for various inks and substrates. One example is flexographic printing of polystyrene applied using the roll-to-roll technology. Several layers need to be applied to achieve hydrophobic channels where width of channels is typically 1 mm [107].

6.1.5 Photolithography-Assisted Methods

Conventional UV lithography can be used to produce pattern on photo-curable polymer that was brought in contact with paper substrate by one or other method. Paper can be impregnated with the photoresist via dipping prior to patterning using a UV lamp through a glass or polymeric mask containing desired patterns. The UV-assisted methods, i.e. FLASH method, was one of the first methods introduced for patterning of paper-based devices [108]. UV lithography can aid very defined and reproducible structures but fabrication costs are high and not every lab has access to photolithography facilities. A variety of hybrid fabrication methods were demonstrated, for example, an interesting method where photolithographic patterning of Parafilm® was subsequently followed by embossing of the film into paper [109].

6.1.6 Plasma Treatment

Definition of hydrophilic and hydrophobic regions can also be achieved by exposure to plasma [110]. Oxygen plasma can also be used to introduce new fluidic functionalities in paper, e.g. multiple-use valves [111].

6.2 Technologies for Assembly of Devices

After paper devices were patterned with hydrophobic materials to form channels and reaction zones, they still need to be assembled. Common methods include cutting, stacking several layers of paper together, shaping of paper by cutting or folding (origami-like technique [71]) with or without use of adhesive tapes [70, 112]. Paper can also be laminated prior to cutting and stacking by methods similar to production of ID cards [113, 114]. Cutting can be done by knife plotter controlled by PC (down to 0.5–0.8 mm with some optimization of cutting parameters), or CO_2 laser can be used for cutting cellulose (down to 1–1.5 mm). Maintaining contact between hydrophilic features in each layer of a 3D paper-based microfluidic device is the key fabrication challenge for these devices. Dry-stored reagents are deposited onto individual layers before assembling the final 3D device [1].

6.3 Technologies for Fabrication of Electrodes in Paper

Many detection methods applied on paper require reproducible fabrication of paper electrodes. Electrodes, typically conductive inks (carbon, Ag/AgCl), can be formed by various techniques, including those discussed above: screen-printing [96, 97], electrospraying [33], painting or dipping paper in conductive ink, or e.g. solution of carbon nanotubes [115, 116], combination of direct printing of Au and Ag stripes and subsequent electrochemical deposition of AgCl layer, and even by using a regular pencil [44]. Combination of fluidic networks in paper printed over electrodes allows for many applications involving measurement of electrical signals [107] such as e.g. digital microfluidics [117]. Functionality of paper can be extended by modifying its properties or hybridly combining it with conductive materials for various electronics applications [118].

7 Detection Methods

7.1 Colourimetric

Colourimetric method has been widely applied in paper for quantification of concentration of analytes [35, 51]. Colourimetric detection may include or not an enzymatic stage. The method is very common for medical laboratories and, as a result, many assays are based on it. It is also a convenient method for multiplexing [119] and can be combined with other detection methods, e.g. electrochemical [120]. Readout of colourimetric signals can be achieved with a smartphone or dedicated readout system.

7.2 Electrochemical

Detection methods which are using change in electric current or potential as a result of biochemical reaction occurring in a paper device are called electrochemical [38, 44, 121, 122]. Cyclic voltammetry is a very common measurement technique

[97, 103], as well as potentiometric [103, 123]. There are also more exotic variations of these such as streaming potential measurements [124].

7.3 Chemiluminescence and Electrochemiluminescence

Detection of electromagnetic radiation as light emitted during chemical reaction is the basis for chemiluminescence [125, 126]. Electrochemiluminescence method is luminescence generated by electrochemical reactions [47, 127].

7.4 Fluorescence

Fluorescence-based assays also have been integrated into paper [128, 129]. Overall, although fluorescence sensing brings new opportunities to paper-based detection, cost reduction and miniaturization of fluorescence readers are still on-going issues [1].

7.5 Nanoparticles

Many detection practices using nanoparticles are coming from the lateral flow assays. Antibody-conjugated gold nanoparticles and monodisperse latex, coupled with fluorescent dyes, are widely utilized in commercial rapid point-of-care tests. In general, colloidal gold particles are a preferred option due to higher colour intensity compared with coloured latex particles; they also can be dispersed in higher density as they typically ten times smaller than monodisperse latex particles [1, 130].

7.6 Other Methods

There are also a few methods, which are less popular and cannot be strictly assigned to any of the methods described above. Paper-based microfluidic calorimeter has been proposed [131]. This method would only suit for assays producing or consuming sufficient amount of heat to be detected as the method is not highly sensitive. Distance-based detection method where distance travelled by analyte producing colorimetric response is used as an analytical signal [50, 132]. Similarly, time to obtain response in the reference area of the device can be used as readout [23, 114].

Various efforts were also made to achieve visual detection just by naked eye which is a very resource-poor settings inspired idea. Specific designs aiming at solely visual determination of test result, e.g. counting bars that turned colorful after analyte has reached it were demonstrated [23]. Electrical measurements were applied for monitoring of bacterial movement through the paper, which both has fundamental interest (measurement of speed of movement of bio-entities through paper network is challenging) and potentially interesting applications [133]. Raman spectroscopy [55] was also applied as a detection method, and although this technique might be very feasible in certain applications, it still remains a sophisticated and expensive technique.

References

1. Yetisen AK, Akram MS, Lowe CR (2013) Paper-based microfluidic point-of-care diagnostic devices. Lab Chip 13:2210–2251. doi:10.1039/c3lc50169h
2. Liana DD, Raguse B, Gooding JJ, Chow E (2012) Recent advances in paper-based sensors. Sensors 12:11505–11526. doi:10.3390/s120911505
3. Martinez AW, Phillips ST, Whitesides GM (2010) Diagnostics for the developing world: microfluidic paper-based analytical devices. Anal Chem 82:3–10
4. Martinez AW, Phillips ST, Butte MJ, Whitesides GM (2013) Patterned paper as a platform for inexpensive, low volume, portable bioassays. Angew Chem Int Ed Engl 46:1318–1320. doi:10.1002/anie.200603817.Patterned
5. Li X, Ballerini DR, Shen W (2014) A perspective on paper-based microfluidics: current status and future trends. Biomicrofluidics 011301:1–13. doi:10.1063/1.3687398
6. Liu B, Du D, Hua X et al (2014) Paper-based electrochemical biosensors: from test strips to paper-based microfluidics. Electroanalysis 26:1214–1223. doi:10.1002/elan.201400036
7. Ballerini DR, Li X, Shen W (2012) Patterned paper and alternative materials as substrates for low-cost microfluidic diagnostics. Microfluid Nanofluid 13:769–787. doi:10.1007/s10404-012-0999-2
8. Santhiago M, Nery EW, Santos GP, Kubota LT (2014) Microfluidic paper-based devices for bioanalytical applications. Bioanalysis 6:89–106. doi:10.4155/bio.13.296
9. Cate DM, Adkins JA, Mettakoonpitak J, Henry CS (2015) Recent developments in paper-based microfluidic devices. Anal Chem 87:19–41. doi:10.1021/ac503968p
10. Ge X, Mohamed A, Du D et al (2014) Nanomaterial-enhanced paper based biosensors. Trends Anal Chem 58:31–39. doi:10.1016/j.trac.2014.03.008
11. Hu J, Wang S, Wang L et al (2014) Advances in paper-based point-of-care diagnostics. Bioelectronics 54:585–597. doi:10.1016/j.bios.2013.10.075
12. Lisowski P, Zarzycki PK (2013) Microfluidic paper-based analytical devices (microPADs) and micro total analysis systems (microTAS): development, applications and future trends. Chromatographia 76:1201–1214. doi:10.1007/s10337-013-2413-y
13. Pelton R (2009) Bioactive paper provides a low-cost platform for diagnostics. Trends Analyt Chem 28:925–942. doi:10.1016/j.trac.2009.05.005
14. Shah P, Zhu X, Li C (2013) Development of paper-based analytical kit for point-of-care testing. Expert Rev Mol Diagn 13:83–91. doi:10.1586/erm.12.130
15. Rozand C (2014) Paper-based analytical devices for point-of-care infectious disease testing. Eur J Clin Microbiol Infect Dis 33:147–156. doi:10.1007/s10096-013-1945-2
16. Smith S, Moodley K, Govender U et al (2015) Paper-based smart microfluidics for education and low-cost diagnostics. S Afr J Sci 111:0–10. doi:10.17159/sajs.2015/20140358
17. Mao X, Jun T (2012) Lab on chip microfluidic diagnostics for the developing world. Lab a chipchip 12:1412–1416. doi:10.1039/c2lc90022j

18. Dickey M, Martinez A, Phillips S, et al. (2014) Paper-based microfluidic systems. US Patent #8,921,118
19. Carrilho E, Martinez AW, Whitesides GM (2012) Methods of micropatterning paper-based microfluidics. US Patent #0,198,684
20. Whitesides GM, Phillips ST, Martinez AW, et al. (2015) Lateral flow and flow-through bioassay devices based on patterned porous media, methods of making same, and methods of using same. US patent #9,193,988
21. Volpatti LR, Yetisen AK (2014) Commercialization of microfluidic devices. Trends Biotechnol 32:347–350. doi:10.1016/j.tibtech.2014.04.010
22. Phillips ST, Lewis GG (2014) The expanding role of paper in point-of-care diagnostics. Expert Rev Mol Diagn 14:123–125. doi:10.1586/14737159.2014.887445
23. Lewis GG, Ditucci MJ, Phillips ST (2012) Quantifying analytes in paper-based microfluidic devices without using external electronic readers. Angew Chem Int Ed 51:12707–12710. doi:10.1002/anie.201207239
24. Zhao C, Thuo MM, Liu X (2015) Corrigendum: A microfluidic paper-based electrochemical biosensor array for multiplexed detection of metabolic biomarkers (2013 Sci. Technol Adv. Mater. 14 054402). Sci Technol Adv Mater 16:049501. doi:10.1088/1468-6996/16/4/049501
25. Nie Z, Deiss F, Liu X et al (2010) Integration of paper-based microfluidic devices with commercial electrochemical readers. Lab Chip 10:3163–3169. doi:10.1039/c0lc00237b
26. Park TS, Yoon J (2015) Smartphone detection of Escherichia coli from field water samples on paper microfluidics. IEEE Sens J 15:1902–1907
27. Chun HJ, Park YM, Han YD et al (2014) Paper-based glucose biosensing system utilizing a smartphone as a signal reader. BioChip J 8:218–226. doi:10.1007/s13206-014-8308-7
28. Park TS, Li W, Mccracken KE, Yoon J (2013) Smartphone quantifies Salmonella from paper microfluidics. Lab Chip 13:4832–4840. doi:10.1039/c3lc50976a
29. Yager P, Edwards T, Fu E et al (2006) Microfluidic diagnostic technologies for global public health. Nature 442:412–418. doi:10.1038/nature05064
30. Su J, Al-Tamimi M, Garnier G (2012) Engineering paper as a substrate for blood typing. Cellulose 19:1749–1758. doi:10.1007/s10570-012-9748-7
31. Chen X, Chen J, Wang F et al (2012) Determination of glucose and uric acid with bienzyme colorimetry on microfluidic paper-based analysis devices. Biosen Bioelectron 35:363–368. doi:10.1016/j.bios.2012.03.018
32. Ge L, Yan J, Song X et al (2012) Three-dimensional paper-based electrochemiluminescence immunodevice for multiplexed measurement of biomarkers and point-of-care testing. Biomaterials 33:1024–1031. doi:10.1016/j.biomaterials.2011.10.065
33. Ruecha N, Rangkupan R, Rodthongkum N (2014) Novel paper-based cholesterol biosensor using graphene/polyvinylpyrrolidone/polyaniline nanocomposite. Biosens Bioelectron 52:13–19. doi:10.1016/j.bios.2013.08.018
34. Klasner SA, Price AK, Hoeman KW et al (2010) Paper-based microfluidic devices for analysis of clinically relevant analytes present in urine and saliva. Anal Bioanal Chem 397:1821–1829. doi:10.1007/s00216-010-3718-4
35. Dungchai W, Chailapakul O, Henry CS (2010) Use of multiple colorimetric indicators for paper-based microfluidic devices. Anal Chim Acta 674:227–233. doi:10.1016/j.aca.2010.06.019
36. Godino N, Vereshchagina E, Gorkin R, Ducree J (2013) Centrifugal automation of a triglyceride bioassay on a low-cost hybrid paper-polymer device. Microfluid Nanofluid 16:895–905. doi:10.1007/s10404-013-1283-9
37. Elisa P, Cheng C, Martinez AW et al (2010) Paper-based ELISA. Angew Chem Int Ed Engl 49:4771–4774. doi:10.1002/anie.201001005
38. Su M, Ge L, Ge S et al (2014) Paper-based electrochemical cyto-device for sensitive detection of cancer cells and in situ anticancer drug screening. Anal Chim Acta 847:1–9

39. Chen C, Lin B-R, Wang H-K et al (2014) Paper-based immunoaffinity devices for accessible isolation and characterization of extracellular vesicles. Microfluid Nanofluid 16:849–856. doi:10.1007/s10404-014-1359-1
40. Li M, Tian J, Al-tamimi M, Shen W (2012) Paper-based blood typing device that reports patient's blood type "in writing.". Angew Chem Int Ed 51:5497–5501. doi:10.1002/anie.201201822
41. Pollock NR, Rolland JP, Kumar S et al (2013) A paper-based multiplexed transaminase test for low-cost, point-of-care liver function testing. Sci Transl Med 4:1521129. doi:10.1126/scitranslmed.3003981.A
42. Warren AD, Kwong GA, Wood DK et al (2014) Point-of-care diagnostics for noncommunicable diseases using synthetic urinary biomarkers and paper microfluidics. Proc Natl Acad Sci U S A 111:3671–3676. doi:10.1073/pnas.1314651111
43. Nery EW, Guimares JA, Kubota LT (2015) Paper-based electronic tongue. Electroanalysis 27:1–7. doi:10.1002/elan.201500054
44. Santhiago M, Henry CS, Kubota LT (2014) Low cost, simple three dimensional electrochemical paper-based analytical device for determination of p-nitrophenol. Electrochim Acta 130:771–777
45. Wang S, Ge L, Li L et al (2013) Molecularly imprinted polymer grafted paper-based multidisk micro-disk plate for chemiluminescence detection of pesticide. Biosen Bioelectron 50:262–268. doi:10.1016/j.bios.2013.07.003
46. Witkowska Nery E, Kubota LT (2016) Integrated, paper-based potentiometric electronic tongue for the analysis of beer and wine. Anal Chim Acta 918:60–68. doi:10.1016/j.aca.2016.03.004
47. Zhang M, Ge L, Ge S et al (2013) Three-dimensional paper-based electrochemiluminescence device for simultaneous detection of Pb 2+ and Hg 2+ based on potential-control technique. Biosens Bioelectron 41:544–550. doi:10.1016/j.bios.2012.09.022
48. Henry CS, Kim Y, Mettakoonpitak J, Guerrero T (2014) Multifunctional paper microfluidic devices for environmental analysis. In: 18th International Conference on Miniaturized Systems for Chemistry and Life Sciences. San Antonio, 26–30 Oct 2014, pp 45–47
49. Cate DM, Nanthasurasak P, Riwkulkajorn P et al (2014) Rapid detection of transition metals in welding fumes using paper-based analytical devices. Ann Occup Hyg 58:413–423. doi:10.1093/annhyg/met078
50. Cate DM, Noblitt SD, Henry CS (2015) Lab on a chip of metals using distance-based detection †. Lab Chip 15:2808–2818. doi:10.1039/C5LC00364D
51. Oktem HA, Senyurt O, Eyidogan FI et al (2012) Development of a laccase based paper biosensor for the detection of phenolic. J Food Agric Environ 10:1030–1034
52. Nguyen TH, Fraiwan A, Choi S (2014) Paper-based batteries: a review. Biosens Bioelectron 54:640–649. doi:10.1016/j.bios.2013.11.007
53. Fraiwan A, Lee H, Choi S (2014) A Multianode paper-based microbial fuel cell: a potential power source for disposable biosensors. IEEE Sens J 14:3385–3390
54. Thom NK, Yeung K, Pillion MB, Phillips ST (2012) 'Fluidic batteries' as low-cost sources of power in paper-based microfluidic devices. Lab Chip 12:1768–1770. doi:10.1039/c2lc40126f
55. Craig D, Mazilu M, Dholakia K (2015) Quantitative detection of pharmaceuticals using a combination of paper microfluidics and wavelength modulated Raman spectroscopy. Plus One 10:1–10. doi:10.1371/journal.pone.0123334
56. Blanes L, Taudte R V, Roux C, Doble P (2014) Using paper-based microfluidics and lab on a chip technologies for the rapid analysis of trinitro aromatic explosives. In: 18th International Conference on Miniaturized Systems for Chemistry and Life Sciences. San Antonio, 26–30 Oct, pp 1581–1582
57. Fu E, Ramsey SA, Kauffman P et al (2011) Transport in two-dimensional paper networks. Microfluid Nanofluid 10:29–35. doi:10.1007/s10404-010-0643-y
58. Elizalde E, Urteaga R, Berli CLA (2015) Rational design of capillary-driven flows for paper-based microfluidics. Lab Chip 15:2173–2180. doi:10.1039/c4lc01487a

59. Li C, Vandenberg K, Prabhulkar S et al (2011) Paper based point-of-care testing disc for multiplex whole cell bacteria analysis. Biosen Bioelectron 26:4342–4348. doi:10.1016/j.bios.2011.04.035
60. Toley BJ, Wang JA, Gupta M et al (2015) A versatile valving toolkit for automating fluidic operations in paper microfluidic devices. Lab Chip 15:1432–1444. doi:10.1039/C4LC01155D
61. Songjaroen T, Dungchai W, Chailapakul O, Henry S (2012) Blood separation on microfluidic paper-based analytical devices. Lab Chip 12:3392–3398. doi:10.1039/c2lc21299d
62. Osborn JL, Lutz B, Fu E et al (2010) Microfluidics without pumps: reinventing the T-sensor and H-filter in paper networks. Lab Chip 10:2659–2665. doi:10.1039/c004821f
63. Fu E, Kauffman P, Lutz B, Yager P (2010) Chemical signal amplification in two-dimensional paper networks. Sensors Actuators B Chem 149:325–328. doi:10.1016/j.snb.2010.06.024
64. Lutz BR, Trinh P, Ball C et al (2011) Two-dimensional paper networks: programmable fluidic disconnects for multi-step processes in shaped paper. Lab Chip 11:4274–4278. doi:10.1039/c1lc20758j
65. Fu E, Lutz B, Kauffman P, Yager P (2010) Controlled reagent transport in disposable 2D paper networks. Lab Chip 10:918–920. doi:10.1039/b919614e
66. Apilux A, Ukita Y, Chikae M et al (2013) Development of automated paper-based devices for sequential multistep sandwich enzyme-linked immunosorbent assays using inkjet printing. Lab Chip 13:126–135. doi:10.1039/c2lc40690j
67. Wu L, Ma C, Zheng X et al (2015) Paper-based electrochemiluminescence origami device for protein detection using assembled cascade DNA—carbon dots nanotags based on rolling circle ampli fi cation. Biosens Bioelectron 68:413–420. doi:10.1016/j.bios.2015.01.034
68. Liu H, Crooks RM (2011) Three-dimensional paper microfluidic devices assembled using the principles of origami. J Am Chem Soc 133:17564–17566
69. Deraney RN, Rolland P, Mace CR (2014) A device architecture for three-dimensional, patterned paper immunoassays. Lab Chip 14:4653–4658. doi:10.1039/C4LC00876F
70. Martinez AW, Phillips ST, Whitesides GM (2008) Three-dimensional microfluidic devices fabricated in layered paper and tape. Proc Natl Acad Sci U S A 105:19606–19611
71. Kirby AE, Wheeler AR (2013) Microfluidic origami: a new device format for in-line reaction monitoring by nanoelectrospray ionization mass spectrometry. Lab Chip 13:2533–2540. doi:10.1039/c3lc41431k
72. Ge L, Wang S, Song X, Yu J (2012) 3D Origami-based multifunction-integrated immunodevice: low-cost and multiplexed sandwich chemiluminescence immunoassay on microfluidic paperbased analytical device. Lab Chip 12:3150–3158. doi:10.1039/c2lc40325k
73. Schilling KM, Jauregui D, Martinez AW (2013) Paper and toner three-dimensional fluidic devices: programming fluid flow to improve point-of-care diagnostics. Lab Chip 13:628–631. doi:10.1039/c2lc40984d
74. Hwang H, Kim S, Kim T et al (2011) Paper on a disc: balancing the capillary-driven flow with a centrifugal force. Lab Chip 11:3404–3406. doi:10.1039/c1lc20445a
75. Vereshchagina E, Bourke K, Meehan L, et al. (2013) Multi-material paper-disc devices for low cost biomedical diagnostics. In: 26th International Conference on Micro Electro Mechanical Systems. pp 1049–1052
76. Shiong C, Lawrence K, Ngin S, Zerda C (2014) A “green” cellulose paper based glucose amperometric biosensor. Sensors Actuators B Chem 193:536–541
77. Godino N, Gorkin R, Bourke K, Ducre J (2012) Fabricating electrodes for amperometric detection in hybrid paper/polymer lab-on-a-chip devices. Lab Chip 12:3281–3284. doi:10.1039/c2lc40223h
78. Fobel R, Kirby AE, Ng AHC et al (2014) Paper microfluidics goes digital. Adv Mater 26:2838–2843. doi:10.1002/adma.201305168
79. Niedl RR, Beta C (2015) Hydrogel-driven paper-based microfluidics. Lab Chip 15:2452–2459. doi:10.1039/c5lc00276a

80. Lutz B, Liang T, Fu E et al (2014) Dissolvable fluidic time delays for programming multi-step assays in instrument-free paper diagnostics. Lab Chip 13:2840–2847. doi:10.1039/c3lc50178g.Dissolvable
81. Alava M, Niskanen K (2006) The physics of paper. Rep Prog Phys 69:669–723. doi:10.1088/0034-4885/69/3/R03
82. Alexander B, Carstens F, Trieb C et al (2014) Engineering microfluidic papers: effect of fiber source and paper sheet properties on capillary-driven fluid flow. Microfluid Nanofluid 16:789–799. doi:10.1007/s10404-013-1324-4
83. Kong F, Hu YF (2012) Biomolecule immobilization techniques for bioactive paper fabrication. Anal Bioanal Chem 403:7–13. doi:10.1007/s00216-012-5821-1
84. Su S, Nutiu R, Filipe CDM et al (2007) Adsorption and covalent coupling of ATP-Binding DNA aptamers onto cellulose. Langmuir 23:1300–1302
85. Wu Y, Xue P, Hui KM, Kang Y (2014) A paper-based microfluidic electrochemical immunodevice integrated with ampli fi cation-by-polymerization for the ultrasensitive multiplexed detection of cancer biomarkers. Biosens Bioelectron 52:180–187. doi:10.1016/j.bios.2013.08.039
86. Hosseini S, Azari P, Farahmand E et al (2015) Polymethacrylate coated electrospun PHB fibers: an exquisite outlook for fabrication of paper-based biosensors. Biosens Bioelectron 69:257–264. doi:10.1016/j.bios.2015.02.034
87. Hansson J, Yasuga H, Haraldsson T, van der Wijngaart W (2016) Synthetic microfluidic paper: high surface area and high porosity micropillar arrays. Lab Chip 16:298–304. doi:10.1039/c5lc01318f
88. Mace CR, Deraney RN (2014) Manufacturing prototypes for paper-based diagnostic devices. Microfluid Nanofluid 16:801–809. doi:10.1007/s10404-013-1314-6
89. Martinez AW, Phillips ST, Butte MJ, Whitesides GM (2007) Patterned paper as a platform for inexpensive, low-volume, portable bioassays. Angew Chem Int Ed 46:1318–1320. doi:10.1002/anie.200603817
90. Wang J, Monton RN, Zhang X et al (2014) Hydrophobic sol–gel channel patterning strategies for paper-based microfluidics. Lab Chip 14:691–695. doi:10.1039/c3lc51313k
91. Lu Y, Shi W, Qin J, Lin B (2010) Fabrication and characterization of paper-based microfluidics prepared in nitrocellulose membrane by wax printing. Anal Chem 82:329–335. doi:10.1021/ac9020193
92. Carrilho E, Martinez AW, Whitesides GM (2009) Understanding wax printing: a simple micropatterning process for paper-based microfluidics. Anal Chem 81:7091–7095. doi:10.1021/ac901071p
93. Renault C, Koehne J, Ricco AJ, Crooks RM (2014) Three-dimensional wax patterning of paper fluidic devices. Langmuir 30:7030–7036
94. Jang I, Song S (2015) Facile and precise flow control for a paper-based. Lab Chip 15:3405–3412. doi:10.1039/C5LC00465A
95. Li X, Liu X (2014) Fabrication of three-dimensional microfluidic channels in a single layer of cellulose paper. Microfluid Nanofluid 16:819–827. doi:10.1007/s10404-014-1340-z
96. Dungchai W, Chailapakul O, Henry CS (2011) Paper microfluidics using wax screen-printing. Analyst 136:77–82. doi:10.1039/c0an00406e
97. Metters JP, Houssein SM, Kampouris DK, Banks CE (2013) Paper-based electroanalytical sensing platforms. Anal Methods 5:103–110. doi:10.1039/c2ay26396c
98. Yafia M, Shukla S, Najjaran H (2015) Fabrication of digital microfluidic devices on flexible paper-based and rigid substrates via screen printing. J Micromech Microeng 25:57001. doi:10.1088/0960-1317/25/5/057001 (11 pp)
99. Mohammadi S, Maeki M, Mohamadi RM et al (2015) An instrument-free, screen-printed paper microfluidic device that enables bio and chemical sensing. Analyst 140:6493–6499. doi:10.1039/c5an00909j

100. Sameenoi Y, Nongkai N, Nouanthavong S, Charles S (2014) One-step polymer screen-printing for microfluidic paper-based analytical device (μPAD) fabrication. Analyst 139:6580–6588. doi:10.1039/C4AN01624F
101. Tao H, Chieffo LR, Brenckle MA et al (2011) Metamaterials on paper as a sensing platform. Adv Mater 23:3197–3201. doi:10.1002/adma.201100163
102. Abe K, Suzuki K, Citterio D (2008) Inkjet-printed microfluidic multianalyte chemical sensing paper. Anal Chem 80:6928–6934
103. Määttänen A, Vanamo U, Ihalainen P et al (2013) A low-cost paper-based inkjet-printed platform for electrochemical analyses. Sensors Actuators B Chem 177:153–162
104. Ko H, Lee J, Kim Y et al (2014) Active digital microfluidic paper chips with inkjet-printed patterned electrodes. Adv Mater 26:2335–2340. doi:10.1002/adma.201305014
105. Le HP (1998) Ink-jet printing technology. J Imaging Sci Technol 42:1998
106. Yamada K, Henares TG, Suzuki K, Citterio D (2015) Paper-based inkjet-printed microfluidic analytical devices. Angew Chem Int Ed 54:5294–5310. doi:10.1002/anie.201411508
107. Määttänen A, Fors D, Wang S et al (2011) Paper-based planar reaction arrays for printed diagnostics. Sensors Actuators B Chem 160:1404–1412. doi:10.1016/j.snb.2011.09.086
108. Martinez AW, Phillips ST, Wiley BJ et al (2008) FLASH: a rapid method for prototyping paper-based microfluidic devices †‡. Lab Chip 8:2146–2150. doi:10.1039/b811135a
109. Yu SL, Shi ZZ (2015) Microfluidic paper-based analytical devices fabricated by low-cost photolithography and embossing of Parafilm®. Lab Chip 15:1642–1645. doi:10.1039/C5LC00044K
110. Li X, Tian J, Nguyen T, Shen W (2008) Paper-based microfluidic devices by plasma treatment. Anal Chem 80:9131–9134. doi:10.1021/ac801729t
111. Salentijn GIJ, Hamidon NN, Verpoorte E (2016) Solvent-dependent on/off valving using selectively permeable barriers in paper microfluidics. Lab Chip 16:1013–1021. doi:10.1039/C5LC01355K
112. Siegel BAC, Phillips ST, Dickey MD et al (2010) Foldable printed circuit boards on paper substrates. Adv Funct Mater 20:28–35. doi:10.1002/adfm.200901363
113. Cassano CL, Fan ZH (2013) Laminated paper-based analytical devices (LPAD): fabrication, characterization, and assays. Microfluid Nanofluid 15:173–181. doi:10.1007/s10404-013-1140-x
114. Lewis GG, Robbins JS, Phillips ST (2013) Point-of-care assay platform for quantifying active enzymes to Femtomolar levels using measurements of time as the readout. Anal Chem 85:10432–10439
115. Pozuelo M, Blondeau P, Novell M et al (2013) Paper-based chemiresistor for detection of ultralow concentrations of protein. Biosens Bioelectron 49:462–465. doi:10.1016/j.bios.2013.06.007
116. Fong K, Lee K, Yang S (2012) Fabrication of carbon nanotube-based pH sensor for paper-based microfluidics. Microelectron Eng 100:1–5. doi:10.1016/j.mee.2012.07.113
117. Abadian A, Jafarabadi-Ashtiani S (2014) Paper-based digital microfluidics. Microfluid Nanofluid 16:989–995. doi:10.1007/s10404-014-1345-7
118. Tobjörk D, Österbacka R (2011) Paper electronics. Adv Mater 23:1935–1961. doi:10.1002/adma.201004692
119. Peters KL, Corbin I, Kaufman LM et al (2014) Simultaneous colorimetric detection of improvised explosive compounds using micro fl uidic paper- based analytical devices (m PADs). Anal Methods 7:63–70. doi:10.1039/C4AY01677G
120. Rattanarat P, Dungchai W, Cate D et al (2014) Multilayer paper-based device for colorimetric and electrochemical quantification of metals. Anal Chem 86:3555–3562. doi:10.1021/ac5000224
121. Nie Z, Nijhuis CA, Gong J et al (2010) Electrochemical sensing in paper-based microfluidic devices †. Lab Chip 10:477–483. doi:10.1039/b917150a

122. Cunningham JC, Kogan MR, Tsai Y-J et al (2016) Paper-based sensor for electrochemical detection of silver nanoparticle labels by galvanic exchange. ACS Sensors 1:40–47. doi:10.1021/acssensors.5b00051
123. Szucs J, Gyurcsanyi RI (2012) Towards protein assays on paper platforms with potentiometric detection. Electroanalysis 24:146–152. doi:10.1002/elan.201100522
124. Leung V, Shehata AM, Filipe CDM, Pelton R (2010) Streaming potential sensing in paper-based microfluidic channels. Colloids Surfaces A Physicochem Eng Asp 364:16–18. doi:10.1016/j.colsurfa.2010.04.008
125. Wang S, Ge L, Song X et al (2012) Paper-based chemiluminescence ELISA: Lab-on-paper based on chitosan modified paper device and wax-screen-printing. Biosens Bioelectron 31:212–218. doi:10.1016/j.bios.2011.10.019
126. Yu J, Ge L, Huang J et al (2011) Microfluidic paper-based chemiluminescence biosensor for simultaneous determination of glucose and uric acid. Lab Chip 11:1286–1291. doi:10.1039/c0lc00524j
127. Liu R, Zhang C, Liu M (2015) Open bipolar electrode-electrochemiluminescence imaging sensing using paper-based microfluidics. Sensors Actuators B Chem 216:255–262
128. He M, Liu Z (2013) Paper-based microfluidic device with upconversion fluorescence assay. Anal Chem 85:11691–11694. doi:10.1021/ac403693g
129. Xu S, Dong B, Zhou D et al (2016) Paper-based upconversion fluorescence resonance energy transfer biosensor for sensitive detection of multiple cancer biomarkers. Sci Rep 2:1–9. doi:10.1038/srep23406
130. Kumar A, Hens A, Arun K, Chatterjee M (2015) A paper based microfluidic device for easy detection of uric acid using positively charged gold nanoparticles. Analyst 140:1817–1821. doi:10.1039/C4AN02333A
131. Davaji B, Hoon C (2014) A paper-based calorimetric micro fl uidics platform for bio-chemical sensing. Biosens Bioelectron 59:120–126. doi:10.1016/j.bios.2014.03.022
132. Cate DM, Dungchai W, Cunningham JC, Henry CS (2013) Simple, distance-based measurement for paper analytical devices. Lab Chip 13:2397–2404. doi:10.1039/c3lc50072a
133. Choi G, Choi S (2015) Bacterial cell transportation in paper-based microfluidics. In: Transducers. Anchorage, 21–25 June 2015, pp 3–6

Chapter 8
Biological Applications of Microfluidics System

Shipra Solanki and Chandra Mouli Pandey

1 Introduction

In the last decade, the applications for microfluidic (MF) devices have proliferated at an explosive rate similar to the revolution brought in the field of microelectronics by the invention of the integrated circuit [1]. These MF devices have shown enormous potential in diverse fields of biological applications, including cell sorting, enzymatic assays, immuno-hybridization reactions, and polymer chain reaction (PCR). The advances in MF technology are revolutionizing the areas of cell biology, molecular biology and diagnostics [2]. The similarity in the dimensions of cells and microchannels (10–100 μm widths and depths) plays a crucial role in modifying the procedures of molecular biology for enzymatic analysis, DNA analysis, and proteomics [3]. MF systems offer many advantages over traditional macroscale laboratory techniques such as:

- The small size of the MF chip requires very fewer sample volumes (typically in nanolitres as compared to the hundreds of microliters needed for microlitre plate based assays). This reduction in the quantity of the sample, as well as the reagents, decreases the cost of the device. Nowadays low-cost paper-based MF devices are being explored as a tentative solution for clinical diagnostics in developing countries.

S. Solanki
Biomedical Instrumentation Section, CSIR-National Physical Laboratory,
New Delhi-110012, India

C.M. Pandey (✉)
Biomedical Instrumentation Section, CSIR-National Physical Laboratory,
New Delhi-110012, India

Department of Applied Chemistry, Delhi Technological University, New Delhi-110042, India
e-mail: cmp.npl@gmail.com

C.K. Dixit, A. Kaushik (eds.), *Microfluidics for Biologists*,
DOI 10.1007/978-3-319-40036-5_8

- The most exciting aspect of MF systems is probably the system integration with different biochemical operations such as sample pre-treatment, sample preparation and detection on one single platform. These devices are referred to as micro total analysis system (μTAS) or lab-on-a-chip (LoC). Technologies need to be developed in biological research to simplify complex assay protocols; to maximize information gathered from precious samples; to conduct analogous experiments together and to provide the investigator with more control and predictability of the spatio-temporal processes of the cellular microenvironment.
- Micro sized components enable researchers to perform experiments which are not possible or practical to execute with other bench top traditional techniques. MF platforms are extensively being used in resource-rich research laboratories to provide innovative solutions to non-feasible problems which may help in understanding the fundamental insight into human biology.
- With the fluidic components that approach the cellular scale, MF circuits offer superb liquid handling capabilities that allow the manipulation of single cells and even individual molecules *in vitro*. Moreover, the small device size makes this technology an ideal platform for portable Point of Care (PoC) diagnostic devices.

The number of research articles published in the field of MF has been exponentially increasing year-on-year, reflecting the growing importance of this field and some MF based devices are also available in the market. In this chapter, we will discuss an overview of how and why MF is changing the scenario of biological analysis and diagnostics. The advances made by MF technology in academic research being conducted in the fields of stem cell research, neurology, tissue culture, drug discovery and synthetic biology will also be presented. Finally, some examples of commercial MF devices that are being successfully used for biological and life sciences applications will be summarized.

2 Microfluidics in Cell Biology

Cells are the building blocks of all living organisms. The knowledge of cells and their functions is thus important in several areas including cell biology, human physiology, and tissue engineering. The basic cellular study involves the three most important steps-isolation, culture, and analysis [2, 3]. But conventional techniques are inadequate for extending the present knowledge of these processes. Traditional petri dish based cell culture systems are not able to mimic exactly in vivo cellular microenvironments and as a result, many cellular phenotypes are lost. Natural suitability of MF technology with all three cell handling processes has given new insights into the study of cells in various contexts [4]. The integrated systems fabricated have revolutionized the techniques of cell sorting, cell culture, and cellular analysis. μTAS or LoC systems are the devices which integrate analytical procedures of pretreatment, separation and analysis of samples in a single MF device [4, 5]. These systems play a significant role in research and discovery of cell biology and tissue engineering. Improvements have been made in the basic requirements of cell culture in terms of required sample volume, the supply of nutrients, control of physical factors, and continuous monitoring and manipulation of cellular processes. The realization of techniques utilized for maintenance and growth of cells *in vitro* is an important milestone in the field of biological sciences [6]. Apart from the often-cited advantages of MF like faster response times, lower reagent volumes, and the potential for integration; the most important benefit of using the technology for biology is the ability to tailor the cellular microenvironment. Latest organ-on-a-chip applications of MF allow the fabrication of smallest functional units of a single organ or multiple organs [7]. These kinds of devices present a more physiologically relevant *in vitro* model compared to cells cultured in dishes. The unprecedented control over spatial and temporal gradients and patterns can be attained while moving from macro- to microscale, which is otherwise not possible with traditional Petri dishes. So far, MF applications have been used in many experimental parts of cell manipulation and analysis, such as cell trapping/sorting, cell culture, cytotoxicity, PCR, DNA sequencing, and gene analysis. In the next section, we will discuss on the advancement made by LoC technologies in stem cell research, neurology, drug discovery, latest fabricated organs on a chip and single cell analysis.

2.1 Stem Cell Research

Stem cells are the cells that are capable of continued self-renewal through replication and becoming precursor cells of a specific tissue types. It offers a steady supply of physiologically relevant cells from pathogen-free sources that both *in vivo* and *in vitro* can differentiate into mature somatic cells [8]. These cells are being successfully used in replacement of cells lost due to degenerative disease and repair of damaged tissue [9]. Other examples of stem cell application are in bone healing, repair of the suspensory ligament, and injury to the nervous system, as well as inflammatory, relieve in arthritis and immune suppression [10]. Regenerative medicine through stem cell therapy is becoming popular and is considered to be the future milestone in the therapeutics. The conventional techniques involved in stem cell experiments are not appropriate as they poorly mimic the physiological cellular environment, and suffer from inaccurate spatial and temporal control. They also give reduced throughput, lack of scalability and reproducibility. On the other hand novel LoC platforms can much better mimic the complexity of *in vivo* tissue and provide a more precise control of different parameters. This capability can be very helpful for understanding the biology and improving the clinical potential of stem cell-based therapies [8, 10].

A lot of research has been conducted, where stem cells have been used to study various life processes. Klein et al. developed a high-throughput droplet-MF approach for barcoding the RNA from thousands of individual cells for subsequent analysis by next-generation sequencing and analyzed mouse embryonic stem cells [11]. They used the analysis to infer gene expression relationships which can be useful for various biological studies. The MF device fabricated was 80 mm deep with four inlets and one outlet. Four events of cell capturing in different droplets, cell lysis, reverse transcription reaction and gene amplification integrated into one single chip made possible the event of cell capturing at the rate of 4000–12,000 cells per hour for their analysis. The device generates monodisperse droplets that can be varied in the range of 1–5 nL at a rate of 10–100 drops per second, simultaneously mixing aliquots from the inlets. Flow rates were 100 μLh^{-1} for cell suspension, 100 μLh^{-1} for mixing of reagents, 10–20 μLh^{-1} for mixing of barcodes, and 90 μLh^{-1} for carrier oil to produce 4 nL drops. The carrier oil used was HFE-7500 fluorinated fluid with EA-surfactant. Jung et al. performed flow-based sorting of human mesenchymal cells by using optimally designed MF chips based on the principle of hydrodynamic filtration (HMD) [12]. Human bone marrow-derived mesenchymal stem cells (hMSCs) were sorted into three subpopulations by focusing the cells (with a proper ratio) between primary and side flows and analyzing the surface marker expressions of cells from each outlet. The specially designed MF chip consisted of a rectangular main channel of length 16 mm with 55 branches and three outlets to achieve the trimodal separation, as shown in Fig. 8.1A. The heights and the widths of primary as well as branch channels, total lengths of main and branch channels, and inter-distances between each branch channel were set by applying equations of steady state laminar flow for the Newtonian fluid. The fluid circulation was done using syringe pumps at a flow rate of 30 $\mu L\ min^{-1}$. Particles of certain size can escape from the main channel at a specific branch where its ratio of the flow fraction is optimized for the right particle size. Thus based on this hydrodynamic filtration method the cells are sorted into three subpopulations: small ($<$25 μm), medium (25–40 μm), and large ($>$40 μm) cells. This process leads to the possibility of sorting stem cells rapidly without damage. Kang et al. developed an efficient on-chip cell culture MF device capable of repeated, temporal delivery of molecules into a population of cells tool [13]. The design of the on-chip localized electroporation device (LEPD) consists of microchannels, a cell culture chamber, built-in electrodes, and a porous substrate containing micro- or nanochannels (Fig. 8.1B). The width and height of the microchannel are 200 μm and ~20 μm, respectively whereas the cell culture chamber is 3 mm in diameter. The working electrode was built-in on a fabricated glass cover slide while the second electrode (Ag/AgCl wire) was submerged into the media in the cell culture chamber. On-chip cell culture was maintained by the continuous flow of culture media through the circulation microchannels located beneath the perforated substrate. For intercellular delivery by electroporation, a solution containing biomolecules to be delivered into the cells was loaded into the circulation microchannels. To elicit the formation of nanopores on the cell membrane voltage was applied between the two microelectrodes due to which molecules are transported into the cell by diffusion. Thus, utilizing this LEPD configuration, localized electroporation was achieved that provides exceptional capabilities such as the transfection of various molecules into primary cells, maintenance of consistent pH levels by continuous media circulation, and application of a focused electric field to a small portion of the cell membrane to minimize stress. Also, the device is simple to use and cost effective. Further, an impedance sensors integrated

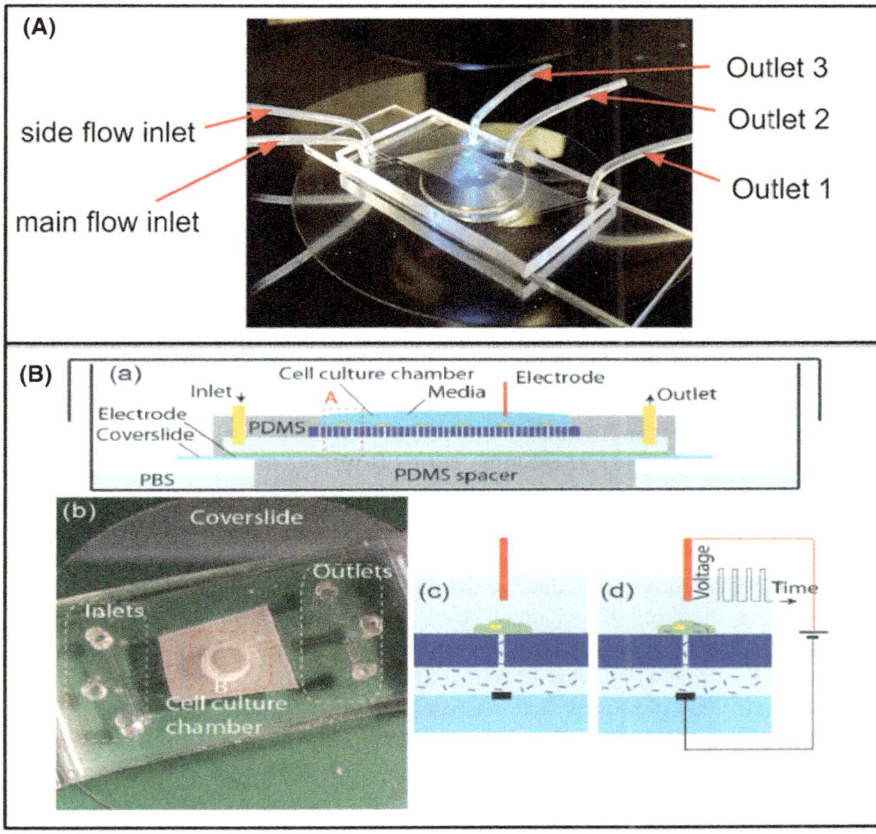

Fig. 8.1 (**A**) HDF microfluidic chip for sorting of particles and hMSCs where the number of branch channels is 55 and the total length of the main channel is 16 mm [12]. (**B**) (*a*) schematic of the LEPD enclosed within a Petri dish, (*b*) an optical image of the LEPD, made of PDMS, with three inlets and three outlets connected to microchannels, and (*c*) and (*d*) magnified views of area A in the schematic drawing depicting an adhered cell covering a micropore, and the mechanism of delivery of molecules by localized electroporation, respectively [13]

microwell platform was used by Lepperdinger research group to culture bone-marrow-derived stem cells for several weeks [14]. The authors reported that the developed LoC platform controls the quality of stem cells in medical cell production platform. Thus, significant attention is given to stem cell research owing to high therapeutic potential and their use as *in vitro* models for drug screening and understanding the developmental mechanisms.

2.2 Neurology

Neurology is another field which is being explored for MF applications. MF is employed for both *in vivo* deliveries of drug solutions from on-chip reservoirs situated on neural implants as well as *in vitro* studies of neuronal cells via highly precise delivery growth and inhibitory factors by the use of gradient-generating devices [15]. It is tough to probe the complex interactions that actually occur among neural cells using conventional methods of analysis. In this context, MF has proved to be the most suitable technique for neurology experiments. For instance, rapid, highly sensitive determination of trans-membrane potential has been possible in MF devices utilizing charged membrane-permeable, potential- sensitive dyes, with minimal use of reagents [16]. The LoC system consists of a quartz sipper chip containing microchannels connected to a sipper capillary, a fluorescence reader, a vacuum pump, and a computer. The flow rate in the detection channel is controlled by pressure driven pumps and could be varied from 2 to 10 nLs^{-1} by varying the applied pressure from 21 to 25 psig. Cells and samples were first mixed on the chip and then the dyes are added to the mixture. After a short incubation in a detection channel, the fluorescence of individual cells was detected. The membrane potential was determined by monitoring the dye uptake rate of the cells and modulated either by opening or blocking K^+ channels by varying cytoplasmic free Ca^{2+}. NMR (nuclear magnetic resonance) micro coils have been efficiently used to study single non-perfused neurons [17], where NMR probes have been micro-fabricated on the glass substrate MF platforms and sucrose solutions are used for their testing [18]. Besides this MF principle and techniques have been applied in the isolation of brain tissue culture studies. The separation of brain tissue specimens under *in vitro* conditions is a very complicated task as it requires exquisite control over experimental conditions and access to neural networks and synapses [19]. Scott et al. designed a MF chip for simultaneous recording of electrical signals and optical characterization of brain tissue slice preparations [20]. The device was utilized to record waves of spontaneous activity in developing cortical slices and to perform multisite extracellular recordings during simultaneous calcium imaging of activity. The device consists of an array of MF channels and a perfusion chamber (Fig. 8.2A). Each channel consists a well at one end, an aperture which is in contact with the perfusion chamber in the middle, and pressure and electronic controls at the other end is connected to port. The apertures are 20 or 50 μm in diameter with a spacing of 300 μm and are designed to probe an array of anatomically relevant sites. The relatively large aperture size and spacing were chosen because they are appropriate for the anatomical features in the particular study. The work paved a way to develop devices of different geometric configurations for other studies also. The use of microchips may overcome the limitations of reduced oxygen and nutrient supply while using conventional interface and submerged slice chambers to the brain slices allow the neuroscientists to design complex experiments to have a better insight into neuroprocesses [15]. Mauleon et al. developed an MF system that allows diffusion of oxygen throughout a thin membrane and directly to the brain

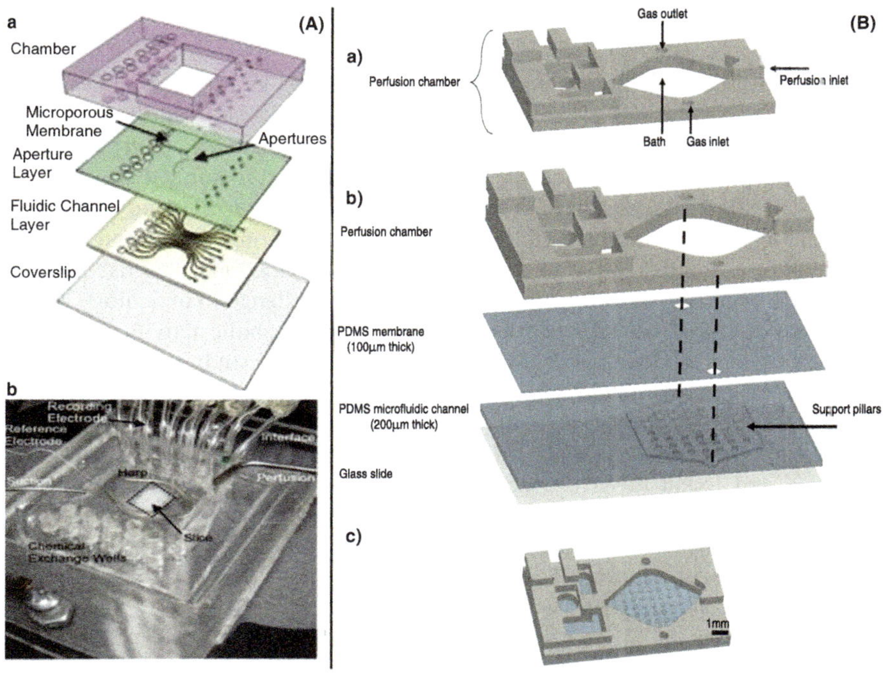

Fig. 8.2 (**A**) (*a*) Schematic of microfluidic multi-electrode array (MMEA) device showing the different layers. (b) Photograph of MMEA during operation. [20]. (**B**) Schematic of device developed by Mauleon et al. (*a*) perfusion chamber having gas inlet and outlet. (*b*) different parts of the device: the perfusion chamber, the PDMS membrane, the PDMS microfluidic channel, and a glass slide and the alignment marks show how the gas is supplied to the device (*c*) *top view* of device [21]

slice via MF gas channels [21]. The device consists of four independent parts: the perfusion chamber, the PDMS layer, the PDMS MF channel, and a glass slide (Fig. 8.2B). The designed microchannel provides rapid and efficient control of oxygen and can be further modified accordingly to allow the various regions of the slice to experience different oxygen conditions. Using this novel device, stable and homogeneous oxygen environment throughout the brain slice has been obtained, and the oxygen tension in a hippocampal slice can be altered rapidly. It was observed that the device allows more complete temporal control and can reach greater differences in oxygen concentration as compared to the standard perfusion method. Although a lot of research is being conducted for exploring the neurological processes seeking the help of MF technologies, still efforts are in their initial stage and scope of MF in neuroscience is endless.

2.3 Drug Development

The drug development process comprises of two phases i.e., drug discovery and drug testing. The first phase includes the target selection, lead identification, and preclinical studies, while the development stage includes clinical trials, manufacturing, and product lifecycle management [22]. Miniaturized MF devices being small sized in nature are emerging as useful tools for the study of target selection, lead identification and optimization and preclinical test and dosage development [23]. Recreation of complex pharmacological and pharmacokinetic interactions occurring in living organisms is a major challenge in toxicology as the toxic effect of drugs in one tissue often depends on the metabolic activity of another tissue. Caviglia et al. developed a MF cytotoxicity assay for studying the impact of anticancer drugs doxorubicin and oxaliplatin (Fig. 8.3A) [24]. The targeted drug delivery tested the cytotoxicity and was evaluated using real-time impedance monitoring. The time-dependent effect of doxorubicin on the HeLa cells was monitored and found to have a delayed onset of cytotoxicity in MF compared with static culture conditions. Although, the cell-based *in-vitro* tests, can provide useful preliminary data but animal tests are still required to verify the pharmacological properties of the drug. To mimic these interactions, organ-on-chip devices i.e. MF system containing a network of interconnected chambers have proved to be very useful. Sung et al. developed a device using hydrogel cell culture for pharmacokinetics–pharmacodynamics (PBPK) studies of the three cell lines that represents the liver, tumor and marrow for testing drug toxicity [25]. These cell lines were grown in the three-chambers of device to test the toxicity of an anticancer drug, 5-fluorouracil. The device is assembled by sandwiching a cell culture chamber layer and a fluidic channel layer between a bottom aluminum frame and a Plexiglass top frame (Fig. 8.3B). Gravity-induced flow achieved the recirculation of cell culture medium and tilting of the device results in the flow of liquid from one well to the other well, where the flow rate through the conduits was calculated. The rocking platform changes the angle, and the medium flows in the opposite direction in almost 3 min. The gravity induced circulation offers several advantages over pump circulation like reduction of the total space occupied by the device, lowering of cost, elimination of the possibility of unwanted binding, and prevention of air bubble formation. Prabhakarpandian et al. designed a synthetic tumor to test in vivo delivery efficiencies of the drug vehicles [23]. A pre-digitized MF-based artificial vasculature assay was used to mimic the tumor microenvironment for in vivo observation. The MF device recreates the in vivo tumor microenvironment of circulatory flow in the vessels, transport across the leaky vessel walls between the vascular and the tumor cells, and delivery to the 3D culture of tumor cells across the interstitial space. The combination of these features distinguishes the present synthetic tumor network model from other in vitro models. Several organs on chips devices are reported to be utilized for drug development purpose (discussed in the preceding subsection).

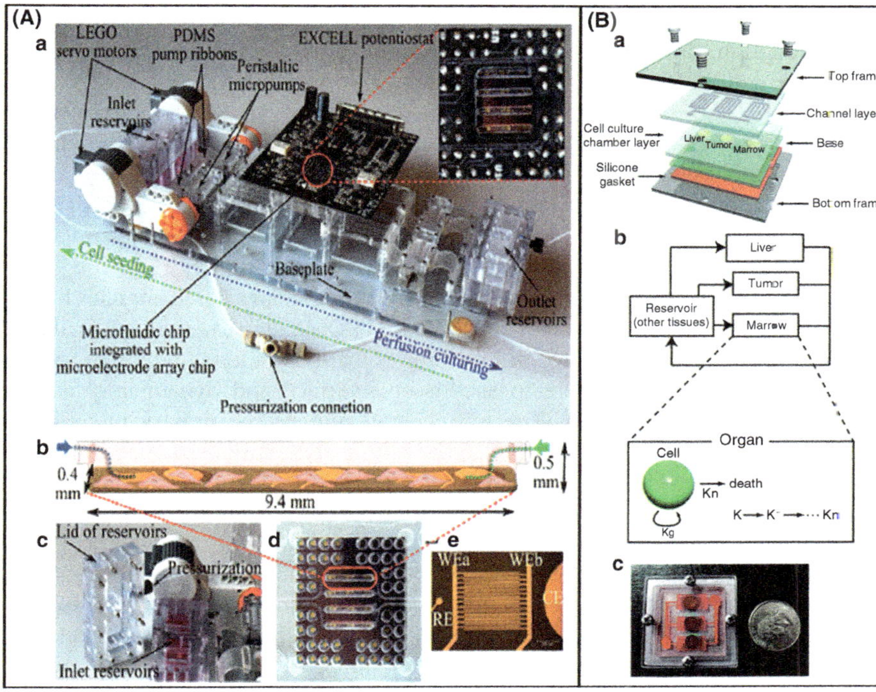

Fig. 8.3 (**A**) (*a*) Photograph of the fully integrated platform showing its main components, including the EXCELL potentiostat with a window for microscopic observations (inset). (*b*) Schematic representation of an individual microfluidic chamber in the cell culture/detection unit containing three electrode arrays. (*c*) Photograph of the reservoirs, showing the lid embedding the gas connection used for pressurization. (*d*) Photograph of the microfluidic cell culture/detection unit (*e*) Microscopic image of a single electrode array showing the counter (CE), reference (RE), and working (WEa and WEb) electrodes [24]. (**B**) PBPK model of Sung et al. consisting of (*a*) schematic of device fabrication (*b*) device operation steps (*c*) photograph of the device [25]

2.4 *Organs on Chips*

These are the MF devices which are used for culturing living cells in micrometer-sized chambers having a continuous supply of nutrients that mimic the physiological functions of tissues and organs. Minimal functional organs that can recapitulate tissue or organ level features are built so that 'near to real' conditions are available for various experiments. Simplest system is a single, perfused MF chamber containing only one kind of cultured cells exhibiting functions of one type of tissue, e.g., hepatocytes or kidney tubular epithelial cells [26, 27]. These systems are utilized to study the organ-specific responses, different chemical stimuli (drugs, toxins or other chemicals) and physical stimuli (physiologically relevant levels of fluid shear stress, cyclic strain, and mechanical compression) [7, 27]. Attempts have been made to micro-fabricate models of blood vessels, muscle, bones, airways, liver, brain, gut, kidney, lung, and heart [28]. For analysis of kidney transport barrier function, a simplified kidney model was created by stacking two microfabricated PDMS chambers, separated by a porous thin membrane [29]. The apical channel is separated from a bottom reservoir by an ECM-coated porous membrane. The presence of the physiological level of apical fluid shear stress helps in culturing the primary human proximal tubule epithelial cells on the membrane. The second compartment (basolateral compartment) is readily accessible for fluid sampling and addition of test compounds to study the active and passive epithelial transport. The dimensions of the MF channel were 1 mm wide, 1 cm long and 100 μm high. The bottom structure beneath the MF channel is rectangular 1 mm wide and 0.6 cm long which is made from a cured PDMS slab to serve as a medium reservoir directly. This fabricated design mimics the living kidney proximal tubule in terms of natural architecture, tissue–tissue interface and dynamically active mechanical microenvironment. The top channel mimics the urinary lumen and has fluid flow, whereas the bottom chamber mimics interstitial space and is filled with media. Primary human proximal tubular epithelial cells were cultured under flow conditions in the MF chip device. On reaching the confluence, the cells are exposed to circulating culture medium at a fluid shear stress of 0.2 dyne cm^{-2} for 3 days using a syringe pump. After 3 days of culture under both fluidic and static conditions the immunofluorescence, microscopic analysis of the proximal tubular epithelial cells revealed a well defined confluent epithelial monolayers which are continuously lined by a linear distribution of the tight junction (Fig. 8.4A). In contrast, exposure of cells to physiological fluid shear stress (0.2 dyne cm^{-2}) results in restoring the cells to normal columnar form and the height is increased by almost twofold (Fig. 8.4B). In a typical MF "alveoli-on-a-chip" setup there is a cyclic propagation of a meniscus over a flexible PDMS membrane which recreates a combined stresses and provides a more physiologic recreation of the stresses, including cyclic distribution of air-liquid interface and wall stretch. During the process, the alveolar chamber can be partially filled with fluid and positioned in the vertical configuration to establish a meniscus at the interface of fluid and air. By withdrawing fluid from the "actuation channel", the membrane can be forced to deform and relax stretching cells and propagating the meniscus over a specified cell region (Fig. 8.4C) [30]. The MF system provides the first in vitro technique to study the role of both solid and fluid mechanical forces in ventilator-induced lung injury systematically.

In another example of lung-on-a-chip a microfabricated lung mimic device was created which uses compartmentalized PDMS microchannels to form an alveolar-capillary barrier on a thin, porous, flexible PDMS membrane coated with ECM [31]. Soft lithography techniques were used to fabricate these hollow microchannels which are conjugated with a new method that uses chemical etching of PDMS to form the vacuum chambers. The fabrication starts with the alignment and permanent bonding of a 10-μm thick porous PDMS membrane (10-μm wide; pentagonal pores) and two PDMS layers containing recessed microchannels (Fig. 8.4D). The entire integrated device was only 1–2 cm in length, with the central channels only millimeters in width, thus making it entirely amenable to

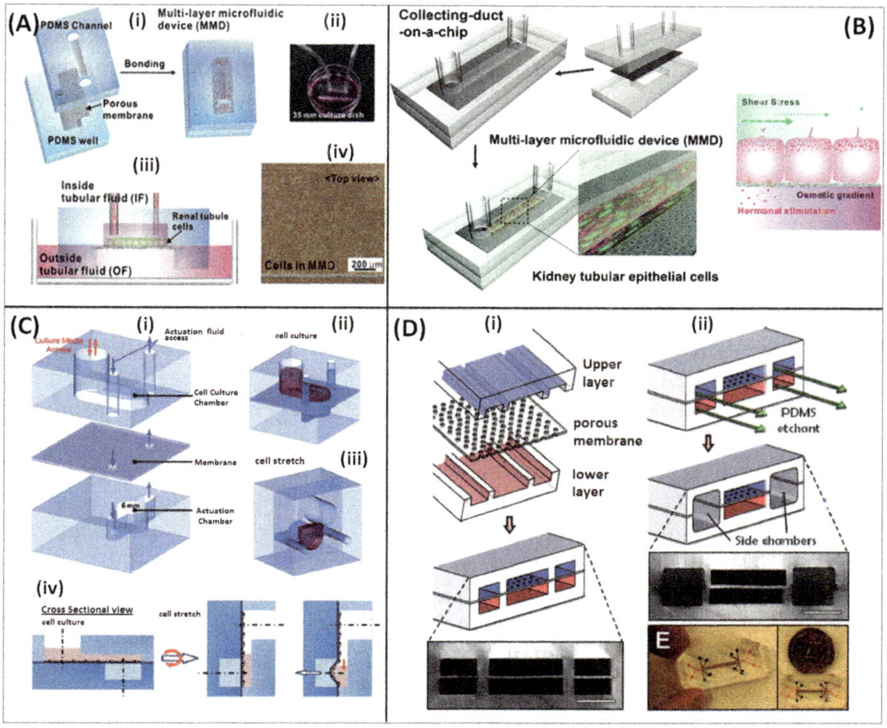

Fig. 8.4 (**A**) and (**B**) Fabrication and operation of a multi-layer microfluidic device (MMD) for kidney on chip application (*i*) fabrication steps (*ii*) image of *top view* of device (*iii*) schematic showing *lateral view* of device (*iv*) SEM image of the cultured cells. Kidney on chip application (*i*) fabrication steps (*ii*) image of *top view* of device (*iii*) schematic showing *lateral view* of device (*iv*) SEM image of the cultured cells. [29]. (**C**) Microfluidic "alveoli-on-a-chip". (*i*) assembly of the components (*ii*) *horizontal* tubules show direction of culture media F-12 K (*iii*) liquid–air interface formation at the alveolar surface (*iv*) A cross-sectional view of the microfluidic device shows the *horizontal* orientation for cell culture and the vertical configuration for experimentation [30]. (**D**) Lung on chip device. (*i*) schematic of fabrication of device. Three PDMS layers are aligned and irreversibly bonded to form two sets of three parallel microchannels (*ii*) After permanent bonding, PDMS etchant is flowed through the side channels. Selective etching of the membrane layers in these channels produces two large side chambers to which vacuum is applied to cause mechanical stretching. (**E**) Images of an actual lung on-a-chip microfluidic device viewed from above [31]

high-density integration with a highly multiplexed microdevice. The microchannels were separated by a 10 μm-thick PDMS membrane containing an array of through-holes with an active diameter of 10 μm. The human alveolar epithelial cells and microvascular endothelial cells are attached opposite to surface of the 3-D extracellular matrix coated membrane (Fig. 8.4D). The device recreates physiological breathing movements by applying vacuum to the side chambers and causing mechanical stretching of the PDMS membrane forming the alveolar-capillary barrier. Epithelial and endothelial cells are grown on either side of a porous thin membrane that is exerted by airflow and fluidic flow respectively via the two channels separated by the membrane. The mechanical forces are operated by two channels beside the main channel. Though complex geometries have been applied to MF system, the models are still small and simple. This device provides a comprehensive tool for lung disease research and related drug screening.

Recently brain on a chip has been reconstructed on MF chip using co-cultured endothelial cells and astrocytes by combining the fluidic shear stress and thin membrane (Fig. 8.5A). This fabricated blood–brain barrier (μBBB) model addresses the previous limitations of fluidic shear stress (static transwell models) and thin dual cell layer interface (conventional dynamic in vitro BBB model). The results indicate that this proposed model might be a valid prototype for simulating the function of μBBB. The fabricated multi-layered microfluidic μBBB device consists of four PDMS substrates, two glass layers, and in the center between the PDMS layers there is a porous polycarbonate membrane (Fig. 8.5A). To ensure the laminar flow the height of the channel is 200 μm, and the width at the cell culture interface is 2 mm (luminal) or 5 mm (abluminal). The porous membrane located at the channel junction has an area of 10 mm^2. To introduce dynamic flow the assembled device houses two perpendicularly crossing channels with a porous membrane at the intersection of the flow channels for cell culture, and to monitor trans-endothelial electrical resistance (TEER) multiple embedded electrodes are introduced across the barrier. Opposite to the membrane on each side are two sets of two AgCl thin-film TEER electrode pairs forming a four-point sensing structure. The fabricated μBBB platform was sterilized and adhesion seeded by steadily perfusing for up to 7 days. The developed μBBB successfully mimicked the dynamic cerebrovascular environment with fluid shear stress. Further, this μBBB model can be effectively used to monitor the changes in the barrier function such as barrier-enhancing or barrier opening drugs [31, 32].

Researchers have also fabricated an MF cell culture device mimicking the microscopic structure in liver tissue (hepatic cords) (Fig. 8.5B). The MF device consists of a medium flow channel with a width of 100 μm and a height of 30 μm, with a cell loading channel (200 μm width and 30 μm height), a cell culture area (37 μm width and 30 μm height), and an endothelial-like barrier [35]. For the smooth alignment of the hepatocytes in two lines, the tip of the cell culture area was asymmetrical design with a width of 37 μm which can accommodate two cells side-by-side. To avoid the deformation of the cells the cross-sectional area of the slits was minimize, (2 μm wide and 2 μm high). For efficient simulation of the flow in the device COMSOL MULTIPHYSICS COMSOL, Inc. was used (Fig. 8.5B). The obtained simulation results show that the flow velocity in the medium flow channel is 1 mm s^{-1}, which is nearly same as the velocity of blood flow in vivo. Under perfusion condition, the flow velocity in the cell culture area is 0 mm/s whereas the flow rate determined is 0.1 μL min^{-1}. To further, avoid the greater fluidic resistance of the endothelial-like barrier in comparison to the medium flow channel both the cell inlet and cell outlet were sealed [33].

A recent human gut-on-a-chip (GoC) model tried to build a more physiologically relevant in vitro model of the human intestine that undergoes peristalsis,

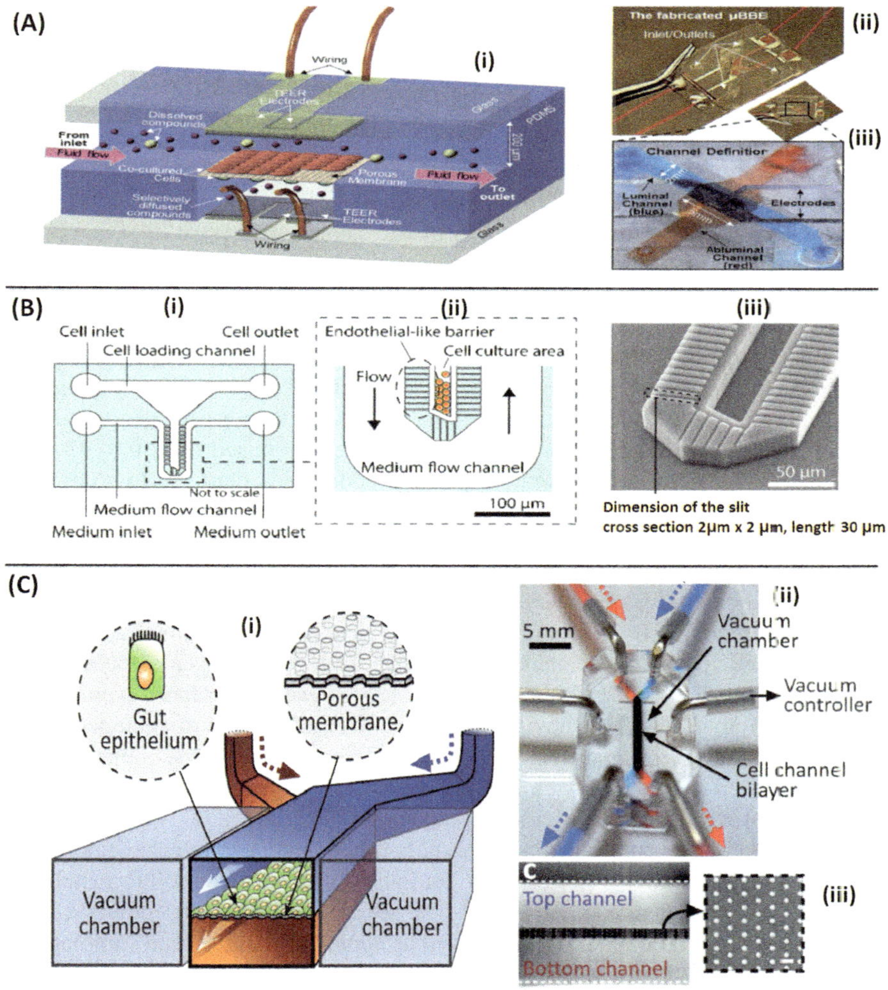

Fig. 8.5 (**A**) Structure and design of the developed µBBB. (*i*) The µBBB system comprises two perpendicular flow channels. (*ii*) The fully fabricated µBBB chip. (*iii*) Close-up view [32]. (**B**) (*i*) and (*ii*) Design of a MF device mimicking the structure of a hepatic cord. (*iii*) SEM image of the device [33]. (**C**) A schematic of the GoC device. (*i*) A schematic of the gut-on-a-chip device showing the flexible porous ECM-coated membrane lined by gut epithelial cells cross horizontally through the middle of the central microchannel, and full height vacuum chambers on both sides. (*ii*) A photographic image of the device. (*iii*) A cross-sectional view of the top and bottom channels (both 150 µm high) of the gut-on-a-chip; square inset shows a top view of a portion of the porous membrane [34]

experiences fluid flow and supports the growth of microbial flora without compromising human cell viability [34, 36]. The GoC' microdevice composed of two MF channels that are separated by a flexible porous membrane coated with extracellular matrix (ECM) and are lined by human intestinal epithelial (Caco-2) mimicking the physiology of human intestine. To create, the gut microenvironment the fluid is flown at a lower rate (30 μL h^{-1}) that results in low shear stress (0.02 dyne cm^{-2}) over the microchannels. These conditions lead to the change in columnar epithelium that polarizes rapidly and spontaneously into folds resulting in the recapitulation of the structure of intestinal villi and the formation of high integrity barrier to small molecules that mimics whole intestine than cells in cultured static Transwell models [36]. A cyclic suction controlled by a computer vacuum manifold was used on both sides of the microchannels that repeatedly stretch and relax the elastic ECM-coated porous membrane. The created rhythmic mechanical deformations of the epithelial cell monolayer are similar to those caused by peristaltic motions of the human intestine, (Fig. 8.5C). Under this condition, the phase contrast microscopic analysis of cell shape in human intestinal epithelial monolayers increased linearly (0–30 %) for both substrate distortion and cell deformation when the level of suction pressure was raised from 0 to 45 kPa. Thus, the GoC microdevice provides a micro controlled- platform to study and perturb critical gut functions in the presence of relevant physiological cues, including cyclic mechanical strain, fluid flow, and coexistence of microbial flora.

An attempt has also been made to constitute a four organ chip system for studying the homeostasis between the organs recapitulated [37]. The authors tried to reconstruct kidney, liver, small intestine and skin equivalents on a multi-organ chip and conducted studies related to barrier integrity, continuous molecular transport against gradients and metabolic activity [28]. Researchers have even microfabricated 'Body-on-a-Chip' devices that contain multiple micro-chambers each containing cultured cells of different origin (e.g., liver epithelium vs. brain neurons) that is connected to a network of micro-channels which permit the recirculation and exchange in a physiologically relevant manner [38, 39].

The technologies promise to have the potential to be more relevant functional models for testing toxicity and efficacy, and allow a better insight into metabolic activities at the tissue and organ levels. The several organ on chip devices, their fabrication and operation parameters as well as applications have been summarized in Tables 8.1 and 8.2.

2.5 Single Cell Analysis

Considerable biochemical heterogeneity exists among cells of the same type caused by many mechanisms including damage, mutations, stages in the cell cycle, and differential exposure to external stimuli [50]. To better understand how and why these differences arise is crucial in cell biology and early detection of disease conditions. Many fascinating examples have been reported in the literature showing the cell-cell variability and its significance to biological phenomena. For example, the transcription events in mammalian cells are affected by random fluctuations, leading to significant variations in mRNA copy numbers. In a clonal population of

Table 8.1 Different organ on chip devices and their applications

Organ on chip	MF Platform used	Application	References
Kidney, brain, heart, lung and liver	Physiologically-based pharmacokinetic (PBPK) model	ADME profiling and quantification of the amount of drugs in different parts of the body	[25, 40]
Gastrointestinal tract and liver	Microscale cell culture analog (μCCA)	Evaluating nanoparticle toxicity and interactions with tissues	[41]
Liver, tumor and marrow	Pharmacokinetic-pharmacodynamic (PK-PD) model combined with a μCCA	Testing drug toxicity and improve insights into the drug's mechanism of action	[25]
Gastrointestinal tract and liver	Gut-parallel tube model	Investigating paracetamol first pass metabolism in intestine and liver	[42]
Intestine, liver, skin and Kidney	Four-Organ-Chip	ADME profiling and toxicity testing	[37]
Liver, colorectal tissues	96-well format-based microfluidic platform	Testing effects of different concentrations drug in several tissues	[43, 44]
Lung, gut	PDMS-based organs-on-chip	Prediction of clinical responses in humans	[28, 45]

mouse multipotent progenitor cells, the cell-cell heterogeneity is found to be connected with cell-fate decisions [51, 52].

The analysis of single cells, however, presents a variety of challenges including micrometer sizes of cells, presence of macromolecules of interest (mRNA and proteins) in low copy numbers, etc. Therefore, there is an urgency for techniques which can handle such small volumes without significant dilution and can be directly integrated with ultrasensitive detection schemes [53]. Moreover, cells must be treated gently before analysis so as not to perturb the biochemical pathways or molecules of interest. Finally, although one wants to examine only one cell at a time, many individual cells need to be analyzed rapidly to understand the statistical distribution of a particular analyte in the cell population, so that the potential disease biomarkers can be identified.

Analysis of single cell can be of two types: whole cell analysis and analysis of cell lysates. The most critical step in both is the development of techniques that can reproducibly transport cells to precise locations for further analysis [29]. Various on-chip culture, immobilization methods, and cell trapping techniques were applied for the purpose. Ideally, these methods should be amenable to parallel formats, have high throughput capability, limit dilution, and be robust. The reported single-cell manipulation techniques include: (a) hydrodynamic flow and focusing [54]; (b) use of on-chip valves and pumps to direct cell transport [55]; (c) incorporation of cells into MF droplets [56]; (d) optical and optoelectronic trapping of cells [57]; (e) dielectrophoretic trapping of cells [58], and; (f) geometrical trapping of cells [59–61] (Fig. 8.6A) [62]. The hydrodynamic focusing takes advantage of the fact that the Reynolds number for fluid flow in MF devices is generally <0.1, so the fluid flow is laminar [64]. Thus, through careful control of flow rates by multiple pumps

Table 8.2 List of different organ on chip devices with few fabrication and operation parameters

Organ on chip	Type of cells cultured	Organ level specific function mimicked	Platform	Rate of fluid flow	Shear stress	References
Heart	Cardiac myocytes	Contractility and tissue structure	Mascular thin film based single channel fluidic chip with metallic base	1 mL min^{-1}	–	[46]
Liver	Hepatocytes, hepatic stellate cells	Improved cell-cell interaction, controlled fluid flow rate	Bichambered MF device with fluid flow controlled by osmotic pump	94.7–304.7 μL per 12 h	–	[47]
Kidney	Primary epithelial cells	Low shear stress conditions, tubular morphology	MF device consisting of two compartments separated by porous polyester membrane		0.2 dyne cm^{-2}	[48]
Lung	Alveolar epithelial cells, pulmonary endothelial cells	Pressure driven stretching occurring in alveolar capillary interface during breathing	Flow stretch strip having two closely apposed microchannels separated by a 10 μm porous PDMS membrane	20 μL h^{-1}	–	[31]
Intestine	Intestinal epithelial cells	Peristaltic movements of intestine, microbial growth in stomach	MF device consisting of two compartments separated by porous polyester membrane	10–100 μL h^{-1}	0.002–0.08 dyne cm^{-2}	[34]
Brain	Brain endothelial cells	Blood brain barrier function of brain	Bilayered membrane based MF device	2.5 mL h^{-1}	0.58 Pa	[49]

and pressure sources, one can focus the cells to the desired location. Single cells were isolated in high-density arrays composed of two channel height levels (Fig. 8.6B). The larger 40-μm channel height served as the main fluid conduits for cell solutions while the 2-μm-height regions were used to form elevated trapping regions. Having a 2-μm gap allows a fraction of fluid streamlines carrying cells to enter a trap. Once the cell enters a trap and partially occludes the 2-μm gap, then there is a reduction in the fraction of fluid streamlines (and cells) entering the trap region which leads to a high quantity of single-cell isolation [59]. The probability of trapping was found to be dependent on the number of cells previously entrapped. For on-chip valves based focusing PDMS-based valves and pumps configured within the device a pressure-driven, feedback-based dynamic trap using on-chip sieve valves that are positioned at both ends of the sensing channel and can selectively capture a cell without replacing the solution (Fig. 8.6C) [63]. The small channel dimensions and laminar flow in MF channels also enable a variety of immiscible droplet-based generation technologies to be developed [65]. Droplets are generated by shearing a dispersed aqueous phase with a constant

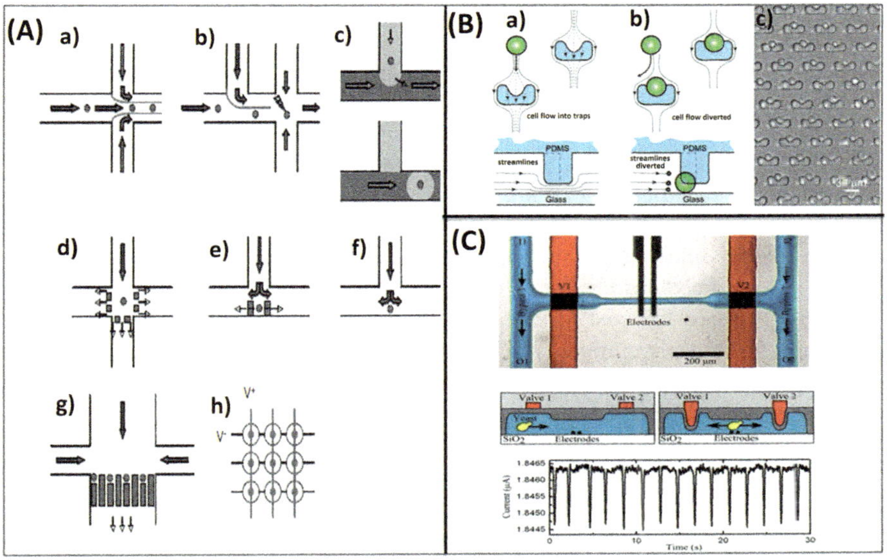

Fig. 8.6 (**A**) Schematic of the various types of methods reported for transporting and manipulating cells. (*a*) Cells can be hydrodynamically focused at an intersection so that they all flow past an interrogation region in single file. (*b*) Cells can be hydrodynamically transported to an intersection, where they can be lysed either electrically or chemically and the lysate injected into a separation channel. (*c*) Cells can be incorporated into aqueous droplets within an immiscible fluid and hydrodynamically transported to an interrogation region. Cells can be trapped in an interrogation region geometrically using pillars (*d*), weirs (*e*), or hydrodynamically balanced flows (*f*). Arrays of cells can also be trapped using filters (*g*). Dielectrophoretic trap for cells (*h*). The gray circles represent cells and the arrows indicate the direction of fluid and cell flow [62]. (**B**) (*a–c*) Hydrodynamic cell sorting method [59] and (**C**) cell sorting based on pumps and valves on chip [63]

oil phase. In optical and electrical sorting of cells, high-frequency electrical fields create forces which move cells into potential wells specified by the placement of the electrodes or the use of light patterns that change the conductivity of a conducting surface. Similarly, the electrical forces can be used in dielectrophoresis to direct the cell to a particular location. After single cell sorting, the analysis of cell components can be done using cell analysis techniques.

Single cell analysis (SCA) reduces the biological noise, and it provides fundamental improvements in experimental design and data analysis for applications predicated on single cells. Stem cells, for example, hold great potential for regenerative medicine as they are capable of self-renew and can divide along different lineages. However, the other cells like the embryonic stem cells, adult stem cells, and induced pluripotent stem cells are all heterogeneous populations. SCA can target specific populations and, therefore, elucidate signaling pathways and networks for self-renewal and for differentiation [66].

3 Microfluidics in Diagnostics

Biomedical diagnostics has been an important application area of MF technologies. The unique features of MF technology make it naturally suitable for the fabrication of PoC testing devices. Till date, some prior DNA separation techniques and diagnostics have been successfully miniaturized [67, 68]. MF seeks to overcome the difficulties or challenges in traditional assays in medical diagnoses such as cell-based assays, disease screens and drug screens. Many MF systems are integrated with sensing modules or sample-pre-treatment modules, which increases the efficiency of the assays and reduces cross-contamination. These advantages have potentially lowered the cost of the assays as well as provided a faster diagnosis. Integration of multiple components/functionalities (sample preparation, detection, data processing) on a simple to use single device has enabled their use by non-specialized person. Moreover, the parallel analysis allows multiple tests to be run simultaneously, either in the same sample or multiple samples. A variety of samples including blood, urine, saliva, stool, and plasma, amniotic and cerebral fluids have been used for diagnosis in PoC devices [69]. Figure 8.7 demonstrates different steps of a general analytical procedure involved in a MF diagnostics device.

Various materials like glass, silicon, and polymers have been used for the fabrication of various MF devices, but recently, the application of paper in MF device development is emerging as a promising format for PoC devices. MF paper-based analytical devices or μPADs as coined by Whitesides are thin, small, cheap, flexible and can easily be disposed of thus, making it an ideal platform for healthcare applications [70]. The PoC devices are designed to detect specific biomolecules like DNA, RNA, whole cell or biomarkers, etc. specific to a particular disease. The paper-based assay is suitable for measuring multiple samples in

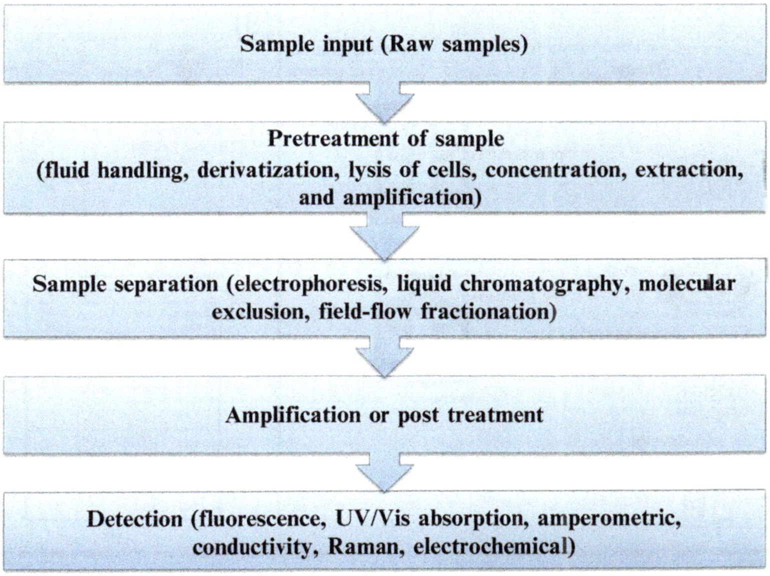

Fig. 8.7 Scheme showing different steps involved in a general procedure for operation of a MF diagnostic device

parallel and in a relatively short period. In one run the researchers were able to run 20 different samples (glucose and BSA) within 7.5 min (another 10.5 min for the color to fully develop). Thus, an 18 min assay is enough for measuring two analytes in 20 different samples (Fig. 8.8A) [71]. In another work Bercovici et al. demonstrated an experimental and analytical study of a novel paper-based analytical device for sample focusing using isotachophoresis (ITP) [72]. The authors used 2.5 mm wide channels and described the processing of a 30 μL sample in several minutes. It was observed that the dispersion was much significant in the paper as compared to glass, and the peak enhancement on the order of 1000-fold in several minute by substantial sample focusing (Fig. 8.8B). Thus, obtaining high sample concentrations in paper results in enhanced reaction kinetics and creating low-cost devices with much-enhanced sensitivity. Further, this paper-based ITP can process large sample volumes, but the small dimension of the microchannels limits their application to the analysis or processing of small sample volumes. The realization of ITP in larger channels or larger diameter capillaries is challenging due to hydrodynamic instabilities and excessive Joule heating. Therefore, these papers based ITP can be helpful in detection of extremely dilute samples (e.g. detection of bacteria at 10–100 copies per mL). In the next section we will discuss the potential application of PoC devices for molecular diagnosis and immunodiagnostics.

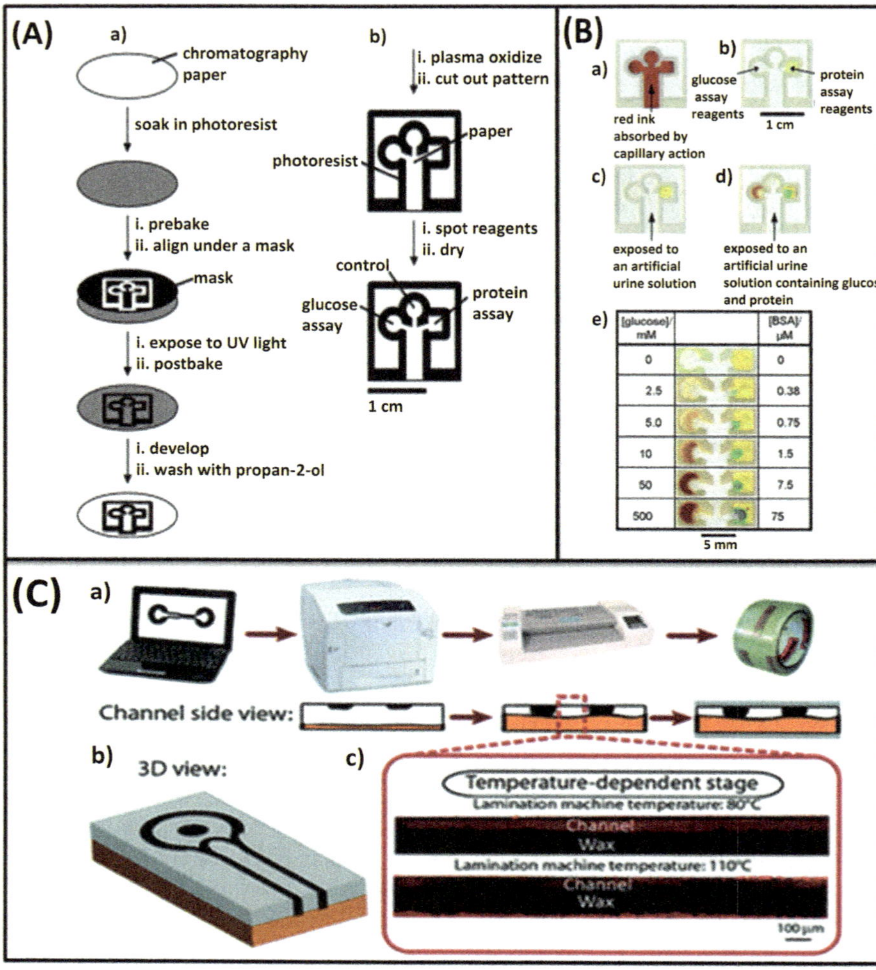

Fig. 8.8 (**A**) Diagram depicting the method for patterning paper into millimeter-sized channels: (*a*) Use of photolithography to pattern SU-8 photoresist embedded into paper; (*b*) modification of patterned paper for bioassays. (**B**) Images of patterned paper (*a*) After absorbing Waterman red ink (5 mL) by capillary action. (*b*) After spotting the reagents. (*c*) Negative control for glucose (*left*) and protein (*right*) by using an artificial urine solution (5 mL). (*d*) Positive assay for glucose (*left*) and protein (*right*) by using a solution that contained 550 mm glucose and 75 mm BSA in an artificial urine solution (5 mL). (*e*) Glucose and protein detection assays by using varying concentrations of glucose and BSA [71]. (**C**) (*a*) Schematic illustration of the multistep fabrication process. (*b*) 3D illustration of the resulting structure. (*c*) Raw fluorescence images of the paper cross section showing the effect of lamination temperature on the penetration of wax into the paper, resulting in control of the channel depth [72]

3.1 Molecular Diagnostics

Numerous MF systems have been used for the analysis of various molecules including DNA, RNA and other chemicals for general purpose as well as for disease detection. A large number of devices used for diagnosis of disease including pathogen detection have reported MF as the general theme of fabrication [73]. Most of the currently designed devices focus on single molecule or single disease identification but the future lies in the fabrication of multiplexed MF devices that can screen hundreds of diseases simultaneously. For detection of bacteria, and virus two methods are commonly used i.e. immunosensing (discussed in next subsection) and nucleic acid based detection. The later detection method employed four fundamental steps (1) designing of a MF platform, (2) lysis of the target bacterial or viral cells, (3) target DNA and RNA purification and amplification, and (4) detection of the target analyte using various transducers. Malhotra et al. has fabricated impedimetric microfluidic–based nucleic acid sensor for quantification of DNA sequences specific to cancer. The MF chip was prepared by patterning an indium–tin–oxide coated glass substrate followed by sealing with PDMS microchannel (Fig. 8.9) [74]. An integrated MF device used by Dimov et al. for tmRNA purification and nucleic acid sequence-based amplification [75]. The device consists of two separate chambers one for RNA purification and other for nucleic acid amplification. The device efficacy was demonstrated by integrating on-chip purification, amplification, and real-time detection of 100 *Escherichia coli (E. coli)* bacteria in 100 mL of crude lysate. The device took less than 30 min for on-chip purification, amplification, and real-time detection of 100 *E. coli* bacteria in 100 mL of crude lysate, thereby, demonstrating the device efficacy. Recently, Chang et al. detected live bacteria from the human joint using an MF platform based on ethidium monoazide (EMA) [76]. The detection process was based on utilization of labeled gold nanoparticles. The limit of detection was 10^2–10^4 CFU for typing bacteria by an on-chip PCR. The system overcomes the problem of human contamination by use of an integrated MF system replacing the usage of RNA sample by EMA to distinguish live and dead bacteria. Lee et al. reported 3D-printed MF device for rapid and facile detection of *E. coli* in milk [77]. They used stereo-lithography to fabricate a vertically designed helical micro-channel around a cylindrical chamber. UV–Vis spectroscopy was used as the detection method and the sensitivity achieved was 100 CFU mL^{-1}. Boehm et al. developed a simple and rapid biosensor with immobilized monoclonal antibodies for the identification and detection of bacteria using an MF LoC [78]. The MF chip utilizes impedance-based measurement and showed that the sensor could detect 9×10^5 CFU mL^{-1} *E. coli* in the solution. The sensitivity of the chip with immobilized bacteria is governed by the height of the sensing chamber. The selectivity of the sensor to different bacterial strains was demonstrated by the identification of *E. coli* in a suspension of *E. coli* and *M. catarrhalis*. Specific extraction of DNA of Hepatitis B virus has been done and several other reports are there for MF based detection of different viruses and bacteria [79].

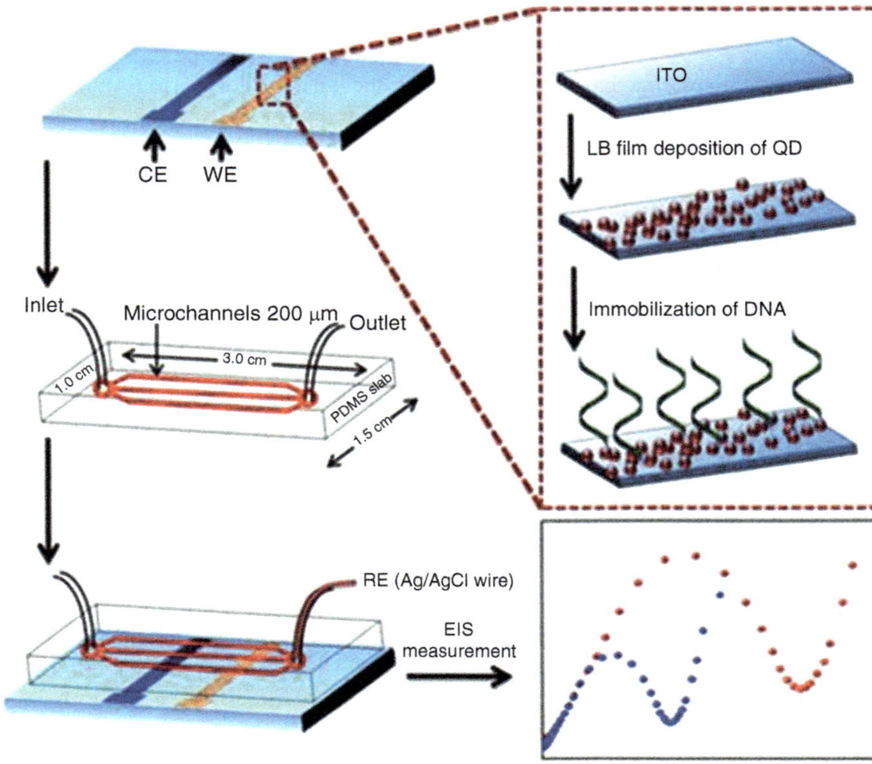

Fig. 8.9 Different steps showing the fabrication of MF biochip for electrochemical detection of DNA hybridization. *WE* working electrode, *CE* counter electrode, *RE* reference electrode [74]

Other than pathogen detection the MF PoC devices include health care devices for monitoring human health from analyzing different molecules from human blood. Blood chemistry analysis is perhaps the most successful application of these kinds of devices and the examples include i-STAT, Bioanalyzer 2100 from agilent Technologies, epoc from Epocal etc. [73, 80]. These devices provide information about various ions like Na^+, K^+, Cl^-, blood pH, nitrogen content and oxygen analysis in very short time duration [81].

3.2 *Immunodiagnostics*

Immunoassays are probably the most important protein analysis technique. These are based on detection of either antigen or cytokines. Antibody-antigen based detection is utilized for the detection of various diseases like hepatitis and autoimmune disorders whereas cytokines based assays are utilized for the studies of diseases like Alzheimer's and cancer. They have proven to be a breakthrough technology that has rendered the analysis of proteins after the work of Yalow and Berson, who employed radioimmunoassays to determine peptide hormones in the 1950s [82]. Immunoassays have since then evolved considerably. The sensitivity limits have been decreased to picomolar concentrations using monoclonal antibodies, labeling techniques, and signal transduction and data acquisition devices. The immunoassays are increasingly ported onto MF formats. MF bioassays have been used in routine analysis of clinically observed symptoms, frequent monitoring of disease progression and continuous assessment of therapeutic efficacy. Two major technologies that are being currently used to fabricate biarrays are (1) microspot array and (2) mosaic array. The most commonly developed immunoassay method is the spot array. It is a simple method whereby many probe spots are printed onto a chip substrate. This bioarray is then tested against a single sample, which flows on top of the array. With this method, a large amount of data can be obtained from using a single sample [83]. The mosaic array is an immunoassay method that allows for multiple probes to react with various samples [84]. Rather than having a single sample react with multiple probes, this method has several samples tested against multiple probes simultaneously. Such robust analysis has the potential of early; possibly even presymptomatic, diagnosis of a disease. MF immunoassays can be classified into two homogeneous or heterogeneous assays. In homogeneous assays, the probes and the samples will be mixed in liquid phase; this method requires a separation step to detect the presence of the reaction products. In heterogeneous assays, the samples in the liquid phase interact with the probes immobilized on a solid phase. While both of these assays are in practical use, heterogeneous assays are much more widespread due to the ease of washing away unbound particles from the solid substrate. Heterogeneous immunoassays usually take the form of the ELISA [83]. Two major areas where immunoassays are now playing a vital role for analysis are cancer detection and cardiovascular biomarker detection. In these fields MF immunoassays have now enabled early diagnosis of disease, resulting in the continuous monitoring of the progress of disease and response to therapy. Legendre et al. reported work on the design and development of an MF device for diagnosis of T-cell lymphoma [85]. The system accepts a whole blood sample as the input, extracts the DNA, amplifies target sequences of the T-cell receptor gene, and electrophoretically resolves the products for detection of a signature consistent with monoclonality. Diercks et al. fabricated a MF device that measures multiple proteins (tumor necrosis factor, CXC chemokine ligand 2, interleukin 6 and interleukin at $pgmL^{-1}$ concentrations in nanoliter volumes [86]. Antibody-coupled polystyrene microspheres labeled with embedded fluorophores were used to detect the analyte (proteins). Optical detection of captured analyte was performed off-chip using a confocal microscope, which proved to be a disadvantage in terms of device portability.

3.3 Commercial Diagnostics

As the global market for PoC diagnostics is expected to grow a value of US$ 24 Billion by 2018 the potential application of MF towards the production of practical devices raises enormous excitement among the researchers [87]. Present diagnostic techniques mainly focus on glucose monitoring, blood chemistry, and electrolyte, pregnancy and fertility, drug and alcohol, cholesterol, hemoglobin/hemostasis, urine chemistry and to some extent infectious disease surveillance. The major areas of growth include HIV testing, drugs of abuse, cardiac and tumor markers detection and diagnostics for infectious diseases. There is a long list of companies involved in the fabrication of MF based devices for both PoC analysis as well as general laboratory tools. Some of which along with the products are listed in Table 8.3.

The i-STAT handheld system was among the first commercially successful LoC based products. i-STAT is a portable handheld system result of an integration of miniature fluidics and electrochemical detection designed for analysis of electrolyte

Table 8.3 List of various PoC devices in the market from different companies [81]

Product	Company	Application and highlights of technology
i-STAT	Abbott	Blood chemistry analysis system, Portable analyzer capillary-driven microfluidics, thin-film electrodes for detection
Bioanalyzer 2100	Agilent Technologies	Cellular and molecular analysis, chip based device utilizing principles of flow cytometry and electro-phoresis based assay for analysis
Piccolo xpress®	Abaxis	Blood chemistry analyser, compact analyzer, injection-molded plastic discs, no sample pre-processing
ACIX 100	Achira	Analysis of hormones, multiplexed assay based device utilizing fluorescence detection
cobas® Liat System	Roche	Infectious diseases and genotyping, Lab-in-a-tube platform for automated analysis in compact benctop instrument
epoc™	Epocal	Blood chemistry analysis, Self-contained cards, patterned electrodes for sensing, wireless data transmission
3M™ integrated cycler	Focus diagnostics	Flu, intestinal pathogens, Portable detector, discs with on-board extraction
GeneXpert	Cepheid	Respiratory infections and cancer detection, disposable cards with benchtop analyzer (GeneXpert1), on-card sample processing (sputum)
Proxima	Sphere medical	Blood chemistry analysis, microanalyser with silicon chips and functionalized electrodes for rapid electrochemical detection
Micronics (Sony)	PanNAT® Molecular Diagnostic System	Bacterial infections and drug susceptibility testing Disposable cards with integrated heating, detection, sample processing in a portable instrument

levels, glucose, cardiac markers and limited immunoassays [81]. The other noticeable blood chemistry analysis LoC based systems are being developed by Abaxis, Epocal, and Sphere Medical. Industries are currently focusing on development of microfluidic CD4 + T-cell detectors for monitoring HIV/AIDS, a disease of high prevalence in developing countries. But a majority of commercial MF is focused on manipulation of DNA and RNA signatures for PoC clinical diagnosis and monitoring of patients. Nucleic acid tests are probably some of the most difficult assays to develop because of additional steps required for sample pre-treatment, signal amplification and target contamination, and instability. The company which initiated the integration of multiple MF procedures for nucleic acid detection was Handylab (founded in 2000 and acquired by BD in 2009). The company developed disposable platforms with onboard dry reagents provided in combination with a benchtop instrument combining heating, mechanical valves for fluid control, and fluorescence detection using molecular beacons. Besides this, several devices aiming at replacing traditional ELISA-based diagnostics with MF technology are also on their way to commercialization.

4 Present Challenges and Future Perspectives

After the recognition of MF potential in diagnosis, the realization of this field has been very slow. In fact, thousands of research publications are there, but the outcome as successful devices is very less. Some commercially available LoC products for DNA analysis, protein crystallization, and performing simple chemical reactions are available but still there is need for so-called "killer application" in the field of clinical diagnostics [88]. The PoC diagnostics have not yet lived up to their forecasted potential. One of the reasons may be the complexity of the systems. Many complex biochemical processes have been demonstrated on-chip for diagnostic application. However, the majority of them require the support of bulky and expensive external parts (pumps, valves, and switches) for the manipulation of the fluid and a variety of optical detectors and sensors for signal measurement. Thus, for practical use, simple approach with more applicable procedures should be devised to make the technology more field application. The other challenge may be the reluctance to the adoption of new technology. As the market is user driven and not technology driven i.e. the users are habitual to the traditional methods of analysis any new technology introduced has to be simple and must be easily operated by non-experts also. Most of the PoC diagnostics used in hospitals or laboratories are not suitable to be used by common people. User-friendly diagnostic concepts should be employed in the devices such as the simple indicator symbols to indicate the presence of antigens, antibodies, viruses, or other biological targets to be analyzed.

Analysis of real samples like blood and saliva in an MF device, however, is more complex and problematic than the purified samples usually used in general laboratories. Therefore, new devices need to be designed which are operable in the virtual environments rather than in laboratory conditions. Lack of funds may be another reason for the slow pace of transformation of academic research to practical devices. Manufacturing of MF devices is quite expensive and uses costly instruments that are not available in all laboratories. These costly MF platforms that are being used for research are not suited for mass production of practical devices. The majority of manufacturing methods published on the LoC devices have been micromachining on glass or silicon, and soft lithography on PDMS, which is again expensive [89]. However, for commercial PoC applications, mass production methods are required. Presently, plastic and paper or membranes are the two popular materials used for achieving high-volume and low-cost production. As the industry is influenced by the market cost of manufactured product more research should be directed towards the designing of low-cost platforms. To justify the switch for the consumers from current products the MF technology must significantly outperform or cost less than the present products. An emphasis on global health has increased the demand for low cost, high through output and integrated PoC devices which are likely to become common in the years ahead. Currently, most MF devices have one diagnostic target, but device targeting 100 s or 1,000 s of diseases will likely be designed and commercialized in the next decade, presenting MF a solution for major disease diagnostics.

MF is an emerging technology in the field of commercial diagnostics as far as the realization of technology from the laboratory to the real world is concerned, and its future holds enormous potential. The MF devices are destined to replace conventional techniques, and the innate advantages of the technology are too hard to ignore. The continuous development of MF applications in manufacturing methods, including platform technologies that can be customized easily for each diagnostic test, will be the drivers of success. Commercial success will result in the expansion of the field from biological domain to other areas also.

5 Conclusion

Here we have summarized the advantages of small fluidic circuit components which are capable of realizing quantitatively more and qualitatively new measurements of biological procedures. Microfluidics can be thus called as a technology with an aim to improve the end products. The microfluidics modified procedures are leading to new discoveries in the laboratory and new devices fabricated based on these discoveries are changing the landscape of biological systems. A lot of work has been done in this direction but still there is enormous scope for future development.

Acknowledgements SS and CMP thank Prof. B. D. Malhotra (DTU, Delhi) and Dr. G. Sumana (NPL, New Delhi) for interesting discussions. Shipra Solanki is thankful to UGC, India, for the award of SRF. C. M. Pandey acknowledges the Department of Science & Technology, Govt of India for awarding the DST-INSPIRE Fellowship [DST/INSPIRE/04/2015/000932].

References

1. Mark D, Haeberle S, Roth G, von Stetten F, Zengerle R (2010) Microfluidic lab-on-a-chip platforms: requirements, characteristics and applications. Chem Soc Rev 39(3):1153–1182
2. Valones MAA, Guimarães RL, Brandão LAC, de Souza PRE, de Albuquerque Tavares Carvalho A, Crovela S (2009) Principles and applications of polymerase chain reaction in medical diagnostic fields: a review. Braz J Microbiol 40(1):1–11
3. Barbulovic-Nad I, Yang H, Park PS, Wheeler AR (2008) Digital microfluidics for cell-based assays. Lab Chip 8(4):519–526
4. Manz A, Miyahara Y, Miura J, Watanabe Y, Miyagi H, Sato K (1990) Design of an open-tubular column liquid chromatograph using silicon chip technology. Sens Actuators B Chem 1 (1):249–255
5. Aa M, Graber N, Widmer HÃM (1990) Miniaturized total chemical analysis systems: a novel concept for chemical sensing. Sens Actuators B Chem 1(1):244–248
6. Mehling M, Tay S (2014) Microfluidic cell culture. Curr Opin Biotechnol 25:95–102
7. Bhatia SN, Ingber DE (2014) Microfluidic organs-on-chips. Nat Biotechnol 32(8):760–772
8. Fernandes TG, Diogo MM, Clark DS, Dordick JS, Cabral JMS (2009) High-throughput cellular microarray platforms: applications in drug discovery, toxicology and stem cell research. Trends Biotechnol 27(6):342–349
9. Griffith LG, Naughton G (2002) Tissue engineering—current challenges and expanding opportunities. Science 295(5557):1009–1014
10. Gupta K, Kim D-H, Ellison D, Smith C, Kundu A, Tuan J, Suh K-Y, Levchenko A (2010) Lab-on-a-chip devices as an emerging platform for stem cell biology. Lab Chip 10 (16):2019–2031
11. Klein AM, Mazutis L, Akartuna I, Tallapragada N, Veres A, Li V, Peshkin L, Weitz DA, Kirschner MW (2015) Droplet barcoding for single-cell transcriptomics applied to embryonic stem cells. Cell 161(5):1187–1201
12. Jung H, Chun M-S, Chang M-S (2015) Sorting of human mesenchymal stem cells by applying optimally designed microfluidic chip filtration. Analyst 140(4):1265–1274
13. Kang W, Giraldo-Vela JP, Nathamgari SSP, McGuire T, McNaughton RL, Kessler JA, Espinosa HD (2014) Microfluidic device for stem cell differentiation and localized electroporation of postmitotic neurons. Lab Chip 14(23):4486–4495
14. Reitinger S, Jr W, Kapferer W, Heer R, Gn L (2012) Electric impedance sensing in cell-substrates for rapid and selective multipotential differentiation capacity monitoring of human mesenchymal stem cells. Biosens Bioelectron 34(1):63–69
15. Gross PG, Kartalov EP, Scherer A, Weiner LP (2007) Applications of microfluidics for neuronal studies. J Neurol Sci 252(2):135–143
16. Farinas J, Chow AW, Wada HG (2001) A microfluidic device for measuring cellular membrane potential. Anal Biochem 295(2):138–142
17. Grant SC, Aiken NR, Plant HD, Gibbs S, Mareci TH, Webb AG, Blackband SJ (2000) NMR spectroscopy of single neurons. Magn Reson Med 44(1):19–22
18. Massin C, Vincent F, Homsy A, Ehrmann K, Boero G, Besse PA, Daridon A, Verpoorte E, De Rooij NF, Popovic RS (2003) Planar microcoil-based microfluidic NMR probes. J Magn Reson 164(2):242–255
19. Huang Y, Williams JC, Johnson SM (2012) Brain slice on a chip: opportunities and challenges of applying microfluidic technology to intact tissues. Lab Chip 12(12):2103–2117
20. Scott A, Weir K, Easton C, Huynh W, Moody WJ, Folch A (2013) A microfluidic microelectrode array for simultaneous electrophysiology, chemical stimulation, and imaging of brain slices. Lab Chip 13(4):527–535
21. Mauleon G, Fall CP, Eddington DT (2012) Precise spatial and temporal control of oxygen within *in vitro* brain slices via microfluidic Gas channels. PLoS One 7(8):e43309

22. Kang L, Chung BG, Langer R, Khademhosseini A (2008) Microfluidics for drug discovery and development: from target selection to product lifecycle management. Drug Discov Today 13 (1):1–13
23. Psaltis D, Quake SR, Yang C (2006) Developing optofluidic technology through the fusion of microfluidics and optics. Nature 442(7101):381–386
24. Caviglia C, Zór K, Montini L, Tilli V, Canepa S, Melander F, Muhammad HB, Carminati M, Ferrari G, Raiteri R (2015) Impedimetric toxicity assay in microfluidics using free and liposome-encapsulated anticancer drugs. Anal Chem 87(4):2204–2212
25. Sung JH, Kam C, Shuler ML (2010) A microfluidic device for a pharmacokinetic-pharmacodynamic (PK-PD) model on a chip. Lab Chip 10(4):446–455
26. Sakolish CM, Esch MB, Hickman JJ, Shuler ML, Mahler GJ (2016) Modeling barrier tissues in vitro: methods, achievements, and challenges. EBioMedicine 5:30–39
27. Huh D, Hamilton GA, Ingber DE (2011) From three-dimensional cell culture to organs-on-chips. Trends Cell Biol 21(12):745–754
28. Huh D, Kim HJ, Fraser JP, Shea DE, Khan M, Bahinski A, Hamilton GA, Ingber DE (2013) Microfabrication of human organs-on-chips. Nat Protoc 8(11):2135–2157
29. Jang K-J, Suh K-Y (2010) A multi-layer microfluidic device for efficient culture and analysis of renal tubular cells. Lab Chip 10(1):36–42
30. Douville NJ, Zamankhan P, Tung Y-C, Li R, Vaughan BL, Tai C-F, White J, Christensen PJ, Grotberg JB, Takayama S (2011) Combination of fluid and solid mechanical stresses contribute to cell death and detachment in a microfluidic alveolar model. Lab Chip 11(4):609–619
31. Huh D, Matthews BD, Mammoto A, Montoya-Zavala M, Hsin HY, Ingber DE (2010) Reconstituting organ-level lung functions on a chip. Science 328(5986):1662–1668
32. Booth R, Kim H (2012) Characterization of a microfluidic in vitro model of the blood-brain barrier (μBBB). Lab Chip 12(10):1784–1792
33. Nakao Y, Kimura H, Sakai Y, Fujii T (2011) Bile canaliculi formation by aligning rat primary hepatocytes in a microfluidic device. Biomicrofluidics 5(2):022212
34. Kim HJ, Huh D, Hamilton G, Ingber DE (2012) Human gut-on-a-chip inhabited by microbial flora that experiences intestinal peristalsis-like motions and flow. Lab Chip 12(12):2165–2174
35. Lee PJ, Hung PJ, Lee LP (2007) An artificial liver sinusoid with a microfluidic endothelial-like barrier for primary hepatocyte culture. Biotechnol Bioeng 97(5):1340–1346
36. Jiang B, Zheng W, Zhang W, Jiang X (2013) Organs on microfluidic chips: a mini review. Sci China Chem 57(3):356–364
37. Maschmeyer I, Lorenz AK, Schimek K, Hasenberg T, Ramme AP, Hübner J, Lindner M, Drewell C, Bauer S, Thomas A (2015) A four-organ-chip for interconnected long-term co-culture of human intestine, liver, skin and kidney equivalents. Lab Chip 15(12):2688–2699
38. Huh D, Y-s T, Hamilton GA, Kim HJ, Ingber DE (2012) Microengineered physiological biomimicry: organs-on-chips. Lab Chip 12(12):2156–2164
39. Esch MB, King TL, Shuler ML (2011) The role of body-on-a-chip devices in drug and toxicity studies. Annu Rev Biomed Eng 13:55–72
40. Sung JH, Srinivasan B, Esch MB, McLamb WT, Bernabini C, Shuler ML, Hickman JJ (2014) Using physiologically-based pharmacokinetic-guided "body-on-a-chip" systems to predict mammalian response to drug and chemical exposure. Exp Biol Med 239(9):1225–1239
41. Esch MB, Mahler GJ, Stokol T, Shuler ML (2014) Body-on-a-chip simulation with gastrointestinal tract and liver tissues suggests that ingested nanoparticles have the potential to cause liver injury. Lab Chip 14(16):3081–3092
42. Prot JM, Maciel L, Bricks T, Merlier F, Cotton J, Paullier P, Bois FY, Leclerc E (2014) First pass intestinal and liver metabolism of paracetamol in a microfluidic platform coupled with a mathematical modeling as a means of evaluating ADME processes in humans. Biotechnol Bioeng 111(10):2027–2040
43. Kim J-Y, Fluri DA, Kelm JM, Hierlemann A, Frey O (2015) 96-well format-based microfluidic platform for parallel interconnection of multiple multicellular spheroids. J Lab Autom 20(3):274–282

44. Kim J-Y, Fluri DA, Marchan R, Boonen K, Mohanty S, Singh P, Hammad S, Landuyt B, Hengstler JG, Kelm JM, Hierlemann A, Frey O (2015) 3D spherical microtissues and microfluidic technology for multi-tissue experiments and analysis. J Biotechnol 205:24–35
45. Huh D, Hamilton GA, Ingber DE (2011) From 3D cell culture to organs-on-chips. Tren Cell Biol 21(12):745–754
46. Agarwal A, Goss JA, Cho A, McCain ML, Parker KK (2013) Microfluidic heart on a chip for higher throughput pharmacological studies. Lab Chip 13(18):3599–3608
47. Lee SA, da No Y, Kang E, Ju J, Kim DS, Lee SH (2013) Spheroid-based three-dimensional liver-on-a-chip to investigate hepatocyte-hepatic stellate cell interactions and flow effects. Lab Chip 13(18):3529–3537
48. Jang K-J, Mehr AP, Hamilton GA, McPartlin LA, Chung S, Suh K-Y, Ingber DE (2013) Human kidney proximal tubule-on-a-chip for drug transport and nephrotoxicity assessment. Integr Biol 5(9):1119–1129
49. Griep LM, Wolbers F, De Wagenaar B, Ter Braak PM, Weksler BB, Romero IA, Couraud PO, Vermes I, Van Der Meer AD, Van den Berg A (2013) BBB on chip: microfluidic platform to mechanically and biochemically modulate blood-brain barrier function. Biomed Microdevices 15(1):145–150
50. Shen F, Li X, Li PCH (2014) Study of flow behaviors on single-cell manipulation and shear stress reduction in microfluidic chips using computational fluid dynamics simulations. Biomicrofluidics 8(1):014109. doi:10.1063/1.4866358
51. Raj A, van Oudenaarden A (2008) Stochastic gene expression and its consequences. Cell 135 (2):216–226
52. Singh A (2014) Transient changes in intercellular protein variability identify sources of noise in gene expression. Biophys J 107(9):2214–2220
53. Yin H, Marshall D (2012) Microfluidics for single cell analysis. Curr Opin Biotechnol 23 (1):110–119
54. Poulsen CR, Culbertson CT, Jacobson SC, Ramsey JM (2005) Static and dynamic acute cytotoxicity assays on microfluidic devices. Anal Chem 77(2):667–672
55. Balagaddé FK, You L, Hansen CL, Arnold FH, Quake SR (2005) Long-term monitoring of bacteria undergoing programmed population control in a microchemostat. Science 309 (5731):137–140
56. He M, Edgar JS, Jeffries GDM, Lorenz RM, Shelby JP, Chiu DT (2005) Selective encapsulation of single cells and subcellular organelles into picoliter- and femtoliter-volume droplets. Anal Chem 77(6):1539–1544
57. Chiou PY, Ohta AT, Wu MC (2005) Massively parallel manipulation of single cells and microparticles using optical images. Nature 436(7049):370–372
58. Taff BM, Voldman J (2005) A scalable addressable positive-dielectrophoretic cell-sorting array. Anal Chem 77(24):7976–7983
59. Di Carlo D, Aghdam N, Lee LP (2006) Single-cell enzyme concentrations, kinetics, and inhibition analysis using high-density hydrodynamic cell isolation arrays. Anal Chem 78 (14):4925–4930
60. Wheeler AR, Throndset WR, Whelan RJ, Leach AM, Zare RN, Liao YH, Farrell K, Manger ID, Daridon A (2003) Microfluidic device for single-cell analysis. Anal Chem 75 (14):3581–3586
61. Peng XY (2011) A micro surface tension pump (MISPU) in a glass microchip. Lab Chip 11 (1):132–138
62. Roman GT, Chen Y, Viberg P, Culbertson AH, Culbertson CT (2006) Single-cell manipulation and analysis using microfluidic devices. Anal Bioanal Chem 387(1):9–12
63. Riordon J, Nash M, Jing W, Godin M (2014) Quantifying the volume of single cells continuously using a microfluidic pressure-driven trap with media exchange. Biomicrofluidics 8 (1):011101

64. McClain MA, Culbertson CT, Jacobson SC, Allbritton NL, Sims CE, Ramsey JM (2003) Microfluidic devices for the high-throughput chemical analysis of cells. Anal Chem 75 (21):5646–5655
65. Clausell-Tormos J, Lieber D, Baret J-C, El-Harrak A, Miller OJ, Frenz L, Blouwolff J, Humphry KJ, Köster S, Duan H, Holtze C, Weitz DA, Griffiths AD, Merten CA (2008) Droplet-based microfluidic platforms for the encapsulation and screening of mammalian cells and multicellular organisms. Chem Biol 15(5):427–437
66. Tang F, Barbacioru C, Wang Y, Nordman E, Lee C, Xu N, Wang X, Bodeau J, Tuch BB, Siddiqui A (2009) mRNA-Seq whole-transcriptome analysis of a single cell. Nat Methods 6 (5):377–382
67. Wu J, Kodzius R, Cao W, Wen W (2014) Extraction, amplification and detection of DNA in microfluidic chip-based assays. Microchim Acta 181(13–14):1611–1631
68. Yager P, Edwards T, Fu E, Helton K, Nelson K, Tam MR, Weigl BH (2006) Microfluidic diagnostic technologies for global public health. Nature 442(7101):412–418
69. Hu S, Loo JA, Wong DT (2006) Human body fluid proteome analysis. Proteomics 6 (23):6326–6353
70. Martinez AW, Phillips ST, Whitesides GM, Carrilho E (2009) Diagnostics for the developing world: microfluidic paper-based analytical devices. Anal Chem 82(1):3–10
71. Martinez AW, Phillips ST, Butte MJ, Whitesides GM (2007) Patterned paper as a platform for inexpensive, low volume, portable bioassays. Angew Chem Int Ed 46(8):1318–1320
72. Rosenfeld T, Bercovici M (2014) 1000-fold sample focusing on paper-based microfluidic devices. Lab Chip 14(23):4465–4474
73. Foudeh AM, Fatanat Didar T, Veres T, Tabrizian M (2012) Microfluidic designs and techniques using lab-on-a-chip devices for pathogen detection for point-of-care diagnostics. Lab Chip 12(18):3249–3266
74. Ghrera AS, Pandey CM, Ali MA, Malhotra BD (2015) Quantum dot-based microfluidic biosensor for cancer detection. Appl Phys Lett 106(19):193703
75. Dimov IK, Garcia-Cordero JL, O'Grady J, Poulsen CR, Viguier C, Kent L, Daly P, Lincoln B, Maher M, O'Kennedy R (2008) Integrated microfluidic tmRNA purification and real-time NASBA device for molecular diagnostics. Lab Chip 8(12):2071–2078
76. Chang W-H, Wang C-H, Lin C-L, Wu J-J, Lee MS, Lee G-B (2015) Rapid detection and typing of live bacteria from human joint fluid samples by utilizing an integrated microfluidic system. Biosens Bioelectron 66:148–154
77. Lee W, Kwon D, Choi W, Jung GY, Jeon S (2015) 3D-printed microfluidic device for the detection of pathogenic bacteria using size-based separation in helical channel with trapezoid cross-section. Sci Rep 5:7717
78. Boehm DA, Gottlieb PA, Hua SZ (2007) On-chip microfluidic biosensor for bacterial detection and identification. Sens Actuators B Chem 126(2):508–514
79. Cho Y-K, Lee J-G, Park J-M, Lee B-S, Lee Y, Ko C (2007) One-step pathogen specific DNA extraction from whole blood on a centrifugal microfluidic device. Lab Chip 7(5):565–573
80. Manini TM, Vincent KR, Leeuwenburgh CL, Lees HA, Kavazis AN, Borst SE, Clark BC (2011) Myogenic and proteolytic mRNA expression following blood flow restricted exercise. Acta Physiol (Oxf) 201(2):255–263
81. Chin CD, Linder V, Sia SK (2012) Commercialization of microfluidic point-of-care diagnostic devices. Lab Chip 12(12):2118–2134
82. Gervais L, De Rooij N, Delamarche E (2011) Microfluidic chips for point-of-care immunodiagnostics. Adv Mater 23(24):H151–H176
83. Kim M, Choi J-C, Jung H-R, Katz JS, Kim M-G, Doh J (2010) Addressable micropatterning of multiple proteins and cells by microscope projection photolithography based on a protein friendly photoresist. Langmuir 26(14):12112–12118
84. B-H C, Huh D, Kyrtsos CR, Houssin T, Futai N, Takayama S (2007) Leakage-free bonding of porous membranes into layered microfluidic array systems. Anal Chem 79(9):3504–3508

85. Legendre LA, Morris CJ, Bienvenue JM, Barron A, McClure R, Landers JP (2008) Toward a simplified microfluidic device for ultra-fast genetic analysis with sample-in/answer-out capability: application to T-cell lymphoma diagnosis. J Lab Autom 13(6):351–360
86. Diercks AH, Ozinsky A, Hansen CL, Spotts JM, Rodriguez DJ, Aderem A (2009) A microfluidic device for multiplexed protein detection in nano-liter volumes. Anal Biochem 386(1):30–35
87. Global Point-of-Care Diagnostics Market Outlook (2018) http://www.rncos.com
88. Blow N (2007) Microfluidics: in search of a killer application. Nat Methods 4(8):665–672
89. Sia SK, Whitesides GM (2003) Microfluidic devices fabricated in poly (dimethylsiloxane) for biological studies. Electrophoresis 24(21):3563–3576

Chapter 9
RETRACTED CHAPTER: On-Chip Immunoassay for Molecular Analysis

Andy Ng

1 Introduction

1.1 Immunoassays

The specificity and affinity of antibody-antigen interactions has long been exploited in immunoassays for the detection of biomolecular analytes. In particular, immunoassays are used in a wide spectrum of applications in clinical diagnostics, molecular biology, proteomics and biosensing, for the measurement of many proteins and, to a lesser extent, small molecules due to their relatively smaller exposed surface for recognition. Immunoassays can be categorized into two main formats: homogeneous and heterogeneous assays. In homogeneous format, antibodies and antigens are in solution, and their interactions that lead to the formation of the antibody-antigen immunocomplex also occur in solution. On the other hand, in heterogeneous format, the antibodies or antigens are immobilized on a solid support, and their interactions take place at the boundary layer, and the unbound, or "non-captured" antibodies or antigens, as well as other reagents can be easily removed. Both immunoassay formats have been extensively studied and widely implemented in microfluidic platforms.

Immunoassays can be further classified into competitive mode and non-competitive modes. In competitive mode, known amount of labeled antigens are introduced into the assay, and the non-labeled antigen competes with the labeled antigens for a limited number of binding sites on the antibodies. As the amount of non-labeled antigen in a

The original version of this chapter was revised: The chapter was retracted as it contains significant parts plagiarizing another publication.
An erratum to this chapter can be found at 10.1007/978-3-319-40036-5_11

A. Ng (✉)
Biomedical Engineering Department, McGill University, Montreal, QC, Canada
e-mail: andy.ng@mcgill.ca

C.K. Dixit, A. Kaushik (eds.), *Microfluidics for Biologists*,
DOI 10.1007/978-3-319-40036-5_9

sample increases, the amount of labeled antigens that remain bound to the antibody decreases. The competitive immunoassay generates a signal that is inversely proportional to the amount of the antigen, if the labeled antigen-antibody complexes are detected, or an increase in signal. In non-competitive mode, there is an excess of antibodies. The antigen are first "captured", and subsequently detected by a second set of labeled antibodies, *via* recognition of a distinct recognition site (epitope) of the antigen. The formation of this antibody-antigen-antibody complex is described as a "sandwich" immunoassay, where the signal is proportional to the antigen concentration.

Conventional immunoassays, in particular heterogeneous immunoassays, are performed in microtiter plates with 96 or more sample wells, and typically involve a series of sample/reagent introduction, washing, mixing and incubation steps. The process often requires at least several hours and could last for up to days to perform a single experiment. The lengthy analysis time is mainly due to the long incubation step, attributed to the limitation in mass transport for all the reagents to migrate from within the solution to the antibody-coated surface, as the association rate of the antigen to the antibody is relatively rapid [1]. Robotic systems can be used to carry out repetitive fluid handling, thereby reducing hands-on time and improving the throughput, but require significant investment in infrastructure, as well as high maintenance efforts.

Miniaturization of the immunoassay within microfluidic systems represents an attractive solution to shortcomings and limitations in conventional immunoassays. Common microfluidic systems are built by networks of channels with dimensions in the μm range. Fluids flow in a laminar manner at these scales, exhibiting neighbouring parallel fluidic streams where mixing occurs only through diffusion [2]. Due to the much smaller scale, mass transport is more efficient, and the increased surface to volume ratio further accelerates the antibody-antigen complex formation. The smaller dimensions of the channel significantly reduce the amount of reagent and sample required. Fluid handling are mostly automated through the design of channel networks, the use of valves and other features, thereby simplifying experimental procedures, improving reproducibility, reducing analysis time, increasing throughput and lowering the operating cost.

This chapter focuses on a number of main developments in microfluidic immunoassays in since 2000. Representative works are highlighted, presented and categorized in three areas: microfluidic immunoassay formats, fluid driving and handling technologies, and multiplexing approaches.

2 Immunoassay Formats in Microfluidic Systems

Both homogeneous and heterogeneous immunoassay formats (discussed above) have been developed into microfluidic-based assays. In homogeneous formats, analyte bound and free antibodies are both in solution, and can be distinguished by their mobility in the microfluidic channels; whereas in heterogeneous formats, all antibodies are immobilized on the surface of the microfluidic device, or on beads introduced in the device. This section will discuss a number of microfluidic implementations in immunoassays based on electrophoresis-based homogeneous format, and surface- and bead-based heterogeneous formats.

2.1 Homogeneous Immunoassays in Microfluidic Systems

Homogeneous microfluidic immunoassays rely on the discrimination of analyte-bound and free antibodies for quantitative measurement. Formation of the immunocomplex creates differences between the analyte-bound and free antibodies in diffusion characteristics, [3] isoelectric point, [4] and electrophoretic mobilities. In particular, gel electrophoresis separates molecules based on electrophoretic mobility due to differences molecular weight and conformation in the case of non-denaturing gel. Microfluidic immunoassays based on gel-electrophoresis that separate bound and unbound analytes have recently gained interests [5–7]. Under non-denaturing electrophoresis conditions, the immunocomplex remains intact in the gel matrix, usually made of polyacrylamide [8]. Since the reagents in homogeneous immunoassays, especially the antibody, are not immobilized and is free in the microfluidic channel, a concentration step is usually required to improve the detection sensitivity. Strategies such as the use of a preconcentration membrane to enrich the analyte have been employed. A microfluidic system integrating analyte preconcentration, reagent loading and discrimination of analyte-bound and non-bound antibody has been shown by Meagher et al. [6]. A protein analyte was first loaded into large pore-size polyacrylamide gel, which then migrated under an electric field onto a preconcentration membrane and trapped. A labeled antibody was then loaded and concentrated in the same area. Analyte-bound and non-bound antibodies were then separated in the separation channel containing polyacrylamide gel with small pore size that is suitable to discriminate the two antibody species. The assay can be carried out in less than 20 min with 10 μL sample and achieved a detection limit of 10 pM for SEB detection and metalloproteinase-8 using saliva samples [9].

Isotachophoresis (ITP) is an electrokinetic method that has been applied in microfluidic chip as a separation and concentration technique in microchip-based electrophoresis that eliminates the use of a preconcentration membrane. ITP has been used as a promising sample pretreatment for a wide variety of analytes such as protein biomarkers, [10] and in a number of microfluidic-based immunoassays. In particular, ITP has been used to preconcentrate the immunocomplex before the electrophoretic separation step, resulting in enhanced sensitivity and reproducibility of the assay [7, 11]. An example is a microchip known as the "electrokinetic analyte transport assay" by Kawabata et al., [12] in which concentration, mixing and incubation of the reagents and analyte were integrated on-chip using ITP prior to electrophoretic separation, completing the entire analysis in about 2 min.

Due to its enhanced compatibility with electric force-based fluid manipulation and the rapid and straightforward separation, capillary electrophoresis (CE) has been used in microfluidic systems for carrying out immunoassays. Since the first report of CE-based microfluidic chip for detection of monoclonal antibodies and a small molecule drug, [13] the Harrison group has developed advanced CE-based microfluidic systems comprising of six independent channels for mixing, reaction, separation and detection within a single wafer. This CE-based chip can perform simultaneous direct immunoassays for ovalbumin and anti-estradiol using fluorescence detection, completing the assay within 1 min [14]. CE-based chips have also been integrated with electrochemical measurement [15] for sensitive detection, as well as high throughput immunoassays in competitive mode [16]. In addition, CE-based chip can be adapted for continuous flow, replacing used reagents and buffer with fresh ones, thereby prolonging CE to 24 h for monitoring insulin secretion from islet [17, 18].

2.2 *Heterogeneous Immunoassays in Microfluidic Systems*

In heterogeneous immunoassays, antibodies are immobilized on the surface of a solid support. In general, the solid support can either be the surface of the microfluidic device or on beads typically of μm-dimension that are embedded in the channel or chamber within the device. Whereas antibody immobilization is a key step because it can dramatically influences the specificity and sensitivity of the assay, strategies for analyte delivery and washing can also affect the performance of the assay in terms of analysis time and limit of detection.

2.2.1 Antibody Immobilization

Microfluidic substrates are commonly made from materials including glass, silicon, silicon nitride [19], polydimethylsiloxane (PDMS)[20, 21], polymethyl methacrylate (PMMA) [22] and polystyrene [23]. The simplest method to immobilize antibody onto a microfluidic substrate is by physical adsorption. While being extremely simple, this method often results in significant reduction in antibody activity due to steric hindrance, random and unfavourable orientation and denaturation of the antibody [24]. Due to these shortcomings, more controlled and specific immobilization strategies by bioaffinity and covalent immobilization have been developed.

One strategy is the use of a long covalent linker to "tether" the antibody, which keeps the antibody a defined distance away from the surface, thereby maintaining a chemical environment around the antibody similar to the bulk solution, minimizing problems associated with accessibility and steric effects. Indeed, different strategies for covalent immobilization of antibodies on silicon substrates have been compared by Yakovleva et al., [25, 26] demonstrating that long and flexible linkers such as polyethyleneimine (PEI) [27] and dextran [28] led to more favourable immunoreaction than short linkers such as 3-aminopropyltriethoxysilane (APTES). Antibody functionality can also be maximized by the use of bioaffinity interactions. A novel antibody immobilization strategy on PMMA surface was reported by Wen et al., [29] who used a surface linker biotin-poly(L-lysine)-*graft*-poly(ethylene glycol) (biotin-PLL-*g*-PEG) for immobilization. The PMMA was activated by plasma, followed by coating with poly(acrylic acid) to add functional carboxyl groups, which form electrostatic interactions with biotin-PLL-*g*-PEG. NeutrAvidin labeled-Protein A binds strongly to the biotin, effectively linking Protein A to the surface, which captured antibody in an orientation-specific manner *via* the Fc region of the antibody. In addition, the PEG moiety on the PMMA surface helped reduce non-specific adsorption. This strategy resulted in a higher analyte capture efficiency than direct covalent linkage of Protein to PMMA.

Antibody can also be directly entrapped in the microfluidic substrate material during the fabrication step. Heyries et al. [30] spotted antibody solution on a Teflon mould by a microarrayer, followed by pouring liquid PDMS. After curing, the PDMS was peeled off and the antibody molecules were thereby transferred to the PDMS. The transferred antibody showed efficient analyte capture with the overall assay sensitivity at ng/mL sensitivity.

2.2.2 Analyte Transport and Delivery

Compared to conventional microtitre wells format, microfluidic systems offer more efficient mass transport due to the significantly reduced diffusion distances inside microchannels/microchambers, thus rapid analyses can be performed at small sample volumes. However, at low analyte concentrations, the transport of analyte can still be limited [31]. While there has been significant effort focused on a reduction of the size of the sensing region, [32] there is less attention on the improvement of the design of microfluidic elements. Transport can be enhanced by further lower the dimensions of the microfluidic features, but there is obviously a limit at which the devices can be miniaturized before the resistance to fluidic flow becomes impractical to handle. Because diffusion is the only method for transporting the analyte to the immobilized capture antibody, and that the replenishment of the analyte at the boundary layer on the capture surface is subject to mass transport limitations, [33, 34] one way to improve analyte delivery to the capture antibody is to integrate mixing functionalities in the microfluidic system. Periodically pulsing the analyte solution in simple "forward-backward" flow within serpentine channels led to higher analyte capture [35]. Passive mixers built with micro-dimensioned features such as patterned grooves/herringbone ridges patterned in the top of the microchannel have been shown to increase the sensitivity of immunoassay by more than 26 % [36, 37]. The grooves in the channel induced mixing of the fluid, thereby encouraging the delivery of analyte to the capture region, as well as continuously replenishing analyte molecules at the boundary layer, preventing analyte depletion. An active mixing strategy developed by Jennisen and Zumbrink [38] used a bubble introduced into the channel upstream of the analyte solution, which moved over the sensing surface through the solution, creating a vortex sheet between the fluid layer close to the surface and the bulk solution that achieve good analyte transport to the sensor via effective mixing. Another strategy to create enhanced transport of the analyte to the capture surface has been developed by Hofmann et al. [39] The rapid delivery of analyte to capture antibodies in small volume sample was achieved by a flow confinement method, in

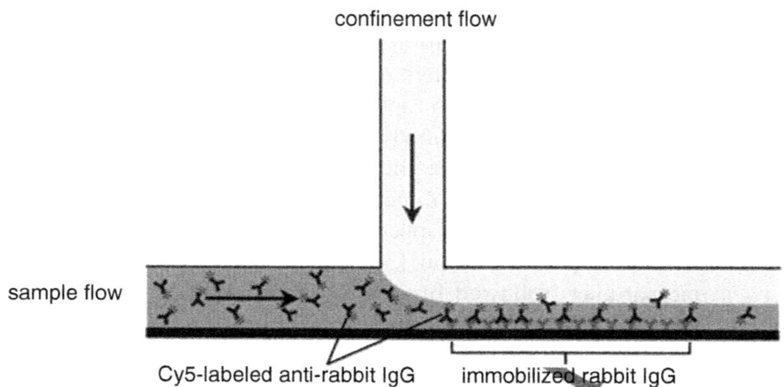

Fig. 9.1 Schematic illustrating the concept of flow confinement. A sample flow is joined by a confinement makeup flow in a perpendicular orientation, leading to a double flux of sample and confinement flow in a planar sheet geometry, effectively confining the sample flow into a thin layer of higher velocity near one surface of the microchannel. Taken from [39]. Copyright © 2002 American Chemical Society

which the flow of the sample is joined perpendicularly by a confinement makeup flow of buffer. Since the flow is laminar, the makeup flow effective confines the sample into a thin layer on the capture zone (Fig. 9.1).

2.3 *Heterogeneous Immunoassays on Microbeads*

2.3.1 Magnetic Microbeads

To significantly increase the surface area to volume ratio, microbeads are often used in microfluidic immunoassays. Microbeads can also serve as a simple vehicle for reproducibly manipulating antibodies and their delivery to specific locations [40]. Microbeads can be magnetic or non-magnetic, which often determines how microbeads are implemented in the microfluidic system.

The use of magnetic microbeads in microfluidic immunoassays is becoming more popular as their manipulation is greatly facilitated by the application of a magnetic field, allowing easy capture of analyte from the sample mixture. Typical magnetic bead immunoassays employ antibody-coated magnetic beads that are immobilized on the surface of the device, [41–43] providing an increased capture or sensing surface. The most efficient use of magnetic beads however is by resuspending and dispersing them in the sample solution to reduce the diffusion distances between the analyte and antibody. For microfluidic systems, collecting and resuspending magnetic beads at specific stages of the assay requires more specialized and complex functionalities such as micropumps, active mixers and microcoils for generating localized magnetic field [44, 45].

A unique magnetic bead retention mechanism has been developed by Lacharme et al., [46] which exploited the self-assembling properties of magnetic beads that led to the formation of a magnetic chain, without the use of sophisticated fluidic handling. The magnetic chain was formed by trapping magnetic beads in microchannels with cross-sections with periodically enlarged size. When a magnetic field is applied, the magnetic beads self-assembled as chains along the length of the channel. This mechanism provided efficient mixing and as a result enhanced the antigen capture in an immunoassay performed completely on-chip with detection limit in the range of a few ng/mL in 30 min, using only nL of reagents and antibody solution.

2.3.2 Non-magnetic Microbeads

Non-magnetic microbeads require physical retention for manipulation, and the removal of unbound analyte. Physical retention of non-magnetic microbeads can be achieved by microstructures fabricated within the microfluidic system. The Kitamori group used a dam structure to trap antibody-coated polystyrene beads [47, 48]. Ko et al. [49] developed an integrated chip for protein sensing using antibody-coated polystyrene beads trapped in a reaction area by micropillars made of PDMS. Sample solution, gold nanoparticles labeling agent, and signal generation solution were then flow through the trapped beads. The gold nanoparticles catalyzed the silver ions in the signal generation solution into metallic silver, which in turn deposited onto the gold nanoparticles and thereby enlarged the particles to 100 nm. The enlarged particles formed an electrical bridge between the electrodes layered on the sensing surface, resulting in a signal due to the decrease in resistance. The entire chip-based assay can be completed in less than an hour, while performing the same assay with a conventional setup requires at least several hours on hands on time.

Employing a configuration similar to solid-phase extraction, Shin et al. [50] packed antibody-conjugated microbeads against a frit structure as a strategy to increase the sensitivity of immunoassay for C-reactive protein (CRP). Captured CRP were captured in the packed bead matrix, and subsequently eluted in an acidic buffer. The extraction also worked successfully to detect CRP in complex sample such as serum and cerebrospinal fluid [51].

Non-magnetic microbeads can also be immobilized by dielectrophoresis [52] or electrostatic forces [53, 54] without the use of microstructures for physical retention of the beads. Holmes et al. [52] developed a microchip-based flow cytometer using commercial methacrylate beads for capture antibody immobilization. The microchip-based flow cytometer used negative dielectrophoresis to focus the microbeads into a stream of particle, and the impedance and fluorescence signals of each bead were measured.

In the work by Sivagananam, [53] positively charged APTES were patterned on a glass substrate, and negatively charged streptavidin-coated microbeads were assembled electrostatically. The assembled beads were then used for conjugation with the capture antibody. Using mouse IgG as a model analyte, this method reached a detection limit of 250 pg/mL when operating in continuous-flow mode.

The majority of microbead-based strategies only employed the outer surface of the beads for antibody immobilization. Unique implementation of the bead-based immunoassay such as the work by Yang et al., [55] in which superporous agarose beads with diameters ranging from 10 to 80 μm were used as a solid support. The superporous agarose beads were covalently conjugated to Protein A, which immobilize the capture antibody in a favourable orientation. Using these superporous beads as immobilization materials, the immunoassay detected goat IgG, as a model analyte, with a detection limit of 100 pg/mL, a 10 times improvement over conventional bead-based microfluidic assays. The large surface area attributed to the inner matrix of the agarose beads, as well as the lowered fluidic resistance due to the porosity of the beads have led to the enhanced sensitivity of the assay.

3 Fluid Driving and Handling Technologies in Microfluidic Systems

Fluid transport plays an integral role in microfluidic immunoassays, and the performance of the fluid driving and handling system directly affect the result of the assays. To miniaturize the fluidic operations analogous to the steps in conventional immunoassays, [56] many strategies for on-chip liquid transport have been reported. The force for fluidic transport modalities can be categorized into electric, pressure and power-free passive forces.

3.1 Electric Forces

Electric forces for fluid flow are generated by electroosmotic pumping based on electroosmotic flow (EOF). EOF is the flow of the bulk fluid resulting from the movement of the solvated ions. A layer of fluid containing a build-up of solvated ions is attracted to the oppositely charged walls. When an electric field is applied, the solvated ions and the waters of hydration are driven toward the oppositely charged electrode. The movement of the solvated ions drags the bulk fluid via viscous forces, forming a uniform plug-like flow [57]. Using this mechanism, fluid handling such as flow initiation, stopping and direction can be easily controlled by the applied electric field, and do not require valves or pumps. Furthermore, controlling fluid flow by electric field alone facilitates automation.

The first application of electric force-driven fluid flow in immunoassays was by Rooij et al. [58, 59] and further developed by Gao et al. [60, 61]. For example, Gao et al. developed a microfluidic chip for detection of multiple microbial antigens based on electroosmotic pumping [60]. The microchip first used a microfluidic network to adsorb the analyte onto a PDMS-coated glass slide. Another PDMS bearing H-shaped microchannels for reagent delivery was then bonded to the slide. The sequential delivery of reagent solutions was performed automatically by a programmable voltage sequencer. This method was able to detect bacterial lysate antigen and achieved a detection limit of 3 μg/mL. Simultaneous detection of human antibodies against two bacterial species in serum was also possible with this automatic system [61].

Electric force-driven immunoassay could have better binding kinetics than pressure-driven assays. Hu et al. [62] built a two-dimensional computation model to simulate the mass transport and binding between the antigen in the bulk fluid flow and the immobilized antibody on the surface of the microchannel. The model showed that, due to the uniform, plug-like fluid flow profile as a result of electro-osmosis, assays using electric force-driven flow indeed exhibited superior reaction kinetics than pressure-driven ones. However, there are strict design constraints for the construction of electric force-based system. The substrate material should be non-conductive, other electrical breakdown or short circuit might occur. Furthermore, the choice of buffer and reagents used in such systems must be conductive. Although most buffers in biochemical assays contain salts and therefore conductive, an ionic strength that is too high might cause Joule heating, which significantly affects the conformation of biomolecules and hence antibody-antigen recognition, as well as causing denaturing and aggregation of other biomolecules present in the sample matrix, particularly in complex biological samples such as blood and urine.

Another electric force-based fluid transport strategy is electrowetting, whereby liquid, in contrast to continuous flow in channels, are handled as discrete droplets, in the presence of voltage sequences applied to an electrode array insulated by a hydrophobic dielectric layer [63]. Electrowetting is often combined with magnetic microbead-based immunoassays for liquid droplet manipulations, separation of bound and unbound analyte and washing steps. Sista et al. [43] developed a microchip containing a sample droplet and a reagent droplet containing capture antibody-conjugated magnetic microbeads and detection antibody. Electrowetting-based manipulation moved and merged the two droplets, and after an incubation of 2 min the droplet containing the sandwich immunocomplex on microbead was delivered over a magnet, followed by capture by the magnetic field. Unbound analyte and other components were then removed by splitting the excess liquid from the beads within the droplet, and a new droplet containing fresh buffer was introduced over the beads for washing, and the process repeated several times to ensure complete washing (Fig. 9.2). Detection was achieved by introducing a droplet containing detection reagents. The entire analysis using electrowetting fluid handling was completed in 7 min.

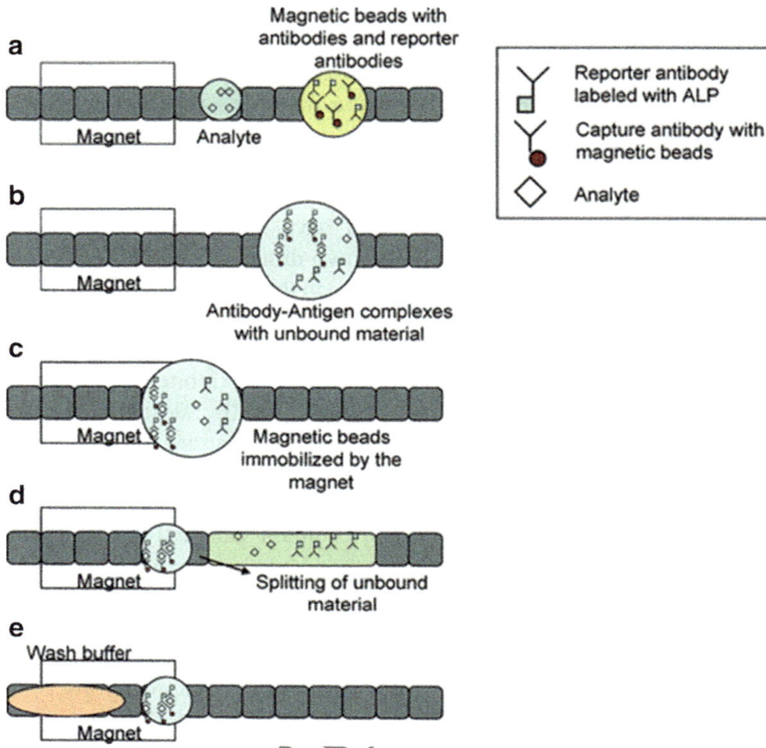

Fig. 9.2 Process flow of the heterogeneous immunoassay using electrowetting for fluid transport. (**a**) Delivery of sample to the immunoassay reagents, (**b**) immunocomplex formation. (**c**) immobilization of magnetic beads, (**d**) removal of unbound analyte and reagents (**e**) addition of fresh wash buffer. Taken from [77]. Copyright © The Royal Society of Chemistry 2008

3.2 *Pressure-Driven Fluid Handling Modalities*

The most popular fluid handling modality for microfluidic immunoassay is by pressure-driven flow. In contrast to electric-force driven flow, pressure-driven modalities do not have strict requirement on the conductivity of the solution and substrates, and therefore are compatible with a range of samples, reagents and device materials. On the other hand, pressure-drive flow exhibits a less uniform, parabolic flow profile, which results in the dispersion of the liquid plug, and therefore is considered less useful for separations [64].

The simplest setup for pressure-driven flow can be created by leaving the inlet open to atmospheric pressure and applying vacuum at the outlet using a vacuum pump, similar to a sucking action. The fluid can also be pushed by applying a positive pressure at the inlet and leaving the outlet to atmospheric pressure. In low-cost microfluidic chip designs, actuation by applying pressure using the hand or thumb have been demonstrated [65, 66].

For more precise control of pressure-driven flow and automation, micropneumatic pumps integrated with the microfluidic chip have been developed and have become popular for immunoassays. The micropneumatic pump is composed of a fluid flow in a microchannel, mechanically flexible membranes such as those made of PDMS and air chambers or valves that pneumatically deflect the PDMS membranes by introduction of compressed air [44, 67, 68]. Liquid is propelled by the peristaltic action driven by the time-phased deflection of PDMS membranes above the fluid flow along the microchannel, and is particularly suitable for sequential liquid delivery. Detection of host-generated antibody against infection pathogens has been demonstrated with micropnuematic pumps in microfluidic immunoassays [67]. In this system, a capture probe was first adsorbed on the surface of the detection zone. Serum sample, washing buffer, detection antibody and signal generation reagents stored in separate reservoirs were sequentially delivered to the detection zone by the micropneumatic pump. The signal generated was then detected optically by absorbance measurement. The peristaltic action required at least three pneumatic valves for controlled sequential deflection of the PDMS membranes [69].

In a more recent design by Schudel et al. [70], the valve structure utilized flexible PDMS membranes with microfabricated barrier features or gates that are lifted away from a substrate by vacuum, thus allowing the fluid to flow. Kim et al. [71] used these lifting gate microvalves and built pumping capabilities within microfluidics. The pump consisted of a linear array of three microvalves in series with cyclic actuation, and was capable of efficient and automated fluid transport. The lifting gate structures offer enhanced and direct integration with patterned solid substrates.

To simplify design and control algorithms, it is desirable to limit the number of pneumatic valves required for pumping. For this, alternative designs for micropneumatic pumps that used air chambers of different sizes interconnected by air channels were developed [72, 73]. Using interconnected air chambers caused the membranes to deflect at different times, allowed for their sequential actuation and introduced the latency required to generate peristaltic pumping action on the PDMS membranes, but only one pneumatic valve was required. This micropneumatic pump was used in bead-based immunoassays with flow cytometry detection [44]. The transport of samples, washing buffer, antibody-conjugated magnetic beads, detection antibody and sheath flow for focusing the sample stream were all provided by the micropneumatic pump, and the entire process including sample incubation and final detection step took 40 min.

3.3 Centrifugal Force-Driven Modalities

Centrifugal force-based microfluidic platforms have been developed and typically built from round substrate similar to the compact disc (CD) form factor and footprint. CD-based microfluidic systems contain networks of channels and chambers that perform many analytical functions, including sample/reagent transport, dilution, mixing and separation, all within a disc format.

The main characteristic of microfluidic systems on a CD is the use of centrifugal force controlled by the rotation speed of the CD. The centrifugal force is to overcome the capillary or surface force which prevents the liquid contained in one chamber from exiting. As the centrifugal force increases from the center towards the outer edge of the CD, sequential pumping of liquid can be achieved. Similar to all pressure-driven flow methods, centrifugal force-driven flow is not affected by the physiochemical properties of the fluid. In addition, because of the CD format, these devices can be easily adapted with optical detectors similar to the mechanism of a CD player.

Centrifugal force-driven microfluidics have attracted attentions and have been implemented for carrying out homogeneous immunoassays for pathogen antigen detection [74]. A number of commercial assays have also been developed. Gyrolab used a packed bed of microbeads and performed multiple reagent delivery steps controlled by centrifugal forces to perform immunoassays [75]. The bed of microbeads formed an assay surface on which reagents flow through for analyte capture and washing, followed by optical detection using a laser directly on the device.

The ability of centrifugal-based microfluidic to implement on-chip blood separation is particularly attractive. Lee et al. [76] developed a system for performing bead-based immunoassay for the detection of host-generated antibody against hepatitis B virus and in particular, the system was capable of plasma separation. The control of flow between chambers was done by an interesting microvalve design involving the use of ferrowax that can form valves that are normally open or closed. Ferrowax is paraffin wax with embedded iron oxide nanoparticles, which upon heating by a low intensity IR laser melts at an accelerated rate, and therefore can be individually actuated. The immunoassay is fully automated started from whole blood, and can be completed in less than 30 min.

A recent centrifugal microfluidic platform reported by Koh et al. performed an innovative sedimentation-based immunoassay for detection of botulinum toxin [77]. This platform first mixed a whole blood sample with capture antibody-conjugated microbeads and fluorescently-labeled detection antibody. Binding of the analyte between the capture and detection antibody formed a sandwich and fluorescent immunocomplex. Subsequently, the centrifugal device separated the beads with the immunocomplex from the blood sample through a density media, which effectively washes the beads and remove all interfering agents. As the bead pellet at the end of the channel located at the out edge of the disk, fluorescent signal can be measured and the analyte quantified (Fig. 9.3).

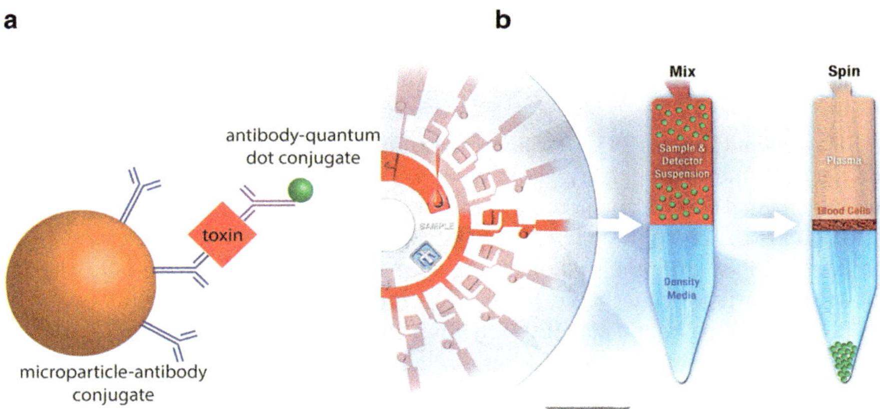

Fig. 9.3 Concept of a centrifugal microfluidic platform incorporating sedimentation-based immunoassay for analyte detection from whole blood sample. (**a**) Schematic of the immunocomplex on microbeads. (**b**) Fluorescent immunocomplex on microbeads were separated from the matrix of whole blood through a density media. Taken from [77]. Copyright © 2014 American Chemical Society

3.4 Passive, Capillary Force-Driven Modalities

Passive fluid handling has been popular because the integration of active actuation of liquids using electric force, pumping by peristaltic action and pressurized valving directly contribute to the complexity and cost of the system. Passive microfluidic systems offer portability, ease of operation and power-free operation. The most common and versatile passive liquid driving force is capillary force, which allow for passive fluid flow, thereby avoid the requirement for external pumping system, tubing and connectors [78–85]. In particular, capillary microfluidic systems can perform autonomous fluid delivery [83, 84] and allows the use of transparent materials including PDMS, PMMA and glass, providing higher flexibility in assay design and detection methods.

Gervais et al. [85] developed a one-step capillary-based microfluidic immunoassay platform comprised of sample collector, delay valves, flow resistors, detection antibody deposition zone and a reaction chamber with immobilized capture antibody. The capillary-force was generated by capillary pumps with vents. The microstructure of the capillary pumps, delay valves and flow resistors determined the overall flow resistance and capillary pressure, which controlled the flow rate of the device [79]. In this system, the delay valves consolidate the liquid flow from the sample collector before reaching the flow resistors, thereby minimizing the problem of air bubble trapping. The flow rate can be modulated by adding flow resistors. Capture of the analyte was achieved by capture antibody patterned on the surface of the reaction chamber. All other reagents were stored on-chip and redissolved as sample solution flowed through during the assay. In this one-step design, the addition of blood sample into the capillary circuit initiated the capillary-driven flow, which then automatically triggered the fluidic operations to complete the immunoassay. However, the system was unable to sequentially deliver multiple reagents required to amplify the signals and enhance the detection limit of the assay.

Safavieh and Juncker [83] introduced two novel capillary elements: trigger valve and retention burst valve. A trigger valve consists of a shallow channel, intersecting another channel that is deeper, covered by a hydrophobic layer. The abrupt increase in the cross-section of the microchannel making it energetically unfavourable for the liquid to flow from the shallow to the deep channel due to the large increase in liquid-air interface at the liquid front. The trigger valve is triggered by flowing a liquid through the larger channel. A retention burst valve is formed by constrictions in the microchannel that produce high capillary pressure. When the pressure in the capillary circuit near the retention burst valve is weaker that the valve's burst pressure, liquid remains pinned by the retention burst valve. When the pressure in the circuit builds up and rise above the valve's burst pressure, the liquid begins draining the channel of the valve until it eventually bursts, and as a result the reservoir downstream of the retention burst valve beings to drain rapidly (Fig. 9.4).

A novel capillary circuit was built by combining a variety of capillary elements, and achieved completely autonomous delivery of multiple liquid reagents according to a defined sequence that followed a preprogrammed and predetermined flow rate and timing. The capillary circuit was able to perform a sandwich immunoassay to measure the concentration of C-reactive protein. The authors defined such circuit as capillarics and introduced symbolic representations reminiscent of those used in electrical circuits.

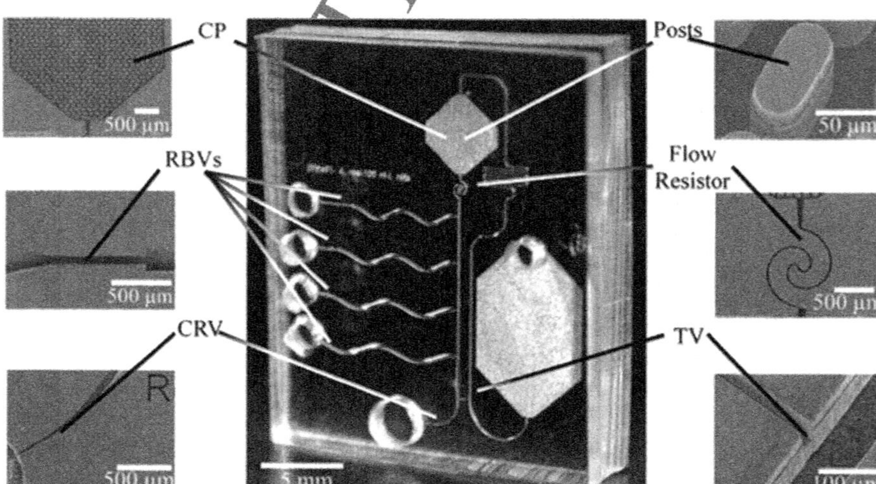

Fig. 9.4 Optical micrograph of a microfluidic capillaric circuit containing 4 side-reservoirs made in PDMS. *CP* capillary pump, *RBV* retention burst valve, *CRV* capillary retention valve, *TV* trigger valve. Taken from [83]. Copyright © The Royal Society of Chemistry 2013

4 Multiplexed Microfluidic Immunoassay Platforms

Multiplex immunoassays facilitate detection of multiple analytes from a single sample, and are of great importance for various fields, such as medical diagnostics, drug discovery, and other biological studies [86, 87]. Multiplex immunoassays are commonly performed on surface microarrays and microbead-based multiplexing. A common characteristic of these two platforms is the reduced volume in which the antibody-antigen recognition takes place, which is facilitated by microfluidic sample and reagent handling.

4.1 Surface Array-Based Multiplexing

Antibody-immobilized microarrays have been developed to achieve high-throughput multiplex immunoassay using chip-based surface microarrays generated by the immobilization of different capture antibodies on different micro compartments (*i.e.*, microspots in microarrays) [88]. Each capture antibody is identified by the spatial encoding (x and y coordinates) on the microarray surface, and can be analyzed using spatially resolved imaging methods such as microscopy and microarray scanner.

Microfluidics has mainly been employed to facilitate fluid handling for reagent delivery to surface microarrays. Microfluidic channels have been developed to create linear channels that connect individual rows or columns of a microarray [89]. Microfluidic flow cells have also been coupled to microarray [90]. As mentioned in the previous section, microfluidics reduce the diffusion distances within the microchannels, thereby enhancing transport of analyte and reagents onto the array surface that leads to faster reaction rate on the surface.

Antibody microarray can be created using microfluidics. A well-known method developed by Delamarche [91] used microfluidic networks (μFN) to pattern antibody on surfaces. First, antibodies were patterned on the surface of the substrate as linear strips using the first μFN. A second μFN, oriented in such a way that the channels are perpendicular to the first set of channels, was used to deliver analytes and reagents for the immunoassays, which subsequently generated signals that can be imaged optically (Fig. 9.5). A number of immunoassays have been developed based on this patterning method, [92, 93] and a more complex system integrating pneumatic valves to control the delivery of reagent solutions for simultaneous detection of 5 different analytes from 10 samples [94]. A modification of the initial design was reported by Ziegler et al. [80] where a stencil was used as an alternative to microchannels for delivering and coating proteins to form microspots [95].

Using the same concept, a double layer microchannel system was developed by Shao et al., [96] which consisted of two layers of μFN that were perpendicular to

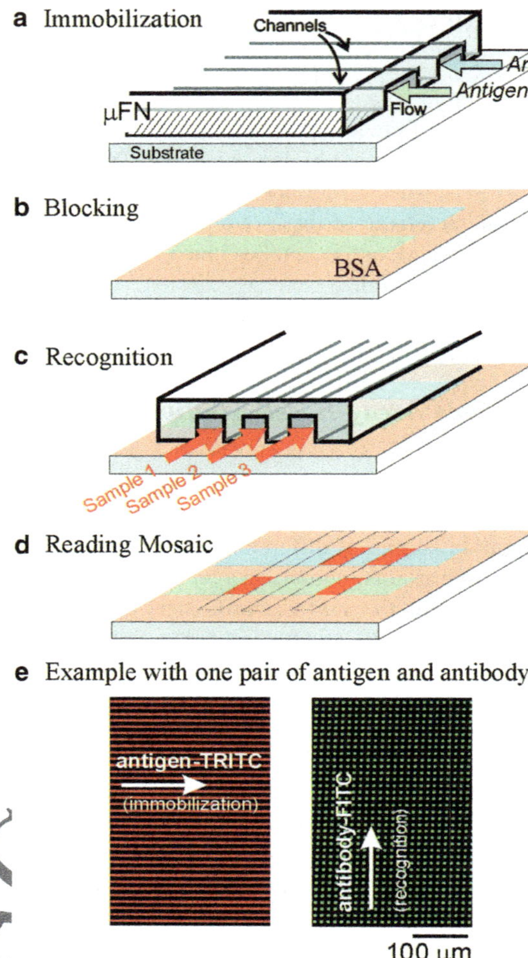

Fig. 9.5 Patterning of surface using microfluidic networks. (**a**) A μFN patterned different protein analytes along lines on a substrate. (**b**) Unpatterned area was blocked with BSA. (**c**) Flourescently-labeled antibodies flowed through the channels of a second μFN and locally bind to the patterned analytes. (**d**) Immunocomplex formation resulted in fluorescent spots. (**e**) Microsopy image of the microspot patterns. Taken from [91]. Copyright © 2001 American Chemical Society

each other, creating a microarray. Microarray has also been fabricated on filter membranes by sandwiching it between two layers of PDMS containing two different μFNs [95]. The major advantage of these systems is the elimination of the need to remove the first μFN and the subsequent introduction of the second onto the pattern.

4.2 *Bead-Based Multiplexing*

The simplest way of multiplexing bead-based immunoassays is to separate beads with different conjugated antibody into sets, and isolate each set of beads in different compartments [20]. A single sample is then delivered and allowed to interact with the antibody on the beads in all compartments. In this format, the number of compartments that can be fabricated and fit within the device is limited, and that a uniform distribution of the sample across all the compartments must occur. On the other hand, in most bead-based multiplexed immunoassays, the microbeads are encoded so that the identity of the antibody conjugated to a particular bead can be determined by reading and decoding the signal from the bead. The use of optically encoded microbeads does not require the precise positioning of the beads because their identity, and hence the identity of the capture antibody, is determined at the end of the assay by imaging. The most common bead encoding method for multiplex immunoassays is fluorescence barcoding, which is mostly analyzed by flow cytometry, whereas in microfluidic implementations, the beads are usually confined to a monolayer to facilitate simultaneous imaging by microscopy. A microfluidic platform was developed by Diercks et al. [97] to simultaneously detect four different analytes with pg/mL sensitivity from a 2.7 nL sample, using antibody-conjugated microbeads encoded by embedded fluorophores. A microfluidic platform was developed by Sasso et al. [98] that fully automated all step of the multiplex immunoassay using magnetic microbeads encoded with the Luminex xMAP technology (www. luminexcorp.com). In this system, magnetic actuation was used to transfer the beads into and out of incubation and washing solutions, which defined the incubation time and also allowed continuous flow operation to perform continuous concentration monitoring. The binding kinetics of the assay was empirically characterized to determine the optimal incubation times, achieving sensitivities in the range 1 pg/mL to 100 ng/mL for detection of IL-6 and TNF-α, respectively.

Another innovative use of microbeads for multiplex immunoassays is the use of DNA for barcoding [41, 99, 100]. In this method, a sandwich immunoassay was carried out on microbeads, where the detection antibody was labeled with gold nanoparticles that were also conjugated with double-stranded DNA (dsDNA) barcodes. After the sandwich immunocomplex was formed at the end of the immunoassay, the dsDNA is denatured, and the dissociated single-stranded DNA (ssDNA) was detected by either DNA assay, PCR amplification to reach aM sensitivity or DNA microarray for multiplex detection. This multiplex immunoassay combined the ease of manipulation of magnetic beads, the straightforward signal amplification by DNA amplification, and the high multiplexing capability of microarrays.

5 Conclusions

In this chapter, representative works on microfluidic immunoassays were categorized and described according to the format of the immunoassay, fluid driving and handling modalities, and multiplexing. Antibody-analyte interaction is accelerated due to enhanced transport within microfluidics, resulting in more rapid analysis and improved sensitivity in comparison to conventional methods. Analysis of complex biological samples such as blood is possible with microfluidics, and is best handled by pressure-driven flow due to the more stringent requirement of electric force-based and capillary-driven systems on the composition and viscosity of the sample. On the other hand, more complex capillary circuits with novel elements have been demonstrated, allowing for precisely controlled sequence of fluid operation performed in full automation with minimal requirement for instrumentation and power. Bead-based heterogeneous assays have experienced significant progress, particularly in the development of novel encoded beads for multiplexing.

Whereas microfluidic immunoassays progress significantly in the last decades, the potential for more advancement remains very high. In particular, multiple fluid handling modalities would be applied in synergy and integrated within a single platform, hence more complex fluid operation can be carried out with enhanced precision and reliability. Integration of novel transducers and detection methods with microfluidic would further improve the specificity and sensitivity of the assay, and facilitate multiplexing.

There are still challenges to be overcome. New strategies for reagents pre-loading and their long-term storage on-chip need to be developed. The issue of reagent stability is particularly important for immunoassays as most biological reagents, such as antibodies, are perishable and require specific storage environments. Whereas polymeric substrates such as PDMS are the preferred materials for prototype fabrication at the laboratory level, materials with similar properties but having capabilities for mass production such as injection molding are in great demand for practical applications and commercialization. The ultimate goal of microfluidic system development is to provide fully integrated, packaged, robust, user-friendly, low-cost and disposable devices. Given the active research and continual progress in the field, and the increasing number of innovations, it is envisioned that this goal can be achieved in the near future.

References

1. Rossier JS, Girault HH (2001) Enzyme linked immunosorbent assay on a microchip with electrochemical detection. Lab Chip 1:153–157
2. Sia SK, Whitesides GM (2003) Microfluidic devices fabricated in poly(dimethylsiloxane) for biological studies. Electrophoresis 24:3563–3576
3. Hatch A, Kamholz AE, Hawkins KR, Munson MS, Schilling EA, Weigl BH, Yager P (2001) A rapid diffusion immunoassay in a T-sensor. Nat Biotechnol 19:461–465
4. Lim TK, Ohta H, Matsunaga T (2003) Microfabricated on-chip-type electrochemical flow immunoassay system for the detection of histamine released in whole blood samples. Anal Chem 75:3316–3321
5. Reichmuth DS, Wang SK, Barrett LM, Throckmorton DJ, Einfeld W, Singh AK (2008) Rapid microchip-based electrophoretic immunoassays for the detection of swine influenza virus. Lab Chip 8:1319–1324
6. Meagher RJ, Hatch AV, Renzi RF, Singh AK (2008) An integrated microfluidic platform for sensitive and rapid detection of biological toxins. Lab Chip 8:2046–2053

7. Mohamadi MR, Kaji N, Tokeshi M, Baba Y (2007) Online preconcentration by transient isotachophoresis in linear polymer on a poly(methyl methacrylate) microchip for separation of human serum albumin immunoassay mixtures. Anal Chem 79:3667–3672
8. Herr AE, Throckmorton DJ, Davenport AA, Singh AK (2005) On-chip native gel electrophoresis-based immunoassays for tetanus antibody and toxin. Anal Chem 77:585–590
9. Herr AE, Hatch AV, Throckmorton DJ, Tran HM, Brennan JS, Giannobile WV, Singh AK (2007) Microfluidic immunoassays as rapid saliva-based clinical diagnostics. Proc Natl Acad Sci 104:5268–5273
10. Bottenus D, Jubery TZ, Ouyang Y, Dong W-J, Dutta P, Ivory CF (2011) 10 000-fold concentration increase of the biomarker cardiac troponin I in a reducing union microfluidic chip using cationic isotachophoresis. Lab Chip 11:890
11. Park CC, Kazakova I, Kawabata T, Spaid M, Chien RL, Wada HG, Satomura S (2008) Controlling data quality and reproducibility of a high-sensitivity immunoassay using isotachophoresis in a microchip. Anal Chem 80:808–814
12. Kawabata T, Wada HG, Watanabe M, Satomura S (2008) Electrokinetic analyte transport assay for a-fetoprotein immunoassay integrates mixing, reaction and separation on-chip. Electrophoresis 29:1399–1406
13. Chiem N, Harrison DJ (1997) Microchip-based capillary electrophoresis for immunoassays: analysis of monoclonal antibodies and theophylline. Anal Chem 69:373–378
14. Cheng SB, Skinner CD, Taylor J, Attiya S, Lee WE, Picelli G, Harrison DJ (2001) Development of a multichannel microfluidic analysis system employing affinity capillary electrophoresis for immunoassay. Anal Chem 73:1472–1479
15. Wang J, Ibanez A, Chatrathi MP (2002) Microchip-based amperometric immunoassays using redox tracers. Electrophoresis 23:3744–3749
16. Bromberg A, Mathies RA (2004) Multichannel homogeneous immunoassay for detection of 2,4,6-trinitrotoluene (TNT) using a microfabricated capillary array electrophoresis chip. Electrophoresis 25:1895–1900
17. Reid KR, Kennedy RT (2009) Continuous operation of microfabricated electrophoresis devices for 24 hours and application to chemical monitoring of living cells. Anal Chem 81:6837–6842
18. Dishinger JF, Reid KR, Kennedy RT (2009) Quantitative monitoring of insulin secretion from single islets of Langerhans in parallel on a microfluidic chip. Anal Chem 81:3119–3127
19. He X, Dandy DS, Henry CS (2008) Microfluidic protein patterning on silicon nitride using solvent-extracted poly(dimethylsiloxane) channels. Sens Actuators B Chem 129:811–817
20. Wen J, Yang X, Wang K, Tan W, Zuo X, Zhang H (2008) Telomerase catalyzed fluorescent probes for sensitive protein profiling based on one-dimensional microfluidic beads array. Biosens Bioelectron 23:1788–1792
21. Jokerst JV, Raamanathan A, Christodoulides N, Floriano PN, Pollard AA, Simmons GW, Wong J, Gage C, Furmaga WB, Redding SW, McDevitt JT (2009) Nano-bio-chips for high performance multiplexed protein detection: determinations of cancer biomarkers in serum and saliva using quantum dot bioconjugate labels. Biosens Bioelectron 24:3622–3629
22. Lucas LJ, Han JH, Yoon JY (2006) Using highly carboxylated microspheres to simplify immunoassays and enhance diffusional mixing in a microfluidic device. Colloids Surf B Biointerfaces 49:106–111
23. Prest JE, Fielden PR, Goddard NJ, Treves Brown BJ (2008) Isotachophoretic analysis using injection-moulded polystyrene chip devices. Meas Sci Technol 19:065801
24. Lee LJ, Yang ST, Lai S, Bai Y, Huang WC, Juang YJ (2006) Microfluidic enzyme-linked immunosorbent assay technology. Adv Clin Chem 42:255–295
25. Yakovleva J, Davidsson R, Lobanova A, Bengtsson M, Eremin S, Laurell T, Emnéus J (2002) Microfluidic enzyme immunoassay using silicon microchip with immobilized antibodies and chemiluminescence detection. Anal Chem 74:2994–3004

26. Yakovleva J, Davidsson R, Bengtsson M, Laurell T, Emnéus J (2003) Microfluidic enzyme immunosensors with immobilised protein A and G using chemiluminescence detection. Biosen Bioelectron 19:21–34
27. Bai Y, Koh CG, Boreman M, Juang YJ, Tang IC, Lee LJ, Yang ST (2006) Surface modification for enhancing antibody binding on polymer-based microfluidic device for enzyme-linked immunosorbent assay. Langmuir 22:9458–9467
28. Jonsson C, Aronsson M, Rundstrom G, Pettersson C, Mendel-Hartvig I, Bakker J, Martinsson E, Liedberg B, MacCraith B, Ohman O, Melin J (2008) Silane-dextran chemistry on lateral flow polymer chips for immunoassays. Lab Chip 8:1191–1197
29. Wen X, He H, Lee LJ (2009) Specific antibody immobilization with biotin-poly(L-lysine)-g-poly(ethylene glycol) and protein A on microfluidic chips. J Immunol Methods 350:97–105
30. Heyries KA, Mandon CA, Ceriotti L, Ponti J, Colpo P, Blum LJ, Marquette CA (2009) "Macromolecules to PDMS transfer" as a general route for PDMS biochips. Biosens Bioelectron 24:1146–1152
31. Parsa H, Chin CD, Mongkolwisetwara P, Lee BW, Wang JJ, Sia SK (2008) Effect of volume- and time-based constraints on capture of analytes in microfluidic heterogeneous immunoassays. Lab Chip 8:2062–2070
32. Kunz RE, Cottier K (2006) Optimizing integrated optical chips for label-free (bio-)chemical sensing. Anal Bioanal Chem 384:180–190
33. Caelen I, Bernard A, Juncker D, Michel B, Heinzelmann H, Delamarche E (2000) Formation of gradients of proteins on surfaces with microfluidic networks. Langmuir 16:9125–9130
34. Goldstein B, Coombs D, He X, Pineda AR, Wofsy C (1999) The influence of transport on the kinetics of binding to surface receptors: application to cells and BIAcore. J Mol Recognit 12:293–299
35. Abrantes M, Magone MT, Boyd LF, Schuck P (2001) Adaptation of a surface plasmon resonance biosensor with miorofluidics for use with small sample volumes and long contact times. Anal Chem 73:2828–2835
36. Stroock AD, Dertinger SK, Ajdari A, Mezic I, Stone HA, Whitesides GM (2002) Chaotic mixer for microchannels. Science 295:647–651
37. Golden JP, Floyd-Smith TM, Mott DR, Ligler FS (2007) Target delivery in a microfluidic immunosensor. Biosens Bioelectron 22:2763–2767
38. Jennissen HP, Zumbrink T (2004) A novel nanolayer biosensor principle. Biosens Bioelectron 19:987–997
39. Hofmann O, Voirin G, Niedermann P, Manz A (2002) Three-dimensional microfluidic confinement for efficient sample delivery to biosensor surfaces. Application to immunoassays on planar optical waveguides. Anal Chem 74:5243–5250
40. Lim CT, Zhang Y (2007) Bead-based microfluidic immunoassays: the next generation. Biosens Bioelectron 22:1197–1204
41. Goluch ED, Nam JM, Georganopoulou DG, Chiesl TN, Shaikh KA, Ryu KS, Barron AE, Mirkin CA, Liu C (2006) A bio-barcode assay for on-chip attomolar-sensitivity protein detection. Lab Chip 6:1293–1299
42. Huang H, Zheng XL, Zheng JS, Pan J, Pu XY (2009) Rapid analysis of alpha-fetoprotein by chemiluminescence microfluidic immunoassay system based on super-paramagnetic microbeads. Biomed Microdevices 11:213–216
43. Sista RS, Eckhardt AE, Srinivasan V, Pollack MG, Palanki S, Pamula VK (2008) Heterogeneous immunoassays using magnetic beads on a digital microfluidic platform. Lab Chip 8:2188–2196
44. Yang SY, Lien KY, Huang KJ, Lei HY, Lee GB (2008) Micro flow cytometry utilizing a magnetic bead-based immunoassay for rapid virus detection. Biosens Bioelectron 24:861–868
45. Hung LY, Chang JC, Tsai YC, Huang CC, Chang CP, Yeh CS, Lee GB (2014) Magnetic nanoparticle-based immunoassay for rapid detection of influenza infections by using an integrated microfluidic system. Nanomed Nanotechnol Biol Med 10:819–829

46. Lacharme F, Vandevyver C, Gijs MAM (2009) Magnetic beads retention device for sandwich immunoassay: comparison of off-chip and on-chip antibody incubation. Microfluid Nanofluid 7:479–487
47. Sato K, Tokeshi M, Kimura H, Kitamori T (2001) Determination of carcinoembryonic antigen in human sera by integrated bead bed immunoassay in a microchip for cancer diagnosis. Anal Chem 73:1213–1218
48. Sato K, Yamanaka M, Hagino T, Tokeshi M, Kimura H, Kitamori T (2004) Microchip-based enzyme-linked immunosorbent assay (microELISA) system with thermal lens detection. Lab Chip 4:570–575
49. Ko YJ, Maeng JH, Ahn Y, Hwang SY, Cho NG, Lee SH (2008) Microchip-based multiplex electro-immunosensing system for the detection of cancer biomarkers. Electrophoresis 29:3466–3476
50. Shin KS, Lee SW, Han KC, Kim SK, Yang EK, Park JH, Ju BK, Kang JY, Kim TS (2007) Amplification of fluorescence with packed beads to enhance the sensitivity of miniaturized detection in microfluidic chip. Biosens Bioelectron 22:2261–2267
51. Peoples MC, Karnes HT (2008) Microfluidic capillary system for immunoaffinity separations of C-reactive protein in human serum and cerebrospinal fluid. Anal Chem 80:3853–3858
52. Holmes D, She JK, Roach PL, Morgan H (2007) Bead-based immunoassays using a microchip flow cytometer. Lab Chip 7:1048–1056
53. Sivagnanam V, Song B, Vandevyver C, Gijs MA (2009) On-chip immunoassay using electrostatic assembly of streptavidin-coated bead micropatterns. Anal Chem 81:6509–6515
54. Sivagnanam V, Bouhmad A, Lacharme F, Vandevyver C, Gijs MAM (2009) Sandwich immunoassay on a microfluidic chip using patterns of electrostatically self-assembled streptavidin-coated beads. Microelectron Eng 86:1404–1406
55. Yang Y, Nam SW, Lee NY, Kim YS, Park S (2008) Superporous agarose beads as a solid support for microfluidic immunoassay. Ultramicroscopy 108:1384–1389
56. Linder V (2007) Microfluidics at the crossroad with point-of-care diagnostics. Analyst 132:1186–1192
57. Hayes MA, Kheterpal I, Ewing AG (1993) Effects of buffer pH on electroosmotic flow control by an applied radial voltage for capillary zone electrophoresis. Anal Chem 65:27–31
58. Dodge A, Fluri K, Verpoorte E, de Rooij NF (2001) Electrokinetically driven microfluidic chips with surface-modified chambers for heterogeneous immunoassays. Anal Chem 73:3400–3409
59. Linder V, Verpoorte E, Thormann W, de Rooij NF, Sigrist H (2001) Surface biopassivation of replicated poly(dimethylsiloxane) microfluidic channels and application to heterogeneous immunoreaction with on-chip fluorescence detection. Anal Chem 73:4181–4189
60. Gao Y, Hu G, Lin FY, Sherman PM, Li D (2005) An electrokinetically-controlled immunoassay for simultaneous detection of multiple microbial antigens. Biomed Microdevices 7:301–312
61. Gao Y, Sherman PM, Sun Y, Li D (2008) Multiplexed high-throughput electrokinetically-controlled immunoassay for the detection of specific bacterial antibodies in human serum. Anal Chim Acta 606:98–107
62. Hu G, Gao Y, Li D (2007) Modeling micropatterned antigen-antibody binding kinetics in a microfluidic chip. Biosens Bioelectron 22:1403–1409
63. Mugele F, Baret JC (2005) Electrowetting: from basics to applications. J Phys Condens Matter 17:R705–R774
64. Henares TG, Mizutani F, Hisamoto H (2008) Current development in microfluidic immunosensing chip. Anal Chim Acta 611:17–30
65. Gong MM, Macdonald BD, Vu Nguyen T, Sinton D (2012) Hand-powered microfluidics: a membrane pump with a patient-to-chip syringe interface. Biomicrofluidics 6:44102
66. Qiu X, Thompson JA, Chen Z, Liu C, Chen D, Ramprasad S, Mauk MG, Ongagna S, Barber C, Abrams WR, Malamud D, Corstjens PL, Bau HH (2009) Finger-actuated, self-contained immunoassay cassettes. Biomed Microdevices 11:1175–1186

67. Wang CH, Lee GB (2005) Automatic bio-sampling chips integrated with micro-pumps and micro-valves for disease detection. Biosens Bioelectron 21:419–425
68. Yang Y-N, Hsiung S-K, Lee G-B (2008) A pneumatic micropump incorporated with a normally closed valve capable of generating a high pumping rate and a high back pressure. Microfluid Nanofluid 6:823–833
69. Quake SR (2000) From Micro- to nanofabrication with soft materials. Science 290:1536–1540
70. Schudel BR, Choi CJ, Cunningham BT, Kenis PJ (2009) Microfluidic chip for combinatorial mixing and screening of assays. Lab Chip 9:1676–1680
71. Kim J, Kang M, Jensen EC, Mathies RA (2012) Lifting gate polydimethylsiloxane microvalves and pumps for microfluidic control. Anal Chem 84:2067–2071
72. Wang CH, Lee GB (2006) Pneumatically driven peristaltic micropumps utilizing serpentine-shape channel. J Micromech Microeng 16:341–348
73. Lai H, Folch A (2011) Design and dynamic characterization of "single-stroke" peristaltic PDMS micropumps. Lab Chip 11:336–342
74. Aeinehvand MM, Ibrahim F, Harun SW, Djordjevic I, Hosseini S, Rothan HA, Yusof R, Madou MJ (2015) Biosensing enhancement of dengue virus using microballoon mixers on centrifugal microfluidic platforms. Biosens Bioelectron 67:424–430
75. Fraley KJ, Abberley L, Hottenstein CS, Ulicne JJ, Citerone DR, Szapacs ME (2013) The Gyrolab immunoassay system: a platform for automated bioanalysis and rapid sample turnaround. Bioanalysis 5:1765–1774
76. Lee BS, Lee JN, Park JM, Lee JG, Kim S, Cho YK, Ko C (2009) A fully automated immunoassay from whole blood on a disc. Lab Chip 9:1548–1555
77. Koh CY, Schaff UY, Piccini ME, Stanker LH, Cheng LW, Ravichandran E, Singh BR, Sommer GJ, Singh AK (2015) Centrifugal microfluidic platform for ultrasensitive detection of botulinum toxin. Anal Chem 87:922–928
78. Cesaro-Tadic S, Dernick G, Juncker D, Buurman G, Kropshofer H, Michel B, Fattinger C, Delamarche E (2004) High-sensitivity miniaturized immunoassays for tumor necrosis factor alpha using microfluidic systems. Lab Chip 4:563–569
79. Zimmermann M, Hunziker P, Delamarche E (2009) Autonomous capillary system for one-step immunoassays. Biomed Microdevices 11:1–8
80. Ziegler J, Zimmermann M, Hunziker P, Delamarche E (2008) High-performance immunoassays based on through-stencil patterned antibodies and capillary systems. Anal Chem 80:1763–1769
81. Wang J, Ahmad H, Ma C, Shi Q, Vermesh O, Vermesh U, Heath J (2010) A self-powered, one-step chip for rapid, quantitative and multiplexed detection of proteins from pinpricks of whole blood. Lab Chip 10:3157–3162
82. Mohammed MI, Desmulliez MP (2013) Planar lens integrated capillary action microfluidic immunoassay device for the optical detection of troponin I. Biomicrofluidics 7:64112
83. Safavieh R, Juncker D (2013) Capillarics: pre-programmed, self-powered microfluidic circuits built from capillary elements. Lab Chip 13:4180–4189
84. Novo P, Chu V, Conde JP (2014) Integrated optical detection of autonomous capillary microfluidic immunoassays:a hand-held point-of-care prototype. Biosens Bioelectron 57:284–291
85. Gervais L, Delamarche E (2009) Toward one-step point-of-care immunodiagnostics using capillary-driven microfluidics and PDMS substrates. Lab Chip 9:3330–3337
86. Joos TO, Stoll D, Templin MF (2002) Miniaturised multiplexed immunoassays. Curr Opin Chem Biol 6:76–80
87. Ling MM, Ricks C, Lea P (2007) Multiplexing molecular diagnostics and immunoassays using emerging microarray technologies. Expert Rev Mol Diagn 7:87–98
88. Angenendt P (2005) Progress in protein and antibody microarray technology. Drug Discov Today 10:503–511

89. Delehanty JB, Ligler FS (2002) A microarray immunoassay for simultaneous detection of proteins and bacteria. Anal Chem 74:5681–5687
90. Cretich M, Di Carlo G, Giudici C, Pokoj S, Lauer I, Scheurer S, Chiari M (2009) Detection of allergen specific immunoglobulins by microarrays coupled to microfluidics. Proteomics 9:2098–2107
91. Bernard A, Michel B, Delamarche E (2001) Micromosaic immunoassays. Anal Chem 73:8–12
92. Murphy BM, He X, Dandy D, Henry CS (2008) Competitive immunoassays for simultaneous detection of metabolites and proteins using micromosaic patterning. Anal Chem 80:444–450
93. Murphy BM, Dandy DS, Henry CS (2009) Analysis of oxidative stress biomarkers using a simultaneous competitive/non-competitive micromosaic immunoassay. Anal Chim Acta 640:1–6
94. Kartalov EP, Zhong JF, Scherer A, Quake SR, Taylor CR, Anderson WF (2006) High-throughput multi-antigen microfluidic fluorescence immunoassays. Biotechniques 40:85–90
95. Liu Y, Yu J, Du M, Wang W, Zhang W, Wang Z, Jiang X (2012) Accelerating microfluidic immunoassays on filter membranes by applying vacuum. Biomed Microdevices 14:17–23
96. Shao G, Wang J, Li Z, Saraf L, Wang W, Lin Y (2011) Poly(dimethylsiloxane) microchip-based immunoassay with multiple reaction zones: toward on-chip multiplex detection platform. Sens Actuators B 159:44–50
97. Diercks AH, Ozinsky A, Hansen CL, Spotts JM, Rodriguez DJ, Aderem A (2009) A microfluidic device for multiplexed protein detection in nano-liter volumes. Anal Biochem 386:30–35
98. Sasso LA, Johnston IH, Zheng M, Gupte RK, Undar A, Zahn JD (2012) Automated microfluidic processing platform for multiplexed magnetic bead immunoassays. Microfluid Nanofluid 13:603–612
99. Nam JM, Thaxton CS, Mirkin CA (2003) Nanoparticle-based bio-bar codes for the ultrasensitive detection of proteins. Science 301:1884–1886
100. Stoeva SI, Lee JS, Smith JE, Rosen ST, Mirkin CA (2006) Multiplexed detection of protein cancer markers with biobarcoded nanoparticle probes. J Am Chem Soc 128:8378–8379

Chapter 10
Challenges and Future

Ajeet Kaushik and Chandra K. Dixit

Since the discovery of microfluidics systems (MEMS), efforts have been used to use these automated architecture widely for health care. The systems are miniaturized and revolutionized the way we deal biological samples and biological matrix. BioMEMS is a choice of system for accurate and precise measurement, wherein a very low sample volume is available to measure. The design and fabrication of an appropriate BioMEMS enable us to understand how a single cell is completely different from multiple cells and spheroids. This happened in practice because BioMEMS technology opened the ways to study physiology of a single cell and to understand the heterogeneity in the cellular population of the same decent using morphological, optical, and electrical analytical tools. Another example is to develop BioMEMS integrated biosensor which can detect target biomarkers using 10 μL of sample volume.

BioMEMS becomes an essential component to develop health care devices for personalized management. However, due to requirement of highly sophisticated fabrication set-up consist of clean room equipped with expensive equipment limits the production of BioMEMS. Beside this, the need of high operation expertise to function and manage a fabrication tool is also a major issue to promote such systems for desired application. Scientists have made considerable efforts to fabricate a BioMEMS of reduced form factor, easy operational, easy to integrate, easy to package, shock proof, no-dead volume, and affordable. Such devices potentially can be useful for field of life sciences, from basics to industrial to diagnostics.

A. Kaushik (✉)
Department of Immunology, Center for Personalized Nanomedicine, Institute of NeuroImmune Pharmacology, Herbert Wertheim College of Medicine, Florida International University, Miami, FL 33199, USA
e-mail: akaushik@fiu.edu

C.K. Dixit
Department of Chemistry, University of Connecticut, Storrs, CT 06269, USA

C.K. Dixit, A. Kaushik (eds.), *Microfluidics for Biologists*,
DOI 10.1007/978-3-319-40036-5_10

However, there is a big communication gap between biologists and micro-technologists; it is because of the lack of training in the fields other than theirs.

Presented book is a very good platform to understand the basic of BioMEMS, fundamental science associated with BioMEMS, materials science in BioMEMS, nanotechnology in BioMEMS, scale-up fabrication of BioMEMS, role of BioMEMS in cell biology, contribution of BioMEMS in health care devices, and development of chip-based approach for diagnosis, especially point-of-care systems.

Overall, this document is call to researchers to explore BioMEMS more to make state-of-the-art more effective and efficient. According to the diagnostics requirement, BioMEMS architecture can be modified via changing design and adopting novel nanomaterials. Such investigated and proposed BioMEMS could be useful in the field of cell and molecular biology, single cell biology, disease diagnostics, and introduced all the commercially available systems that has been either introduced or have the potential of being get used in research and development. To make possible, government agencies, national laboratory, universities and industries should pay more attention and manage more budget to establish BioMEMS fabrication center along with the introduction of more basic coursed to produce engineers. This approach certainly will help to establish BioMEMS based start-up companies. This will reduce the cost of fabrication and increase BioMEMS availability.

The proposed book explored the need of novel microtechnologies, their integration strategies for developing new class of assay systems to retrieve the desired health information of patient in real-time. To achieve accomplishments, the selection of each sensor components and operational parameters of desired property is very crucial to develop an efficient sensing platform to capable of performing at POC. System-on-a-Chip (SoC), Diagnostic-on-a-Chip (DoC), and Lab-on-a-chip (LOC) are the core to the next generation bioanalytical sciences; therefore, this book will help biologists to understand these systems and will allow them to make educated decisions on selecting the nature and type of microtechnologies that suits best to their methods thus enhancing the rate of translational research in the field.

RETRACTION NOTE TO

On-Chip Immunoassay for Molecular Analysis

Andy Ng

C.K. Dixit, A. Kaushik (eds.), *Microfluidics for Biologists*,
DOI 10.1007/978-3-319-40036-5_9

Retraction Note:

"On-Chip Immunoassay for Molecular Analysis," by Andy Ng, published as a chapter in Microfluidics for Biologists, edited by Chandra Dixit and Ajeet K. Kaushik, copyright 2016, ISBN 978-3-319-40035-8, is being retracted due to striking overlap with another published source, "Immunoassays in Microfluidic Systems", published 27 Apr 2010 in Analytical and Bioanalytical Chemistry volume 397 issue 3 on pages 991 to 1007 by Alphonsus H. C. Ng, Uvaraj Uddayasankar, and Aaron R. Wheeler.

The Editor(s) of this volume regret the oversight.

The updated online version of this chapter can be found under
DOI 10.1007/978-3-319-40036-5_9

C.K. Dixit, A. Kaushik (eds.), *Microfluidics for Biologists*,
DOI 10.1007/978-3-319-40036-5_11

Index

C.K. Dixit, A. Kaushik (eds.), *Microfluidics for Biologists*,
DOI 10.1007/978-3-319-40036-5